The Small Farmer's Guide to
RAISING LIVESTOCK AND POULTRY

The Small Farmer's Guide to
RAISING LIVESTOCK AND POULTRY

Edited by Katie Thear
and Dr. Alistair Fraser, MVO, MRCVS

Consultant: Dr John Skinner
*Poultry and Small Animal Specialist,
University of Wisconsin*

Photography Brian Hale

ARCO PUBLISHING, INC.

NEW YORK

Note: References in this book to liquid quantities are given in UK gallons, which are larger than US gallons. (5 UK gallons are equal to approximately 6 US gallons.)

Published 1980 by Arco Publishing Inc,
219 Park Avenue South, New York, N.Y. 10003

© Martin Dunitz Limited 1980

First published in the United Kingdom in 1980
by Martin Dunitz Limited

Library of Congress Cataloging in Publication Data
Main entry under title:
Small farmer's guide to raising livestock and poultry.
1. Livestock.   2. Poultry.   I. Fraser,
Alistair.   II. Thear, Katie.
SF65.2.S62   636   80-13811
ISBN 0-668-04687-2

Designed by Colin Lewis.

Editorial advisor for North America: John Skinner,
Poultry and Small Animal Specialist, University
of Wisconsin.

Printed in Hong Kong

CONTENTS

4. Goats 73

by John and Jill Halliday

5. Sheep 104

by Lawrence Alderson

6. Pigs 140

by Ranken Bushby

PREFACE

There is now a great revival of interest in small-scale, mixed farming as more and more people are starting to discover the pleasures and rewards of rearing their own animals for eggs, meat or milk. Whether they are involved in a small, commercial enterprise or whether they just want to keep a cow and a few chickens to supply their families, the principles of animal care remain the same.

Up until the 1940s there were a great number of small farmers. Even in the mid-fifties it was still common to find country people who kept dairy animals, pigs and poultry for their own use, but the numbers were declining. People whose families had for generations maintained a link with the land and with small-scale livestock husbandry, suddenly found that modern economics and the demand for convenience foods had combined to produce a specialized consumerism in which they could play little part. Small poultry keepers found themselves out of business as large intensive battery and broiler units took over.

This meant that the long heritage of local knowledge and traditional agricultural practices all but disappeared. Nowadays a typical huge commercial farm will often be run by only two men and it is not unusual to meet a dairy farmer who has never hand-milked a cow, or a poultry farmer who cannot diagnose the simplest chicken ailment. If one wants to learn the practicalities of keeping livestock on a small scale, the big commercial farmer is therefore unlikely to be a good source of help. This book is designed to fill that gap by providing all the relevant information,

sound advice and encouragement needed by anyone starting with livestock.

Each chapter has been written by someone who can pass on first-hand experience in looking after the animals concerned and who knows exactly what questions the small-scale farmer is likely to need answered. Clear, step-by-step photographs help demonstrate many of the practices described. All the chapters are structured in such a way that either they can be read straight through for general guidance or individual sections can be consulted when a crisis arises or when the reader wants advice on a specific topic.

People decide to take up small-scale farming for a wide variety of reasons. Certainly the steadily increasing costs of the chemicals used in modern farming practices have strengthened the economic sense behind less intensive systems, but commercial profit is not always the small-scale farmer's greatest or his only priority. Some people decide to keep their own livestock simply in order to supply their families with really good fresh food which they know has been naturally produced, and some in order to escape or to counteract the more materialistic and mechanized aspects of modern life. In the process they are helping to bring back life to our denuded and often sterile countryside.

Whatever the aims and attractions of such an enterprise are for you, *The Complete Book of Raising Livestock and Poultry* will enable you to look after your animals in the best possible way and to gain the maximum amount of satisfaction from it.

Katie Thear

CHICKENS

Background

Traditionally hens were the province of the farmer's wife who fed them on kitchen scraps and spare grain. The rest of their diet was dependent upon free-range foraging, usually around the farmyard. This was a cheap way of keeping hens, but the number of eggs must have been low, although the dual-purpose bird did furnish meat for the table. For a long time the Dorking and the Old English Game were the only recognizable breeds of domestic fowl in Britain amongst the farmyard mongrels, the former reputedly introduced by the Romans, and the latter developed as a fighting cock.

Gradually other breeds were imported from the Continent and from the East. Many of the Asiatic heavy breeds originated as table birds in France and other Continental countries. Some of the best breeds were produced in North America around the turn of the century, perhaps the most notable being the Rhode Island Red and the Leghorn, which was developed from an Italian breed. Since 1947 Australia has had a quarantine embargo on poultry importation which has prevented any new introductions, but commercial and fancy strains have been preserved. White Leghorn cocks mated to Australorp hens produce the most common layer-bird which is referred to as the Crossbred. One commercial breed based on crossing of selected strains of White Leghorn has also been developed and is called a strain-cross or hybrid.

After the First World War, many ex-servicemen in Britain started poultry farms. These were free-range and even in the late 1940s one could still see large fields filled with grazing hens – a sight which has now disappeared from the landscape. Gradually more intensive methods were developed. In the 1950s and early 1960s the deep-litter system came into general use, where many hens were housed indoors on straw, shavings etc. Then in the late sixties and seventies came the system we have today – the intensive battery, where hens spend their lives (one laying season) in small cages. In recent years there has been a reaction against this, and we are again seeing the resurgence of part-time farmers and smallholders who keep chickens less intensively for eggs, meat and pleasure.

Origin and development of breeds

The domestic hen (*Gallus domesticus*) has been with us for a long time. Originally domesticated in prehistoric times, hens were kept by the Egyptians, Greeks, Chinese and Romans.

Their common ancestor is generally believed to have been the Jungle Fowl (*Gallus gallus*), a small Asiatic bird, beautifully adapted to life in the tropical forests of the East. The Jungle Fowl lived in a flock which had a dominant male and a definite social pecking order. Its large feet and sharp claws had evolved to scratch the soft, leaf-littered ground of the forest and also to perch on tree branches for protection at night. Its beak was hard and pointed – ideal for pecking seeds, insects and plant shoots. External parasites were discouraged by periodic dust baths in fine, dry earth. In fact, the Jungle Fowl's characteristics are almost identical with those of today's domestic hen. The species is still to be found in its natural habitat in the East, as well as amongst the stock of dedicated poultry fanciers all over the world.

The original farm hens were dual-purpose, combining the qualities of both egg layers and table birds. Gradually, however, selective breeding developed and improved these characteristics separately. This resulted in a division into light breeds which, on the whole, produced more eggs, and heavy breeds which were bigger and provided more meat. We know little of breeding before the nineteenth century, but by this time many different breeds had been established and the poultry fanciers had emerged.

For the first half of the twentieth century commercial breeders and hatcheries were supplying poultry farmers and smallholders with such memorable utility breeds as the Rhode Island Red, the Light Sussex, the White Leghorn, the White Wyandotte, the Plymouth Rock, and, in Australia, the Australorp. After the Second World War, however, the commercial poultry world moved towards intensive practices, and the division between light and heavy breeds became greater, with the emphasis on hybridization and the crossing of different strains. This division has since been carried to its extreme so that two huge industries exist, one for egg production and one for broiler meat.

The modern egg-laying hybrid is a small, light bird which lays far more eggs than her ancestors did, but the ability to become broody and hatch her own eggs has largely been bred out of her. The cross-bred layer is large in size, lays slightly larger eggs and also does not become broody. The Australorp is a poor layer and does go broody.

In their hybridization programme the commercial poultry interests bought up the best of the pure utility stock, and it is only thanks to the poultry fanciers that some breeds remain in existence today. Unfortunately, breeders who specialize in show varieties may be more interested in appearances than in utilitarian virtues, and there is no doubt that many of the utility breeds have declined in the last few decades. Perhaps, now that the number of smallholders is on the increase, there will be a reversal of this trend.

Buying chickens

Which to choose

When choosing chickens there is a choice between pure breeds, first crosses and hybrids.

White Wyan-
dotte cock

White Wyan-
dotte hen

Plymouth
Rock
(Buff variety)

White Leghorn

Pure breeds

These are the birds (such as the Rhode Island Red, White Leghorn or Australorp) that will breed true to type. The further choice, as has already been indicated, is between a light and a heavy breed. Up until the mid- to late-fifties, it was the pure breeds specially bred for utilitarian purposes which dominated the poultry scene, but since then they have declined in numbers, vigour and performance. There is now a revival of interest in them, and it is to be hoped that new and vigorous strains will emerge as a result of selective breeding by smallholders and poultry keepers. If however you want eggs more than anything else, you should buy in modern hybrids every season.

Bantams

These are small fowl which are either naturally small such as the Sebright, or which have been bred down from larger fowl crossed with other bantams. They are true-breeding miniatures of the larger breeds and varieties. There are nearly 400 combinations of colours, shape and features to be found. The eggs that they lay are small and they therefore tend to be ignored by commercial interests. The interest among poultry fanciers and many other amateur poultry keepers is however considerable. Their utilitarian value lies in the low cost of feeding them, in the relatively small space that they require and in their great capacity for broodiness. The best of the bantam layers are Light Sussex, Rhode Island Reds and Anconas.

The Silkie

This is often mistakenly referred to as a bantam because of its relatively small size, but it is in fact classified as a large fowl. Its particular value to small farmers is its tendency towards broodiness; many Silkies will go broody twice, or even three times in a year. They are often used as foster mothers for game birds and other ornamental fowl. When crossed with other sitting breeds, the Silkie produces first rate broodies. A particularly good cross for broodiness is the Silkie crossed with a Light Sussex.

First crosses

These are produced by the mating of two pure breeds, such as a Rhode Island Red cock and a Light Sussex hen (indicated as RIR x LS). This particular cross is an excellent one for the smallholder because the colour is sex-linked and day-old chicks can be clearly distinguished as male or female. The males are silvery white and the females orange-brown. The most common first cross in Australia is the White Leghorn cock mated to Australorp hens. Generally speaking, first crosses are hardier than pure breeds, but you will not be able to breed from them and retain the breed characteristics. So you will either have to buy in new first crosses periodically or breed your own from a foundation stock of pure breeds.

Hybrids and strain-crosses

If you want a large number of eggs, then the commercial hybrid which has been developed from several different strains and breeds is the most suitable – but there may be snags. She will not be a good table bird, and in order to attain peak performance, she will need to be fed on intensive (commercial) rations. As her breeding will have been for a controlled environment, there is a possibility that she may not be as hardy for outdoor conditions as a first cross. Having said that, I have known hybrids adapt extremely well to outside conditions and a less intensive diet.

Age to buy

There is a choice of buying stock in any of the following ways – as day-olds, as pullets, as point-of-lay pullets or as year-old hens.

Day-old chicks

This is the cheapest way of acquiring stock, but they may be unsexed, so about half of them will be cockerels. The usual practice is to raise the cocks to table weight and then slaughter them.

You will have to provide the chicks with artificial heat in some sort of brooder unless you happen to know someone with an available broody hen. There will be feeding costs, although you will get no eggs for at least twenty weeks. With chicks there is also a higher mortality rate than with older stock.

Pullets

The term pullet is used to describe any female chicken between eight and twenty weeks. If you buy them slightly under age and the weather is cold, they may still require artificial heat, but they will be stronger than chicks and there will be no cockerels among them.

Point-of-lay pullets

Strictly speaking, these are pullets of twenty weeks which are about to start laying. Be sure to check the age with the vendor, for the term is often misused. This is the most expensive way to buy stock, but the birds need less feeding prior to laying, and they will produce more eggs during their first laying season than they will do subsequently.

Year-old hens

Commercially, hens have a productive life of one laying season; after that they are disposed of because the number of eggs declines and they tend to eat more. It is often possible to buy these rejected birds cheaply but, if they are from a battery system, be prepared to rehabilitate them. A year in a cage, often crammed in with others, leaves hens exhausted. They emerge with feathers missing and often have to learn to walk after their long confinement. They will also usually undergo a complete moult when relocated at this stage in life.

Where to buy

Generally speaking, markets are not good places for the beginner to buy poultry, unless he has an experienced

Rhode Island Red pullet

Indian Game

Light Sussex

Ancona

Silkie

Barnevelder

Dorking cock

Cuckoo
Maran cock

Rhode Island
Red cock

Ross Ranger
hybrid

Welsummer
cock

Maran hen

poultry keeper with him to advise. The best place to obtain first crosses is to buy stock from a small poultry keeper who is a reputable breeder. There are not all that many of them around, but they do exist. (In Australia a list of approved chicken hatcheries can be obtained from the State Department of Agriculture.) The larger hatcheries will usually only have hybrids and may only supply in large minimum quantities. They often have local agents however, who will probably supply small numbers, particularly if the customer will collect. Failing this you can always try cooperating with friends to place a bulk order.

Specialist breeders are usually the only ones to have pure breeds, and these are normally bred for show. I have already made the point that if a bird has been bred for its appearance

Free-ranging hens with a house provided in each paddock. Note the drinker made from an adapted oil drum.

it may not necessarily be a good layer, so it is important to check this. Exhibition birds will also command a higher price than first crosses.

If a flock of more than twenty hens is kept in Australia, it must be registered with The Egg Board and a licensed 'quota' must be purchased. Backyard flocks must therefore have twenty hens or fewer.

Poultry-keeping systems

Which method of poultry keeping you adopt will depend on a number of factors – notably your own space, time, finances and inclination, and the needs of the hens themselves. The two extremes are the back-to-nature brigade who let hens wander and scratch wherever they want to around the farm, and the battery-intensive farmers who cram all the birds in cages. Hens kept in the former manner will destroy every vestige of a flower or vegetable garden, lay eggs in inaccessible places and probably be carried off by foxes. The latter to my mind is an obscene practice which is illegal in some countries and which many people condemn. However, since it is undoubtedly true that the hens respond with good production records, large-scale commercial enterprises will no doubt continue to keep poultry in this way. It is interesting that recent research by Dr David Sainsbury at Cambridge University has shown that layers kept in covered straw yards perform just as well as battery birds.

The free-range system

This is suitable for those with a lot of land and fields of pasture. It was once popular with farmers, who would let the chickens graze the stubble after harvest. The birds gleaned much of their food in this way, as well as ridding the field of insect pests and weed seeds. A hen house was provided in the field, and when it was time to move on to a new field the house would be transferred as well.

The system had much to recommend it. For most of the year the birds found their own food, and feeding costs were thereby greatly reduced. Where they followed grazing cattle, they did useful work in scratching up and spreading cow-pats which, together with their own droppings, helped to build up the fertility of the soil. They also got rid of parasites.

In this sort of situation a density of 250 birds per acre (0.5 ha) was the maximum when they were put on fresh pasture. Where no other livestock had been before them, the density was increased to 300–400 birds depending on soil type and drainage. (In the USA zoning now keeps poultry off many home sites of less than 5 acres.)

There were disadvantages of course. The losses by foxes and other predators were high, and eggs were often laid in hidden nests, rather than in the houses. In cold winters many hens died and egg production dwindled to nothing.

The fold method

This method utilizes a fold unit, or a house complete with its own covered-in run. A unit measuring 20 ft × 6 ft (6 m × 2 m) will hold between twenty-five and thirty hens which graze on pasture. As soon as the grazing is exhausted (usually after one day) the unit is moved on the distance of its own length, so that the hens have access to fresh grass. The unit is equipped with carrying poles, which also act as perches inside, but moving it every twenty-four hours is still a time-consuming and laborious task.

The advantages are that grazing is controlled, fresh grass is available and there is no build-up of droppings with the consequent parasites and disease. The hens are also safer from foxes and other predators.

Although this system requires pasture and is more suitable for the farmer with several fields, the poultry keeper with a few hens and limited space can use his own smaller version. Units to house half a dozen hens are easily made at home and the controlled grazing is an effective way of keeping the lawn weeds down. Small units are also much easier to move.

Some form of grass management is needed to ensure that future grazing is preserved. On smaller areas, you can encourage fresh growth which has the required nutrients by regular mowing. Bare patches can be reseeded and a light dressing of lime applied once a year will not only help to make the nutrients available, but will also help to deter parasites. Caked manure should be raked up so that it can be dispersed by rain.

On a large scale most permanent pastures were not originally laid down with poultry in mind, but rather for larger grazing animals. Many of the taller, strong-growing plants are not relished by hens, who prefer dwarf grasses that will form a compact carpet. A useful pasture for

A traditional fold house on wheels.

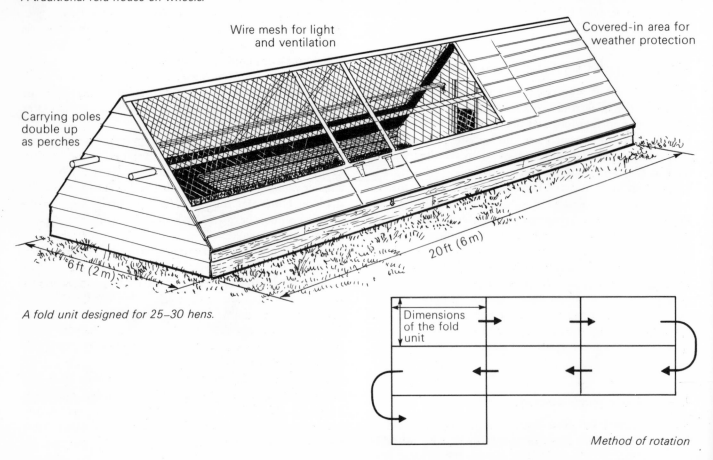

Wire mesh for light
and ventilation

Covered-in area for
weather protection

Carrying poles
double up
as perches

20 ft (6 m)

6 ft (2 m)

A fold unit designed for 25–30 hens.

Dimensions
of the fold
unit

Method of rotation

The semi-intensive system.

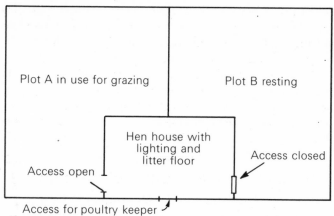

Plot A in use for grazing

Plot B resting

Hen house with lighting and litter floor

Access open

Access closed

Access for poultry keeper

poultry is made up of a mixture of perennial ryegrass, smooth-stalked meadow grass, creeping bent, chewing fescue, crested dogstail and wild white clover.

The semi-intensive system

The semi-intensive system requires a permanent house, often a converted barn or farm building, with access to two plots of grazing area. The hens have access to each plot in turn, while the other is rested, and the number of birds you can keep depends on the amount of land available. In the USA the hen house is often located among the farm buildings and the hens are allowed to range over the farmyard and adjacent fields. This method is particularly suitable for smallholdings where land is usually limited.

The floor of the house is kept covered with straw and, in the event of really bad weather, the hens are not let out but are given their scratch feeds in the straw. Lighting in the house ensures that winter egg production does not decline. This is the system that I use, and in winter I do not let the hens out at all. As soon as the weather improves, I allow them out to graze in the new grazing area while the original plot is rested.

The deep-litter system

This is a more intensive system than the previous one because the hens are not let outside at all. They are housed in a single, large building with a straw- or litter-covered floor, and electric lighting. The litter build-up is removed once a year, and provides valuable compost for vegetables. It is particularly suitable for people with little land, although there are those who would condemn it on the grounds that it is unnatural. It is, however, more humane than the battery system where individual birds are confined in cages. The hens are free to walk around and peck in the litter, while the year-round protection from cold and rain, together with artificial lighting, does ensure a high level of egg production.

The straw-yard system

This is a variation on the deep-litter system, which consists of a straw-covered yard with housing in an adjacent

A deep-litter house. Note the litter-covered floor, nestboxes and perches over a droppings pit which only needs periodic cleaning out.

A small straw yard opening out from a hen house. Note the 'pop hole' with door for closing at night and the ridged supports to stop straw spilling out.

building. Woodshavings also make a good floor covering. The house is equipped with electric lighting and nestboxes, and in the event of really severe weather the hens are kept inside.

The yard is confined, usually by a stout post and wire mesh fence with a gate. If feeders and drinkers are placed outside as well as in the houses, they must be raised above the level of the straw to prevent spoilage. Outside hoppers also need to have weather shelters.

This method uses more straw than does an indoor deep-litter system because rain will dampen the litter. It is therefore not suitable if you live in an area of high rainfall. Roughly two bales of straw a week will be required for twenty-five hens if you want to avoid damp, muddy conditions.

This system comes close to recreating the natural habitat of the original domestic fowl which was adapted to scratch on leaf-littered ground, and is one I would like to see in greater use. The only problem I have discovered with keeping hens on litter is that when they scratch down to the damper, compacted layer underneath the dry straw, some of the litter sticks to their claws and leads to a build-up. I have known a hen to have a solid ball of material stuck to a claw, rather like a convict with a ball and chain, and the only way to remove it without hurting the bird is to soak it in hot water. Even then it may take an hour to remove it completely.

The battery system

This is the system which is used commercially and involves keeping hens indoors in individual cages within a large, controlled environment. Droppings fall straight through the cage floors, and food and water are available automatically on demand. Each time an egg is laid, it rolls onto a collection trough. Smaller versions are available for the small poultry keeper, and there is no doubt that this system produces a high number of eggs.

I, in common with many poultry keepers, dislike this way of keeping hens and would rather have fewer eggs from a more humane system which allows the birds to follow their natural tendencies to peck, scratch and to take dust baths.

The backyard or small scale system

Where space is limited, it is still possible to keep half a dozen hens in a small hen house and run. I have already mentioned the small fold unit where poultry graze on lawns or sometimes a gravel covered run. With a house and run the arrangement is more permanent and one of the disadvantages is that the ground may become over-used and muddy in the winter. You are also likely to get a build-up of parasites.

One of the methods which suburban and allotment poultry keepers in Britain used during the Second World War was to divide the garden into two areas, one for hens and one for vegetables. In winter, they moved the hens onto the vegetable patch and left them to scratch and clear up any insect pests and weeds. Meanwhile they planted spring cabbage on the chicken run. In summer the hens were moved back to their original quarters and the patch which had been scratched up and manured reverted to a vegetable garden. This rotation benefited both hens and vegetables, and gave each portion of land six months' rest from its former use.

Small house with run designed for 6–8 layers.

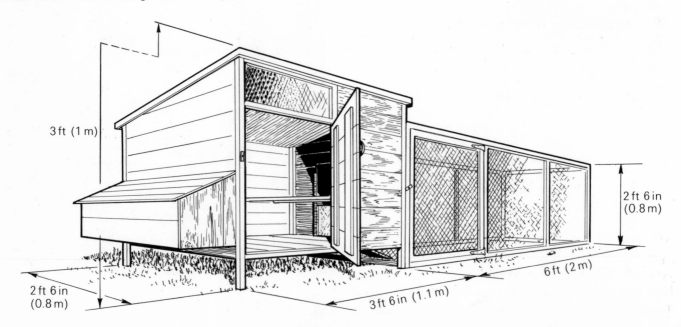

3 ft (1 m)

2 ft 6 in (0.8 m)

2 ft 6 in (0.8 m)

3 ft 6 in (1.1 m)

6 ft (2 m)

Housing and equipment

The hen house

As I have pointed out, the housing depends largely on the poultry-keeping system you are following. A house should provide adequate shelter from the weather, particularly from damp. If you are making the house yourself, you should make sure there is an overhang of at least 3 in (75 mm) on the roof, which should be covered with waterproof material. Corrugated iron tends to be too hot in summer and too cold in winter.

Wood is undoubtedly the best and warmest material for making poultry houses. Although I have seen plastic used in the construction of small houses, I am not convinced that it is healthy for the hens. One of the problems is to provide adequate ventilation, and plastic tends to encourage a build-up of condensation which inevitably leads to damp. Wood will absorb a certain amount of condensation and in that sense is said to 'breathe'. Tongue-and-grooved or match boarding is commonly used in house building, but in exposed areas a construction of overlapping weather boards is more durable.

Where winters are particularly hard some form of extra insulation may be necessary. If you have only a small number of birds they cannot generate enough heat in really low temperatures. The problem is to give them extra insulation without impairing the ventilation. One method which has been used in North America is to provide insulation outside by stacking up straw bales on three sides of the house – straw, like wood, 'breathes'. Ventilation is effectively provided in the ridge or roof of the building, which lessens the likelihood of draughts. However this practice has now been largely discontinued because it encourages rodents.

Perches

These are essential for the birds to roost on at night. As with all perching birds, a hen's foot is adapted in such a way that once her toes have gripped the perch, a muscular system rather like an automatic lock ensures that she does not fall off while she is asleep.

Perches should be 24–30 in (0.7 m) off the floor – no higher or the heavy breeds may suffer foot damage, particularly a condition known as bumblefoot. The perch should be 3 in (75 mm) wide, slightly rounded off and, ideally, removable to make it easier to clean them and check for red mite. Each bird should have at least 8 in (200 mm) of space on the perch, and it is a good idea to have a droppings board underneath.

Nestboxes

There should be one nestbox to every three hens. The most convenient are the ones which are attached to the side of the house, so that the hens can get in from the inside, while the egg collector can lift up the nestbox lid from outside.

A slatted floor in the poultry house allows most of the droppings to fall through.

An ideal perch, 3 in (76 mm) wide and slightly rounded to suit the bird's foot.

A convenient nestbox outside a small house makes egg collecting easy.

A group of hens taking a dust bath in woodshavings litter.

Nestboxes should be placed slightly above the floor level of the house but below the level of the perch, in order to discourage hens from sleeping in them. Straw or wood shavings make satisfactory nesting material, and this needs to be changed frequently to avoid the contamination of eggs by droppings.

Dust baths

It is part of the hen's instinctive behaviour to take a dust bath. This means she will find an area of fine, dry earth or dust and literally wallow in it, allowing the dust to trickle through her feathers and onto her skin. This removes many of the parasites, such as lice, which infest the skin and base of the feathers.

Free-range hens will find their own favourite places, but in confined runs a natural dust bath may be difficult for the hens to make. It is essential, therefore, to provide one for them. A shallow wooden box filled with fine silver sand is ideal, but it will need to be protected from the rain. Fine, dry soil is also suitable, but ashes are not recommended, for the alkaline conditions that they create are conducive to the spread of the mite which causes Scaley Leg, a condition in which the scales of the legs are pushed outwards by the burrowing action of the mites, and can lead to lameness. Insecticides can be put in the dust baths to help control lice and mites.

Feeding and drinking equipment

Hens need access to feed several times a day and to water at all times, so providing the right feeding and drinking equipment is important.

Home-made equipment is usually adequate if you have only a small number of hens but, for larger numbers, purpose-made feeders and drinkers are usually more efficient, less time-consuming and easier to clean. Food should be stored in dry, rat-proof containers and a convenient way of measuring out the grain and mash is to use a purpose-made scoop. A grain grinder is a useful piece of equipment for, set at its coarsest setting, it can be used for the preparation of grains, peas and beans. A mincer for using on raw vegetables is also useful, as well as a large pan with a lid for cooking up scraps and potatoes. An economic way of cooking such scraps is in a hay-box which will cook them slowly over a twenty-four hour period. The pan needs first to be brought to the boil on the stove before being transferred to the insulated box.

For chicks a useful feeder is the type which allows small heads to get in to peck, but prevents larger birds from feeding. This ensures that the mother hen will not eat the chick crumbs (which she does not need anyway).

For adult birds a suspended tube-feed hopper can be used in a building, but it will only take dry mash or pellets; wet mash would merely clog it up. Wet mash is best given in a plain, open trough which will give access to all the pecking beaks and not allow particles to get trapped anywhere.

Outdoors a weatherproof hopper is needed, and a large

A suspended tube feeder for indoor use.

Automatic watering.

A galvanized metal drinker suitable for adult birds.

A small jam jar drinker suitable for chicks.

one can be filled sufficiently to allow you to have the odd weekend away, without worrying about not being there to feed the hens every day. One drawback, however, is that wild birds and rats could also be feeding there at your expense. Grit hoppers are useful for holding limestone grit and oystershell if ranging is limited.

The most convenient system for providing water is an automatic one, situated indoors or outdoors. This is designed for large poultry units and may be beyond the reach of many smaller poultry keepers particularly in areas where there are water restrictions or where a special water rate may be levied.

Chick drinkers made of plastic are easy to keep clean, and the water level can be seen through the plastic, but they are more easily knocked over than metal ones. For adult birds a gravity-fed drinker is suitable for indoor or outdoor use if it is made of galvanized metal. It is best to suspend it a short distance above the ground or to place it on a stand; there will then be less likelihood of contamination by droppings.

Feeding

The hen's digestive system

The hen is a pecker, not a chewer, and is equipped with a hard, horny beak, but no teeth to grind up food particles. Food, which is taken into the gullet, passes down the oesophagus and into the crop which is basically a storage container. Here the food is moistened and held until it is ready to continue its travels. If you pick up a hen just after she has been eating wheat, you can feel the grains in her crop, in the breast area.

Between the crop and the gizzard is the glandual stomach, a short, swollen section of the food pipe, which secretes hydrochloric acid and food-digesting enzymes. Once in the gizzard, which is a strong, muscular bag, the food is crushed and ground into a cream consistency. It passes into the first part of the small intestine, where it is further acted upon by secretions from the pancreas and gall bladder. After this, the food is in a suitable condition to be absorbed through the walls of the small intestine, which is lined with many finger-like projections called villi, each of which is richly endowed with blood capillaries.

The blood transports the absorbed food to all parts of the body, wherever it is required, while the rest moves on to the large intestine. At the junction of the small and large intestine is a valve which prevents matter moving backwards, and also at this point are two caeca or cul-de-sac tubes whose function it is to digest fibre and absorb water. From the cloaca, waste matter is periodically ejected through the anus.

Nutritional requirements

Fresh, clean water is essential at all times, and there is no exception to this rule. Nutritional requirements are essentially the same throughout life, but the quantities and proportions differ, so that, for example, growing chicks need more protein in their first weeks of life than they do subsequently. The vital nutrients, their function and major sources are shown in the table on this page.

How much to feed

As a general guide, a layer will need $3\frac{1}{2}$–$5\frac{1}{2}$ oz (100–150 g) of balanced layer's mash a day, depending upon her size and level of egg production. As this is a completely balanced diet, the only other necessity is water.

If she is not being fed exclusively on proprietary mash, but is also receiving grain, an average daily intake would be $2\frac{1}{2}$–3 oz (70–85 g) of balancer pellets and 1–2 oz (30–60 g) of whole grain. Balancer pellets (available in the UK and USA) are intended to be given with grain, and the concentration of ingredients is specially formulated to balance the grain ration.

You can give each bird $1\frac{1}{2}$ oz (45 g) of kitchen scraps with the balancer pellets but this should be reduced to 1 oz (30 g) if you are feeding them on layer's mash. If hens are fed

Feed requirements

Nutrient	Function	Main sources
Proteins	Body building and repair	Fish meal, blood meal, soya, skim milk
Carbohydrates	Energy	All cereal grains
Fats & oils	Energy	Fish meal, meat and bone meal, groundnuts (peanuts)
Vitamin A	Normal growth, disease resistance	Grass meal, yellow maize meal
Vitamin B complex	Optimum growth rate and production	Fish meal, yeast, skim milk, cereals
Vitamin D	Healthy growth, prevention of rickets, strong egg shell	Sunlight, fish meal, cod liver oil
Vitamin K	Healthy blood	Grass meal
Calcium & phosphorus	Healthy bones and strong egg shells	Meat, fish meal, bonemeal, limestone flour
Zinc	Healthy skin and feather development	Available in supplements
Manganese	Strong egg shells, good hatching rate	Available in supplements
Iodine	Control of metabolism	Seaweed extract, available as a supplement

entirely on kitchen scraps and grain, the level of egg production will fall dramatically, and it is up to the individual poultry keeper to decide on his priorities.

What to feed

Proprietary feeds

These days ready-mixed feeds provide all the essential nutrients in their correct balance, and are available to suit fowl at all stages of their lives. They are available in several forms – as powder, crumbs or pellets – and may be fed dry or wet.

Proprietary feeds have been developed for the intensive battery-egg and broiler meat industries, as easy-to-feed, all-in-one convenience foods, and there is little doubt that hens will lay far more when fed exclusively on layer's mash. What is worrying to many people is that antibiotics are added to many of these feeds. While these may be necessary to counteract infection in the closely confined environment of the large, intensive houses, they are not needed by the small poultry keeper whose hens are kept along more natural lines.

Green feed suspended in a run provides interest as well as nutritional value.

Table of wild and home-grown feeds

Potatoes	Boiled and added to mash – do not use green ones
Jerusalem Artichokes	Cooked or minced raw
Grass mowings	Given dried or green
Brassicas	Given green
Turnips	Tops fed green; roots cooked or minced raw
Carrots	Cooked or minced raw
Sunflower seeds	Dried and cracked
Buckwheat	Fed as grain
Groundsel	Fed green
Fat Hen	Fed green
Acorns	Dried and crushed
Beech masts	Dried and crushed
Nettle tops (young)	Boiled with potatoes or scraps
Parsley	Fed green
Beans (all types)	Dried and ground up
Lettuce	'Gone-to-seed' plants given whole
Linseed	Small quantities only as it is a laxative
Peanuts	Dried and cracked
Fruit	Small quantities given fresh

These feeds are also expensive, and careful costing is needed to ensure that home-produced eggs are not working out more expensive than those you can buy at the supermarket.

Green feed

Green feeds are useful, not only as a source of supplementary nutrients, but also to provide interest for the hens. In a small run, the birds are particularly prone to boredom and this can trigger off abnormal behaviour patterns such as feather and vent pecking or egg eating.

However, do remember that, while greens are a good appetizer, they are not high in energy and should not be substituted for large amounts of the diet. Also, cooking feed is clearly time-consuming and is something to which you should only commit yourself if you are able to go on providing it regularly, every day and at the same time. If you can't, don't start it.

Large vegetables such as cabbage can be suspended in the run. Other green feeds can either be chopped up and added to the mash or given whole for the hens to peck at will. Weeds or grass mowings are also acceptable in small quantities, provided no weedkillers have been used on the lawn. It is important not to leave surplus greens to heat up and decay, but to clear them away promptly.

An old-fashioned practice was to dry grass mowings by spreading them out thinly immediately after they had been cut, and to use them as a winter feed supplement. This kept the yolk colour of the egg a deep yellow. However some people, particularly in North America, object to this dark colour and to the fact that it varies from season to season.

Onion tops, chives and finely chopped garlic are useful disease and worm preventatives, although some of them may impart flavours to the eggs.

Sprouted grains

Sprouted grains provide a good source of winter greens for egg producers. Any grains may be used, but the usual ones are wheat or oats. Traditionally it was always oats, and in particular the black oats.

Allow 1 lb (450 g) of grain for every thirty hens and leave it to soak in lukewarm water for twenty-four hours. Drain off the water, then continue sprinkling morning and evening. As soon as the grains begin to sprout, spread them out on a clean surface to the thickness of an inch (25 mm), and continue sprinkling until the sprouts are about an inch long. At this stage stop sprinkling and allow them to dry off (but watch out for signs of mould!). They can either be chopped up and added to the mash or given with dry grain as a scratch feed.

A hand grinder is invaluable in the preparation of small quantities of home-mixed feed.

Calcium and grit

Any calcium deficiency will lead to poor shell formation. Proprietary foods contain appropriate amounts of both calcium and phosphorous, but hens fed on a more traditional diet may need to have them provided as a supplement.

Crushed oystershell is available commercially, as is limestone grit, and a container of each of these in the run will enable the hens to help themselves when they need it. Another source of calcium is recycled egg shells, but these must be baked to sterilize them and then finely ground up. If any of the pieces of shell are still recognizable, this may trigger off an egg-pecking habit.

Grit in the form of small pieces of flint or gravel is needed to aid digestion in the gizzard. Free-range hens normally find enough for themselves, but birds in confined runs may need it specially provided for them. Oystershell, as well as providing calcium and phosphorous, is a good source of grit.

Feeding practices

In the following notes on feed rations I have included modern and traditional feeding practices. The traditional methods are included not because I recommend them, but because many people are interested in them. They are undoubtedly more time-consuming than modern practices, especially if mixtures need to be made or freshly prepared before each feed. Besides this the casualty rate among chicks will be lower if you use chick crumbs. It is a matter of individual choice whether you use one or another, or a combination of both practices. Only your own situation can be the judge.

Day-old to six weeks

Chick crumbs are a complete food, and will provide everything a growing chick needs from day-old to six or eight weeks, except water. They are about 20 per cent protein, and give chicks a good start in life, building them up during the crucial first weeks. I use them for my chicks, and revert to a more natural diet after eight weeks. The crumbs are fed ad-lib so that the chicks can help themselves when they want to. I also grind up wheat and give them a few of the coarse particles as a scratch feed after one week, as well as finely chopped chives. A traditional practice was to give a special diet for the first week consisting of finely chopped hard-boiled egg and breadcrumbs mixed with milk. The reasoning was that the newly hatched chick retains remnants of the yolk in its abdomen for the first two days after hatching and this was the nearest thing to the yolk of the egg that they could provide. The chicks were fed five times a day which must have been quite a chore. For the next six weeks they were given three meals a day of kibbled grains and seeds, made up as follows:

Kibbled wheat	50%
Groats (husked oats)	25%
Canary seed	10%
Kibbled maize	5%
Millet seed	5%
Hemp seed	5%

In addition, the chicks were given one meal a day of boiled rice and wheat in equal proportions.

After six weeks

Once a chick is six to eight weeks, it needs less protein in relation to other foods, and a ration with approximately 15 per cent is adequate. Proprietary grower's meal is available commercially, and may be fed until the pullet reaches point-of-lay or sexual maturity. A point worth making is that any change, particularly with food, should be gradual rather than abrupt. Start by giving less of the previous ration, and make it up with the new, gradually changing the proportions, until the change-over is complete. A popular and traditional grower's ration was a mixture of:

1 part kibbled maize
3 parts kibbled wheat
1 part whole oats.

This was fed morning and evening, along with grit and plenty of greens, including chopped cabbage, lettuce and onion tops or chives.

At mid-day the hens were given a wet mash, made up as follows:

4 parts wheat meal
4 parts ground oats
1 part maize meal
¼ part meat meal.

My own system of feeding growers is to give them a mixture of available grain, usually wheat and oats, supplemented by kitchen scraps to provide their protein requirements. In addition, they have plenty of chopped greens and access to grazing. I feed them morning and evening.

Twenty weeks onwards

From the age of twenty weeks, the pullet is at point-of-lay and is ready for layer's food which will help her to be as prolific as possible. About 15 per cent protein is required, but it is important not to give too much fat, or carbohydrate which may be converted to surplus fat and affect laying performance. The heavy breeds are more prone to this difficulty and sometimes even have heart attacks as a result of overfeeding.

Proprietary layer's food in the form of dry or wet mash or pellets provides all the hen's nutritional requirements, but I prefer to give a less expensive and intensive ration, and to put up with fewer eggs accordingly. In the morning I give proprietary mash mixed with water to a crumb consistency, to which I add vegetable and meat scraps. If I have a surplus of goat's milk, I use a little of that to mix up the mash instead of water.

In the evening, I give them wheat and they have access to natural grazing during the day. In addition they are given green foods, usually in the form of whole plants such as suspended cabbages. When available, home-grown daffa beans and sunflowers can be kibbled to make a scratch feed.

Some traditional layer's recipes are:

1 part bran
1 part wheat meal
1 part ground oats
1 part maize meal
½ part fish or meat meal.

This can be mixed to a crumb consistency with water and fed at mid-day. In addition a scratch grain mixture can be given morning and evening, made up as follows:

2 parts whole oats
2 parts whole wheat
1 part kibbled maize.

Fattening cockerels

It goes without saying that if you raise your own chicks, half of them will be cockerels, and the best way of using them is to fatten them up for the deep freeze. They undoubtedly fatten up more effectively if they are caponized or made sterile, but for the small poultry keeper who is fattening them for his own use, this is unnecessary.

The traditional way of caponizing involved a certain degree of cruelty and is not to be recommended. It meant making an incision in the side, between the last two ribs nearest the tail and removing the testicles. This operation carried with it the risk of damaging a main artery which, if punctured, led to death in seconds.

The modern way of caponizing is perhaps less painful for the bird but is hardly more responsible with regard to the possible effects it can have on the people who subsequently eat the flesh. The female hormone oestrogen is inserted in pellet form under the skin of the neck, or in some cases is added to the food in another form. The unsuspecting public is then invited to buy the neatly packaged meat at the supermarket. I am glad to be able to raise my own! (In Australia, chemical caponizing is illegal and implanted pellets are likewise no longer approved in the USA.)

When my cockerels are twelve to fourteen weeks old, they are separated from the pullets and confined in a small pen. If they are left to run loose they will not fatten effectively. I feed them three times a day with a mixture of oats, barley and boiled potatoes in equal proportions, mixed with skim milk. 'Chat' potatoes, or the small ones left behind when the big ones are dug up, are ideal for this. There is no need to peel them; just scrub them clean and boil them whole. If you don't grow your own main crop it is often possible to buy these very cheaply as 'stock feed potatoes' from farmers. Including corn in the fattening diet is also a good idea and gives the carcass more yellow colour. The cockerels are given as much as they will take at each meal, and have access to fresh water at all times. Three weeks later they are killed for the freezer.

If you wish to concentrate on meat production rather than on eggs, it is best to buy broiler hybrids which have been selectively bred for this purpose.

Chickens for the table

Culling

Commercially, layers are culled at the end of the first laying season. In the second season egg production declines, the hens eat more and each egg is costing more to produce. Another reason for culling at the end of the first season is that, by starting each year with new stock, you avoid the possibilities of disease being passed on from the older to the younger hens. The hens are also more liable to have laying problems in that second season. There are obvious exceptions and all poultry keepers know of the occasional hen which continues to lay reliably for several seasons. Hens selected for breeding purposes will also of course be kept, as breeding does not usually take place in the first season. If a hen has proved herself to be a good breeder or has a particular characteristic which is valuable, such as reliable or early broodiness, then it is obviously worthwhile to keep her for several seasons.

1. *Killing a chicken by breaking its neck.*

2. *Do not worry if the bird flutters for a few minutes after death.*

The individual poultry keeper is the only one who knows his chickens, and he is therefore the only one who can decide which hens to keep and which to cull. The one characteristic which is no use to anyone is sentimentality. A hen which is a poor layer or which does not have any of the qualities mentioned above should be humanely killed, plucked and eaten, unless you happen to be a vegetarian – in which case, someone else will be pleased to eat it for you. Only sound, healthy birds should be eaten.

Killing and plucking

It is not difficult to kill a chicken, but if you are in any doubt, *don't do it* – get someone experienced to do it for you, or show you how it is done. The traditional way is to grasp the legs in the left hand, hold the hen upside down and, holding the neck at the base of the head, push down and twist sharply to the right. As soon as the neck is broken, stop, otherwise you will pull the head off. The bird will flutter for a few seconds after death, but this is merely a nervous reaction of the muscles. No feed should have been given for twenty-four hours.

Pluck immediately, for as the skin cools it becomes more difficult to get the feathers out. Try to avoid tearing the skin. Scalding it with hot water often can make plucking easier.

Drawing

Cut off the head, using a sharp knife or poultry secateurs for the backbone. Remove the neckbone and insert your hand in the space. With your fingers loosen the connective tissue which is holding the innards. Make a cut between the vent and the tail, taking care not to cut the rectum. Now make a circular cut right round the vent and carefully pull it out with all the attached innards. The gizzard, lungs and heart will follow on, but the crop may be easier to remove from the top end. (See photographs on page 26.)

Eggs

Most people keep chickens for eggs, and there is certainly nothing to beat the taste of one of your own free-range eggs, despite what cholesterol-conscious authorities may say. The modern hybrid can lay up to three hundred eggs a year if adequately housed and fed, and is a far better choice than the traditional pure breeds or first crosses if egg production is the priority.

From the age of twenty weeks onwards, a pullet will begin to lay eggs. The first one will probably be rather small, but her triumphant cacklings when she emerges from the nestbox will not be subdued by that. Her pride is not unjustified as an egg is indeed a small miracle. It contains protein, fat, calcium, magnesium, potassium, phosphorus, sodium, sulphur and iron, as well as vitamins A, B, D and E.

Egg production is controlled by hormones and begins

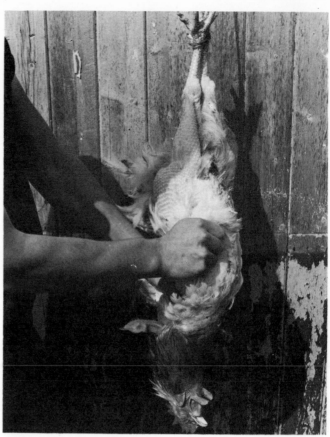

Plucking: 1. *Suspend the carcass and start plucking while it is still warm. Remove the primary wing feathers first.*

2. *Pluck the leg feathers next, taking care not to tear the flesh.*

3. *Now pluck the body and down feathers.*

4. *A nice, cleanly plucked bird.*

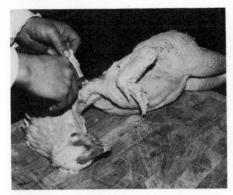

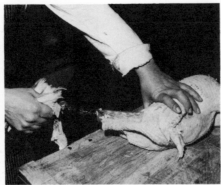

Drawing: 1. *Cut the skin along the back of the neck and up towards the head.*

2. *Cut through the neck bone behind the head and pull the head with the blood vessels away.*

3. *Remove the neck bone, then loosen the innards by moving your hand around inside the carcass.*

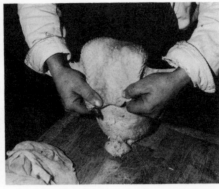

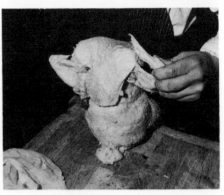

4. *Remove gullet and attached tissues.*

5. *Fold the skin over the hole.*

6. *Twist the wings over the loose skin.*

7. *Cut around the vent, taking care not to pierce the rectum.*

8. *Draw out the intestines, followed by gizzard, liver, lungs, heart and crop.*

9. *Make a cut above the feet, then break the leg bones, but do not sever the tendons.*

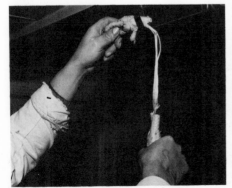

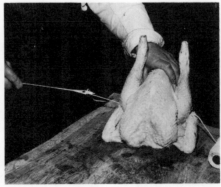

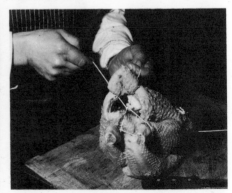

10. *Pull out the leg tendons by suspending the feet on a hook and pulling the legs downwards.*

11. *Pass a needle and thread through the body and wings, then tie the ends together along the back.*

12. *Wind the thread around the legs, cross it behind the hocks, then tie it behind the tail, pulling it tight.*

Checking whether a hen is in lay by the two-finger test.

The vent of a layer is moist and whitish and expands gradually into a conical shape as laying proceeds.

in the ovaries where a number of yolks, each with a small unfertilized egg on one side, begin to develop. As each egg is released from the ovary, it may or may not be fertilized by sperm from the cockerel as the result of an earlier mating. The egg enters the funnel-shaped top of the oviduct, where it is coated with a layer of albumen or egg white. Further down the tube it receives a coating of two membranes, and finally a shell is secreted around it by a gland, before it finally emerges through the cloaca and vent. The whole process takes approximately twenty-four to twenty-six hours.

The onset of lay

Generally speaking, pullets that grow and mature fast tend to make the best layers. One of the more obvious signs of the onset of lay is the bright red, full comb. Another indication is the 'filling out', or the 'body capacity' as it is sometimes called. This is the general shape and size of the hen's body, the width between the pelvic bones and the depth between them and the back of the breastbone.

The 'finger' test is a way of establishing whether a pullet is in lay by measuring the width between the pelvic bones. These can be felt one on each side of the vent. If you can fit two finger-widths comfortably in between them, the chances are that the hen is laying. Later on, when she is really heavily into lay, the width will be sufficient to take three fingers. Another way to establish lay is to examine the vent. That of a non-layer will be small, round and yellowish. That of a pullet approaching lay will be expanding and turning white and after several months' laying, the vent will be conical in shape. The best indication of a hen which is laying today is a moist, enlarged vent.

Before laying her first egg, a pullet will often manifest a restless pattern of behaviour, clucking in a complaining way and going in and out of the hen house. It is important

Young Light Sussex pullets approaching 'point of lay'.

that the nestboxes are ready, in the darkest corner of the house, and with a liberal quantity of nesting material such as straw. Any disturbance or sudden shock should be avoided. It is not unknown for shock to cause egg laying problems such as the yolk failing to drop into the oviduct, and entering the abdominal cavity instead, causing peritonitis.

Occasionally the pullet will lay on the floor of the house, and any straw lying there should be picked up to discourage the habit. The traditional way of encouraging the hens to lay in the nestboxes was to place china or pottery eggs there before the onset of lay.

The first egg will probably be small, and may even be a 'wind' egg, which is merely a portion of white covered by a shell. This is nothing to worry about, provided that the hen does not continue to lay them.

Getting reluctant pullets into lay

There are always some pullets which, while demonstrating all the characteristics of being in lay, have still not produced their first eggs. In this situation, I feed them exclusively on layer's mash, providing enough in the hopper for them to be able to eat at will. Water is of course freely available, but nothing else is given until the first egg has arrived. I find that a few days of this treatment is usually enough to bring them into lay and, once started, they carry on. At this stage, I revert to a less intensive diet, giving them mash or pellets in the mornings and grain in the afternoons, with access to grazing and other green foods. If, despite this treatment, a reluctant pullet has still not come into lay, I would recommend that she go into the pot or the freezer.

It makes a difference, too, whether the mash or pellets are given in the morning or afternoon. I once had a letter from a lady who said that her new pullets were given the best of everything in housing, feeding and care, but they would not lay. She was giving them large quantities of grain in the morning and layer's mash in the afternoon. I wrote back, suggesting that she reversed this pattern and gave mash in the morning and grain in the afternoon. Shortly afterwards, I had a letter to say that this had done the trick.

Hens love grain and will gorge themselves on it, often at the expense of other foods, but if they have too much starch it will not only make them fat, but will also be a cause of reduced egg production.

Eggs in winter

Egg production is intimately linked with daylight hours. The light rays received through the eyes affect the pituitary gland at the base of the brain, releasing hormones into the bloodstream and stimulating the ovaries into action. As the daylight hours shorten, egg production correspondingly decreases, until by mid-winter it is usually non-existent.

To ensure continued production hens must have a minimum of sixteen light hours a day. As the hours of natural daylight decrease, artificial lighting can be gradually introduced for longer and longer periods to make up the difference. Thus, for example, by the time you reach a winter's day lasting only nine hours, you will be providing artificial lighting for an extra seven.

In a poultry house with windows evening lighting, early morning lighting, or a combination of both may be used to prolong natural daylight. When evening lighting is used, it is important to instal a dimming device so that the light goes out gradually and the birds are not suddenly plunged into darkness before they have found their way to the perches. When early morning lighting is used, it is as well to incorporate a time switch, otherwise you will find yourself having to get up in the early hours of a freezing winter's morning to switch on. My hens over-winter in a stable on deep litter, and neon strip lighting is automatically switched on by a time switch to extend the morning light. Natural light comes in through the top of the stable door and through a side window.

For a small number of hens, a 25 watt bulb is adequate. Paraffin or bottled gas lamps may also be used, as long as they are securely installed, out of reach of hens which may knock them over and cause a fire hazard. Any electrical work should be carried out by an experienced electrician.

An important point to remember is that winter layers need a slightly increased ration because they need extra warmth and energy. I find that the easiest way to provide this is to give them a few extra layer's pellets on top of their normal rations. The best birds to use as winter layers are spring-hatched chicks which come into lay in autumn. These are far more reliable than older hens.

Moulting

All adult birds will moult once a year, usually towards the end of the summer or in early autumn. The onset of moulting (or the loss and replacement of feathers) inevitably means no eggs, for the hen's system needs to devote much of its energies to moulting rather than to laying eggs.

Spring-hatched chicks which come into lay in the autumn will continue to lay through the winter and spring if given artificial light, but will then moult when the warm days begin.

Force moulting is a practice which is sometimes used to ensure that hens take a vacation from the egg production business and regrow their new coat of feathers before the late summer, so that egg production can continue by early autumn. The bird is stimulated to moult by reducing her period of light, putting her in a warm place, withdrawing her normal rations for two to three weeks, and feeding her only on grain. My own preference is to let hens moult at whatever time is natural for them.

Egg eating

Most poultry keepers have come across some hens with this irritating tendency to peck and eat eggs. Often it can be remedied by giving extra calcium in the form of oystershell grit or baked, ground egg shells, for it can be a calcium deficiency which triggers off the habit.

It may also be lack of water or shortage of food generally. Sometimes there is no apparent reason, except possibly boredom, and I have on several occasions found that a small pile of gravel, or even some upturned grass turves in the run, will provide sufficient interest to break the habit.

There are always the hardened culprits of course, and for these there is a special treatment. Take an egg and carefully cut a hole on one side. The easiest way to do this is to crack it gently against a hard surface, then slip a scalpel into the crack to widen it. Remove the contents and replace them with freshly made mustard. Seal the hole with a piece of tape and put the egg, tape-side down, in the nest where eggs are normally eaten. I have never known this to fail, but other poultry keepers claim to have failures even with this method.

Debeaking, where the top of the beak is removed, is an effective remedy, but I dislike this practice because it also deprives the hen of her natural ability to peck, and I would rather dispose of her quickly and humanely into the freezer.

Abnormal eggs

Soft-shelled eggs
These are a frequent problem, and the trouble is often a calcium deficiency for which the remedy is oystershell grit or baked, ground egg shells. Sudden shocks may also cause the shell-secreting glands to malfunction, so if you live near a rifle range, your hens may be shell-shocked on two counts.

Eggs with double yolks
These are fairly common and tend to occur more in second-season hens. It is believed that there may also be a hereditary tendency towards the simultaneous release of two ova which combine to form one shelled egg.

Eggs with blood spots
These are also thought to be the result of a hereditary tendency for blood to escape from the ovarian follicle, and become caught up in the albumen. Some commercial egg producers candle eggs (see page 34) and discard spotted ones, but they are quite harmless.

Fertile eggs
These, of course, are not abnormal, but there is a feeling that eggs for eating should be infertile. The fertile ones are perfectly alright to eat, but as there is no reason at all to keep a cock running with the layers they can easily be avoided. It is a complete fallacy that hens will lay more eggs if there is a cock with them, or that this will make them happier. The only reason for keeping a cock is for breeding purposes.

Wind eggs
I have already referred to these small eggs which contain no yolk, and which are quite common in the young pullet whose egg-laying system is just beginning to function. They may also be the result of foreign matter passing down the oviduct and stimulating the shell-secreting glands into action. They also commonly occur as a hen is ceasing production.

Other egg problems

Peritonitis
I have already referred to this condition which occurs when eggs are released into the abdominal cavity as a result of shocks. It may also occur when the production of abnormally large eggs causes a blockage of the oviduct. A small percentage of hens fail to lay eggs as a result of a tipped or malformed oviduct.

If no birds are kept for a second season of lay, the likelihood of disorder in the egg-laying mechanism is increased, and this factor should be taken into consideration when deciding which stock to cull.

Prolapse of the oviduct
This is another possible sequel of straining to lay over-size eggs. The inner walls of the cloaca are pushed out as a red mass of tissue through the vent. An affected hen is often attacked and pecked by other birds, and it is usually better to dispose of her. If the condition is discovered in time, it is sometimes possible to push the tissue gently back, after applying petroleum jelly, and then put the hen in a quiet place on her own to recover. The condition does, however, have a tendency to recur. Hens become more prone to prolapse when they are too fat.

An egg bound condition
If you notice a hen enter and leave a nest several times without producing an egg it may well be that she is egg bound. A teaspoonful of pharmaceutical grade liquid paraffin fed through the beak will deal with constipation which may be causing a blockage by pressing the bowel against the oviduct. Petroleum jelly applied around the vent may help to lubricate the passage of the egg outwards. Where this fails and the hen is in obvious distress, hold her over hot water so that steam bathes and relaxes the vent, and at the same time bathe the vent with water.

Unwanted broodiness

It occasionally happens that a hen will go broody at a time when peak egg production is crucial or a pullet may manifest 'false' broodiness as a result of being in lay before her bodily development is complete. To break this broodiness pattern the bird should be removed from the nest and placed in a slatted coop. The air circulating through the gaps will keep the temperature sufficiently cool to deter her from sitting. Extra rations should be given as well as a dusting of lice and mite powder for she will not be able to get to a dust bath. A few days of this treatment is usually enough to make her change her mind and resume laying.

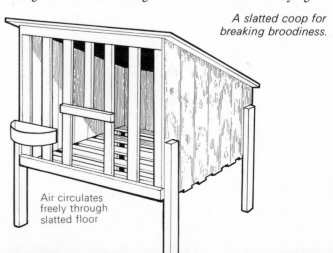

A slatted coop for breaking broodiness.

Air circulates freely through slatted floor

Storing eggs

Freezing

If you have a large surplus of eggs, they will keep for up to six months in the deep freeze, provided you have shelled them. An easy way of storing individual eggs in this way is to break each egg into a separate compartment of an ice-cube tray.

Storing in solution

Another method is to store them in a solution of waterglass or sodium silicate. Dissolve one part of the soluble water-glass in nine parts of water and stir thoroughly. Put the solution into a plastic, enamel or earthenware container with a tightly fitting lid. Choose only clean, fresh and uncracked eggs and place them in the solution, pointed end downwards. They will keep for about nine months like this.

Painting and dipping

It is possible these days to buy proprietary brands of plastic-based solution which can be applied with a paint brush, or by dipping the eggs into the solution, and will preserve them.

Pickling

Eggs can also be pickled by first hard-boiling them and then immersing them in pickling vinegar which has previously been brought to the boil and cooled. 1 oz (30g) of mixed spice added to the vinegar while it is being heated imparts a delicious flavour to the eggs. They will keep for up to two months by this method.

Breeding

If you intend to breed a few replacement pullets for the next season, it is better to stick either to pure breeds or to first crosses, so that you can be reasonably sure of what you will be getting. If you choose first crosses many of them will be sex-linked, enabling you to identify the cockerels at day-old.

It is not a good idea for the amateur to breed from first crosses or to attempt hybridization. This is an extremely skilled and time-consuming business which, if done properly, requires careful recording.

Breeding of course carries its responsibilities, and only the very best stock should be allowed to breed. The perpetuation of hereditary weaknesses is to be condemned, as is the practice of just letting them get on with it on the grounds that this is more natural.

Selecting breeding stock

The characteristics to be borne in mind in the selection of utility breeding stock are as follows:

1. Good laying ability: Ideally the number of eggs laid in a certain period should be recorded before you make this selection. Light breeds such as the White Leghorn tend to produce more but the strain is more important than the breed, and some Leghorns may be poor layers.

2. Egg size and quality: Obviously a hen which lays misshapen eggs should be avoided, and the size and texture of the eggs should also be of acceptable quality. If, for example, you are breeding Barnevelders which lay dark brown, speckled eggs, the colour and pattern will be important.

3. The cock's mother: It is important that she should have come from a good laying strain.

4. The age at which sexual maturity is reached: While the average age is twenty weeks for a pullet, it is by no means certain that egg-laying will commence at this time. Late developers are best avoided – this also applies to the cocks.

5. Fertility: If you are buying in a cock from a breeder, ask for records of his parentage and performance record.

6. Broodiness, or the tendency towards sitting and hatching eggs: If this tendency is required (and it may not be if you are just concerned with numbers of eggs) it is best to use a heavy breed. Bantams and Silkies often have this tendency, and one of the best broody crosses is a Silkie cock crossed with a Light Sussex bantam. Small fowl however are only able to cover a few eggs.

7. Weight: If your aim is to produce table birds, then heavy breeds should be your choice. The Indian Game cock is often crossed with a heavy breed, because of its broad breast and tendency to put on flesh.

8. Resistance to disease: It goes almost without saying that only the really healthy and vigorous birds should be allowed to breed.

Points of the cock.

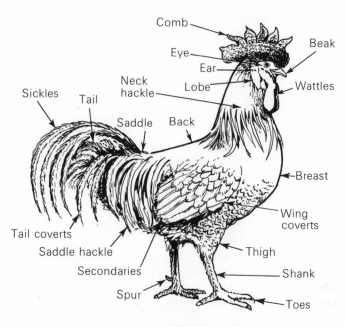

Spur trimming: 1. *Cutting the ends of the spurs.*

2. *After trimming.*

The stud cock

A heavy breed cock should have no more than eight hens at a time, and should be at least ten months old before starting work. A light breed cockerel can have up to ten hens. In appearance he will have bright red wattles and comb, a sleek cape of neck hackles, a glossy coat and a strutting, confident walk. He should be unrelated to the hens and, ideally, should be removed for a rest for a day or two once a week. At this time it is not a bad idea to give him extra rations in the form of chopped-up liver to keep his strength up.

It is always difficult to know how long to keep a cock for breeding, as so much depends on his performance, and you can only gauge this by the abilities of his offspring. If he is a particularly good cock then he may go on for five or six years, but three to four is usually the limit.

Spur-trimming

The cock's spurs should be kept trimmed $\frac{1}{2}$ in (13 mm) in length otherwise he may damage the hens by ripping their sides during mating. There is no truth in the old myth that spur-trimming causes infertility. It is easily carried out by wrapping him up tightly in an old towel, and either trimming the spur with wire-cutters or singeing them. The latter method involves holding the end of the spur in a candle flame, keeping your fingers and thumb on the side of the spur nearest to the leg. It will not hurt him, for his spurs are like our nails. After a minute or two, grasp the blackened end with a damp cloth and twist, when it will come away. File smooth. Some breeders have used soldering irons with great success.

The breeding hens

It is not a good idea to breed from pullets, and hens should normally be in their second year of lay before being considered for breeding, although commercial enterprises breed in the first season.

A trap-nest (see next page). **1.** *As the hen enters she pushes the wires inwards, releasing the door (which has a weighted base) behind her.*

2. *Trap-nest closed. Note the gap for ventilation.*

Fitting an identifying leg ring.

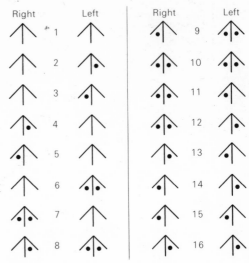

A chart showing possible toe-markings.

Ideally, a record of their egg-laying capacity should have been kept during their first season. This is not as easy as it sounds. Although with a small number of birds it is sometimes possible to identify eggs as having come from a particular hen, the only safe way is to use trap-nests with doors which close automatically on the hen as soon as she is inside. If each hen has a leg-ring with her own number on it, the number and quality of the eggs can be recorded. Needless to say, they need to be frequently inspected, and this can be time-consuming and laborious. There should be at least one trap-nest for every two hens.

Selective breeding

Egg recording table

Leg-ring number	Eggs laid in a year	Result
01	130	No breeding because of poor egg record
02	110	
03	150	
04	142	
05	170	No breeding because of poor shell quality
06	260	CHOSEN FOR BREEDING
07	190	No breeding because of tendency to have colds
08	280	CHOSEN FOR BREEDING

Let us assume that eight pullets in their first egg-laying season had their eggs recorded as shown in the table above. It will be seen that 01, 02, 03 and 04 have been eliminated as possible breeders because they did not produce enough eggs. If a further selection is made, based on the criteria we have already mentioned such as shell quality and disease-resistance, we may find that only two remain, as in our example where 06 and 08 are left. When eggs from these two are incubated it is obviously important to keep them separately identified, and the shells are marked with the hens' numbers.

On hatching the chicks are normally marked on the web of the foot by means of a toe-punch, but you must take care to avoid the bones. The position of the marking identifies the chick, and you will be able to see from the above diagram that there are sixteen possibilities. Indelible ink can be used for additional marks, such as a spot on the head of all the chicks from 08, which has the best egg-laying record. Wing bands are another means of identifying day-old chicks. At eight weeks, you can fit them with leg-rings, but before then it is undesirable as their legs may be damaged.

If several different family lines are established in this way, with a different cock for each line, it will be possible to produce the best hen from one line and mate her with the best unrelated cock from a separate line.

Inbreeding is not generally recommended because of the inherent danger of congenital weakness but it may occasionally be done. For example, a hen with outstanding qualities may be mated with one of her carefully selected sons. This should only be done once, and you should make sure to mate the female progeny of these parents with an unrelated male.

Incubation and rearing

Once the hen has been mated several times, the chances are that she will lay fertile eggs. The sperm from the cock fuses with the egg cell on the surface of the yolk. The fertilized egg and yolk then move down the oviduct whose walls secrete albumen (white) to surround them. The shell is added just before laying, at the lower end of the oviduct.

The temperature in the hen's body is approximately 100°F (38°C), and if division and development of the fertilized egg is to continue, this temperature must be maintained. You can either rely on natural incubation where a broody hen sits on the eggs until they hatch and then rears them, or you can use an artificial incubator.

The broody hen

Broodiness is a bird's natural, inherited instinct to reproduce the species. The ancestor of the domestic fowl laid two clutches of eggs, one in spring and the other in late summer, making a total of about thirty eggs, but commercial hybridization has bred out the tendency to broodiness, so we must look to the older, heavy breeds and their crosses, or to the bantams.

A broody hen is easily recognized because she will have decided to stay on the nest, and on your approach will fluff herself up to nearly twice her size, giving loud, complaining and guttural squawks. If removed, she will walk around in circles, still fluffed up, and clucking loudly, then will make her way back to the eggs.

Housing

It is important to move her to a place of her own, away from the other hens, otherwise some of them will sit with her to lay their eggs, leading to considerable confusion. A dry, draught-free, rat-proof shed is suitable, but best of all is a broody coop which is cosy and warm and can be easily made. Sawdust or woodshavings provide a good nesting material, as does good quality hay or straw, but if this is of poor quality it may harbour mites or fungal spores. Try moving her to her new quarters to see if she will accept the sitting place. If she has only just gone broody, she may refuse and make agitated attempts to get out. There is a danger of her breaking the eggs and her broodiness pattern if this happens, so it is best to remove her and put her back in the original place for another day or two until broodiness has become well established. She will sometimes adjust better to her new surroundings if you move her at night rather than in the daytime. Give her a dusting against lice and mite attack, for she is particularly vulnerable at this time, and it is a good idea to put a little on the nesting material too.

Food and water should be provided nearby, as well as a dust bath. She will usually leave her eggs once a day for food, water and exercise, and a quick examination of the eggs at this time will give you the opportunity to remove any droppings from the nest. If she does not leave the nest of her own accord, she should be lifted off once a day. There is no need to turn the eggs, for she will do this several times a day with her beak. During the last three days before hatching, I usually sprinkle the eggs with warm water to ensure against excessive drying out of the membranes around the embryo.

Hatching

Twenty-one days is the average incubation period for a chick, and any time from the twentieth day onwards it will

A small incubator in use.

begin to 'pip' or peck its way out. The tip of the beak has a specially adapted hard section for this purpose. It is not a good idea to help chicks out of the shell, for if there are any weakly ones the chances are that they would not survive long anyway. It does occasionally happen however that dehydration of the membrane makes it stick to the chick's down, so that although the chick has made a hole in the shell, and may be partly out, it cannot get any further. In this situation, a little warm water may be used to bathe it, before you replace it in the nest to finish emerging from the shell. Take care, however, that water does not trickle into the chick's nostrils and drown it.

Artificial incubation

Broody hens are not always available when they are needed, particularly early in the year, and artificial incubation as a means to overcome this is not a new idea. It was used over 2000 years ago by the Chinese who hatched eggs in large, clay-brick ovens, heated by burning wood. A modern artificial incubator can be a useful investment, and if bought from a reputable manufacturer, will give many years of service. It is also possible to make your own. Whether you buy it or make it yourself the incubator should be sited in a place where it is in a cool, even temperature, out of the sun and away from the possibility of knocks and sudden jarring – an outhouse is ideal.

Temperature, humidity and the position and movement of eggs are crucial factors which influence hatching quality, so it is worth going into these in detail. The optimum temperature for hatching is 100°F (37.7°C) at the centre of the egg. With still-air incubators the temperature at the surface of the eggs should be 103°F (39.4°C) taken with a thermometer suspended 2 in (50 mm) above the eggs. In still-air machines there is a tendency for the temperature to layer and a reading taken too far above the eggs could be quite different from one taken nearer to them. In forced-draught incubators the temperature should be 99.5°–100°F (37.5°–37.8°C). Embryos will withstand temporary cooling, but are very sensitive to increases of temperature above the optimum. The critical temperature above which germs start to develop is about 55°F (12.8°C).

A hygrometer will give the humidity reading and many incubators have these built in, but it is comparatively easy to take a reading using two thermometers. One thermometer has wet, absorbent material wrapped around the bulb, and the other is left as normal. The difference in the reading between the lower, wet bulb and that of the dry thermometer is in direct relation to the relative humidity and is known as the wet bulb depression or WBD.

Wet bulb depression (at 99.5°F, or 37.5°C)	Relative humidity
15.8°F (8.8°C)	50%
11.2°F (6.8°C)	60%
8.8°F (4.9°C)	70%
6.6°F (3.1°C)	80%

Eggs need to be turned at least three times a day, and preferably more. An odd number of turns is better than an even number, otherwise the eggs tend to spend the long night period always on the same side.

Storing the eggs before incubating

Eggs can be stored for up to two weeks before incubating, but after four days hatchability decreases. (To a certain extent, hatchability is a hereditary factor. It is a good idea to have breeding stock blood-tested for diseases like Pullorum which are egg transmitted. In Australia it is illegal to establish groups of breeding fowl unless they have this test.) Suitable eggs are clean, of a good shape and texture, with no chalkiness or hairline cracks. Store them, pointed-end down, in a cool place in egg trays. The temperature should be 55°–60°F (12°C) with a humidity of 75%–80%. You can usually achieve this by placing a container of water nearby.

Introducing the eggs

Before you introduce the eggs, the incubator should already have been warmed up. If you are turning the eggs by hand it is a good idea to mark each one with a nought on one side and a cross on the other, so that you can see easily which ones you have turned. Keep a careful eye on the temperature and humidity and do not neglect the turning of the eggs.

Candling

After seven days the eggs can be candled to detect any infertile ones. A candling box is easily made with a 60 watt bulb and a box with a hole $1\frac{1}{8}$in (28mm) diameter. In a darkened room the egg is held in front of the hole through which the light passes. A developing egg will have a small, dark red spot with veins radiating out in every direction. If you have not done this before it is a good idea to have a fresh egg as a comparison. Any addled eggs (i.e. eggs which have gone bad because, for example, the chick has died or bacteria have infected them) should be disposed of, but the

Mark eggs in an incubator with a cross on one side and a circle on the other so that you can turn them in sequence.

After seven days' incubation, check that the eggs are fertile by candling them.

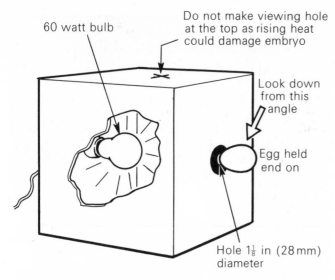

60 watt bulb

Do not make viewing hole at the top as rising heat could damage embryo

Look down from this angle

Egg held end on

Hole 1⅛ in (28 mm) diameter

A home-made candling box.

Candling after seven days.

infertile ones are alright to use for cooking. I must admit however, that I have never been able to bring myself to use them, so I hard boil them and either feed them back to the hens or give them to the pigs. Dark or speckled eggs such as those of Barnevelders or Welsummers are difficult to candle, and it may be necessary to give them the benefit of the doubt.

Hatching
By the nineteenth day, you should stop turning the eggs, and you will soon hear the chicks pipping the shell. Leave them for twenty-four hours to dry off completely before removing them to a brooder. There is no need to worry about food and drink at this stage, because they still retain the remnants of the yolk in their abdomens.

Brooding

Introducing day-old chicks to a broody hen
If you have a broody hen who has been sitting for at least two weeks, it may be possible to introduce the new chicks to her. She may not accept them, but if she does you will be spared the trouble of having to brood them yourself.

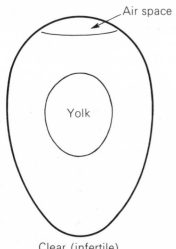

Air space

Yolk

Clear (infertile)

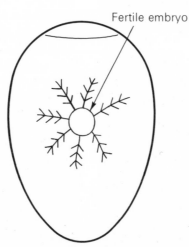

Fertile embryo

Fertile

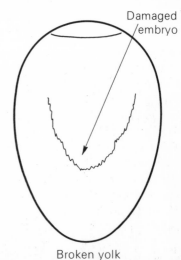

Damaged embryo

Broken yolk

Hatching eggs.

A newly-hatched chick.

Chicks in a brooder. The lamp provides warmth and the corrugated cardboard keeps them confined. Woodshavings provide a warm base and the egg box serves as a container for chick crumbs.

A haybox brooder with run. The end section is insulated with hay and the whole run is vermin-proof.

Introduce the chicks gently, one by one, from behind – for it is vital that you do not let her see them. (Hens do not have a well-developed sense of smell, so they do not recognize their kin by smell as sheep do with new-born lambs) If she has eggs, take them away one by one, and replace each with a chick. A large hen may accept as many as fourteen. If she begins to cluck in response to their cheeping and to spread herself protectively over them, then she has probably accepted them. It is as well, though, to check up on her several times within the next few hours, without disturbing her, to make sure she has not abandoned them.

The artificial brooder

I have successfully reared small numbers of chicks in a cardboard box in an airing cupboard, moving them successively to the kitchen, the verandah and then to a small pen outside for a few hours. For rearing larger numbers however, a brooder is essential. Some incubators are made as dual-purpose incubator/brooders and there are also many other types of brooder on the market, but it is relatively easy to make your own. The one essential is to buy an infra-red lamp – far too many people have used ordinary bulbs or oil lamps and then wondered why their chicks have either died of distress or been burned up.

If the brooder is in an outhouse, it must be rat-proof. One rat will kill dozens of chicks in a single night, by biting them in the throat then dragging them away. Anyone who claims not to have rats on his land is indulging in wishful thinking. Wherever there is livestock with its associated feedstuffs there will be rats, and while we can do a great deal to control them by poisoning and trapping, total eradication is impossible. Building blocks provide useful temporary walls, and close-mesh netting over the top will keep out rats. A concrete floor is cold for young chicks, but I have found that empty paper feedsacks, laid out and covered with a layer of woodshavings, make an excellent base. The infra-red lamp should be suspended above the breeze blocks for at least twenty-four hours before the chicks are put in the brooder. This ensures that all the dampness has been removed.

Initially, the temperature should be around 95°F (35°C), reducing gradually by 6°F (3°C) a week. Experimenting with raising and lowering the lamp will soon establish the right position. If the chicks are too hot, they will spread out to the edges of the brooder, if too cold they will huddle in the centre, but when the temperature is just right they will be somewhere between the two. At about six weeks the chicks can do without artificial heat, unless the weather is particularly cold.

Chick management

As soon as they are put in the brooder, the chicks should be given fresh water in a purpose-built chick drinker. An open container will not do, for there is always the risk that a chick will drown in it, or at least become chilled. Chick crumbs can be left in a container which is low enough for them to reach and they will help themselves from it when they feel like it. I have found that in the absence of a mother hen, it is good practice to teach the chicks about food and water by first dipping the beak of each chick in the water, then taking a handful of crumbs and dropping them on a piece of brown paper. This imitates the action of the hen who, when she finds food, calls the chicks and drops the food on the ground for them to eat.

A grain grinder set at its lowest setting will grind wheat into suitable particles for the chicks, and this provides a good scratch feed for them. Chopped chives are a good source of greens. Detailed information on feeding is given in the section on page 20.

Hygiene is all important and a build-up of droppings should be avoided at all costs, otherwise the risk of disease is greatly increased.

Once the chicks are big enough to do without artificial heat, they should be given their own run, with plenty of grass where they can free-range. It is not a good idea to mix them with older birds because of the risk of transference of disease. For the same reason, they should not have ground which has been used by hens, and which may harbour parasites. The veterinary surgeon or the district poultry officer will give advice on disease protection. There are certain diseases such as Marek's disease which they can be protected against by vaccination.

Health and disease

Prevention is obviously better than cure, and you can avoid a great deal of trouble and disease simply by paying adequate attention to hygiene, feeding and general care. There is no doubt that most of the small poultry keeper's problems arise from the fact that his ground is not properly rotated. When hens spend year after year on the same ground, there is an inevitable build-up of parasites, and the soil becomes literally sick. Although most of these parasites such as gapeworm can be controlled by the use of veterinary preparations, they will inevitably be reintroduced by wild birds, and on over-used ground a gradual concentration will accumulate. If poultry keepers were to apply the same principle of rotation as that used by gardeners, they would have less trouble.

Even when every effort is made to avoid problems, there will always be occasions when you will find yourself with an ailing fowl. Often it is nothing serious, but there is always the chance that it could be. It is important to be able to recognize general and specific symptoms and to develop the 'stockman's eye' which will help you spot any uncharacteristic behaviour in the poultry at an early stage. Any sudden death or a number of deaths should receive immediate attention, either from a veterinary surgeon, or the district poultry officer.

Local conditions vary, and it may be that certain diseases such as fowl cholera, Marek's disease or fowl pox are prevalent in your area. Take the advice of the experts and, if necessary, have your birds vaccinated accordingly. There are certain conditions which are egg transmitted from the hen to the chick. One of these is Pullorum or white diarrhoea, caused by the bacteria Salmonella pullorum. It is a fairly simple procedure to have any breeding birds blood-tested to establish that they are clean.

Preventive measures

Cleanliness

Ensure that all food and drink containers are kept clean. Clean houses regularly and put fresh straw in the nestboxes. The straw and manure makes excellent compost when it has rotted down, and I always use it as a surface mulch around blackcurrant bushes which do well on the extra phosphates. It is a good practice to site compost heaps where poultry cannot gain access to them, because there is a possibility that disease or parasites could be transferred through the droppings.

Action against lice and mites

Houses, perches and nestboxes should have a regular dusting with powder to control infestation. Treat the fowl at the same time, giving particular attention to the areas under the wings, at the back of the head and around the vent. Once a year, the houses should be creosoted inside and out, but birds should not be allowed back inside until the wood is quite dry and the smell has dispersed. Allow at least three weeks for this. Hens may suffer burns to the legs and breast from freshly creosoted perches.

Adequate feeding

One cannot over-emphasize the need for a balanced and adequate diet, and access to fresh, clean water at all times. I have discussed nutritional values and the need for oyster-shell grit as a source of calcium carbonate in the section on feeding. An occasional tonic in the form of chopped, boiled nettle-tops is a useful addition to mash, and fresh-chopped chives are particularly good for chicks. Epsom salts added to the drinking water at the rate of one teaspoonful to each quart of water are another useful stimulus to the digestion.

Details on recognizing a sick bird, together with a list of poultry ailments, are given at the end of the following chapter on page 60.

DUCKS, GEESE & OTHER POULTRY

Buying ducks

There are many poultry keepers who keep ducks for the simple reason that they like them – and who would dispute that this is an excellent reason for keeping any livestock?

Ducks are less susceptible than hens to some poultry diseases and, indeed, are immune to a few of them. Some breeds such as the Khaki Campbell will outlay hens, and the eggs themselves are bigger. A duck will continue laying for a longer period than a hen which, commercially at least, is replaced after one laying season. It is only fair to say that ducks eat more and tend to be messier than hens, but they are also more efficient scavengers and, if they have access to a pond as well as grazing, will find a lot of their own food.

Generally speaking, ducks are less destructive in gardens than hens which will scratch up every vestige of plant growth if allowed to do so. Ducks will investigate plants for insects and slugs, but will not usually cause damage in a herbaceous border or on a lawn, unless there are any particularly delicate plants which may be trampled by their webbed feet. They will, however, cause a certain amount of destruction in a vegetable garden, and young lettuces, peas and brassicas are particularly at risk.

The correct way to carry a duck.

Breeds

It is thought that all breeds, with the exception of the Muscovy, originated from the wild Mallard, and that 'sports' were thrown up at various times which were selectively bred for a particular purpose. For example, the white Aylesbury is thought to have originated as a white mutation of the Mallard, and was subsequently bred for increased weight as a table bird. The Muscovy from South America has come from a different line, although no one is certain of its ancestry. It is in many ways more akin to the goose, and it is thought that the goose world is probably where it belongs. Although it will mate with various duck breeds, the resulting eggs are infertile.

There are now many breeds of ducks, both ornamental and domestic, but as the small farmer is likely to be mainly concerned with getting meat and eggs from his ducks, I shall concentrate on those with utilitarian virtues. That is not to say that functional ducks are not ornamental – some are extremely attractive. One cannot, alas, claim that the opposite is necessarily true of ornamental breeds.

The following is a list of the most common breeds, but popularity varies from one place to another. The Welsh Harlequin, Buff Orpington, Cayuga and Black East Indian for example are not generally bred in Australia although they are well known in other countries.

Khaki Campbell

One of the best breeds for egg production is the Khaki Campbell which was originally bred by a Mrs Campbell from Uley in Britain at the turn of the century. It has Mallard, Indian Runner and Rouen in its make-up and is the most popular breed for eggs in most parts of the world. It is capable of producing over 300 eggs a year if properly fed and housed, but it should be emphasized that the individual strain is more important than the breed. If you have Khaki Campbells, they will not necessarily lay well unless you happen to have a particularly good egg-laying strain.

Indian Runner

Before the Khaki Campbell ousted it from its position as top egg-duck, the Indian Runner was one of the most widely kept breeds. It was introduced into Britain from Malaya around 1870 and spread from there to other countries all over the world. There are several varieties, including the White, the Buff and the Fawn and White. They can lay an average of 180 eggs a year, and are easily recognized by their characteristic upright stance.

Welsh Harlequin

This breed was bred in Britain in 1949 from two 'sports' off Khaki Campbell stock. A good strain can equal the Khaki Campbell in egg production, and it has the added advantage

Khaki Campbell

Welsh Harlequin

Aylesbury drake

Buff Orpington

Pekin drake

Black East Indie

Muscovy drake

of being an excellent table bird with a weight of up to 6½ lb (3 kg). It is docile and placid, with less of a tendency to flap and panic than many other breeds, and also has an attractive plumage. The Welsh Harlequin is my personal favourite.

Aylesbury

The Aylesbury, laying up to 100 eggs a year, is perhaps a more general favourite. It is a snow-white, heavy bird which dominated the table duck trade in Britain for many years. Weighing up to 10 lb (4.5 kg) when adult, it has been recently ousted from the commercial scene – first by the Whalesbury, produced from an Aylesbury drake and Welsh Harlequin duck, and latterly by strains of the Pekin.

Pekin, Rouen and Buff Orpington

The Pekin was introduced from China in the 1870s and is now the premier table duck in North America and Australia. It produces an average of 130 eggs a year. Two other good table birds are the Rouen, a heavy duck bred in France but which has a poor record of about ninety greenish eggs a year, and the Buff Orpington, which was bred in Britain as a dual-purpose duck from strains of Indian Runner,

Rouen and Aylesbury, and lays an average of 240 eggs a year. In the USA this breed is called the American Buff Duck.

Cayuga and Black East Indie

These two are, strictly speaking, ornamental breeds, but they are often kept by gourmets because of their superb game flavour. The Black East Indie originated in North America, and the Cayuga is thought to have been bred from it around 1850 in New York. They are both poor layers as far as quantity is concerned, but on the other hand they have a reputation for being good mothers in hatching and rearing their own eggs.

Crested Duck

The Crested Duck is aptly named for its characteristic top-knot. It originated in Britain, and is a medium-weight bird suitable for the table, but unfortunately it is not as readily available as it perhaps deserves to be.

Muscovy

This South American breed is worth keeping purely for its readiness to hatch not only its own eggs but any others you

Domestic ducks in a small natural pond.

put under it. It also makes a good, if rather gamey, table bird. Muscovies are however great fliers and it may be necessary to clip their wings (see page 47). In the wild they perch on tree branches and their clawed feet are adapted for this purpose.

Duck-keeping systems

Ducks are waterfowl and ideally should have access to their natural environment in the form of a pond. However, it is not essential for them to have a pond, as long as they are able to submerge their heads in water and can periodically splash their feathers. Ducks which are not able to dip their heads in water will quickly become affected by eye or nostril problems. Dust, food or any other matter will clog up their nostrils or beaks. If you do not have a pond, it is no great problem to make a small one, or even to put an old bath or sink into the ground. This will need to be cleaned and refilled frequently or the water will quickly stagnate.

Ducklings should not be allowed to swim until they are eight weeks old – in this way you will save them from any problems associated with chilling. In large ponds there is also a danger from rats, which are not averse to water and will soon decimate a flock of swimming ducklings.

There are basically five ways of keeping ducks, depending on the amount and type of land you have available and on how many you want to keep. Within these categories, however, there is an infinite variety of possibilities and, whichever system you use, you will need to adapt it to your own particular situation so it is difficult for me to generalize. My own system has evolved around the fact that we have a pond close to our house. Our 2-acre site (1 ha) is divided up into four areas – the house, outbuildings and pond; the

vegetable garden; the hen orchard and the paddock. It seemed sensible to let the ducks free range as much as possible around the pond area, without letting them wander anywhere where they could cause damage or interfere with other livestock. Consequently the vegetable garden, hen orchard and paddock are out of bounds to them, but they are allowed to free range around the house and pond, on the lawns and in the flower gardens. They have a wooden hen house with the perch and nestboxes removed, and with its floor covered in clean straw. This housing is placed near the pond and they are confined to it at night which ensures that their eggs are laid inside.

In my experience ducks will cause a fair bit of damage to young vegetables, but will largely ignore flowers. Once the autumn comes, however, I let them go into the vegetable garden so that they can clear up any slugs or other pests. At this time of year there is not much damage they can do.

The free-range system

On a large scale 1 acre (0.5 ha) will support 100 ducks on a permanent basis. A greater concentration may damage the pasture by compaction and puddling of the soil, although heavy clay soils are more prone to this than light sandy soils. Where soil is particularly heavy a stocking rate of seventy-five ducks to the acre should not be exceeded. Relating this to a smaller scale, half a dozen ducks will need about 50 sq ft (4.6 sq m) of land if you want to avoid a quagmire.

Ducks are hardy creatures, but they do need shelters at night and during severe weather. Traditionally farmers have used movable structures made of wooden planks and corrugated iron. The planks provided walls, usually not more than 2 ft 6 in (0.75 m) high and 10 ft (3 m) long. The corrugated iron sheets were bent in a U-shape and over-

A suitable temporary pond which allows the ducks to submerge their heads and splash their feathers.

A hen house adapted for ducks and strategically placed near their pond.

lapped to form a waterproof roof. This was secured by wires stretched over the top and pegged into the ground on either side. The ground under the shelter had to be raked regularly to clear away debris, and clean, dry straw was always provided.

Straw bales have also been used successfully to make houses for ducks on free range. In this case, a satisfactory roof is made of planks resting on bales, and the corrugated iron sheets attached to the planks. It is not a good idea to use corrugated iron on its own – apart from the problems of attachment, it produces a great deal of condensation which drips onto the ducks. Contrary to popular belief they do not like being wet all the time. If you use straw bales, the house will last longer if you peg wire netting on the inside and outside of the walls.

Temporary structures of this kind are cheap to construct and easy to transport, but they do not confine the ducks, so many of them may lay their eggs outside, in dirty or inaccessible places. For this reason many people prefer to have a house where it is possible to lock up the ducks overnight. They then lay their eggs in the clean straw and are also safe from predators such as foxes or rodents.

A poultry house of the type that was traditionally used for free-range hens is ideal, although the perches and nest-boxes are not used by the ducks and could therefore be removed. It is important to provide a ramp for them to get in and out, as a duck's legs are easily damaged if they have to scramble up or drop down.

The Dutch system

In the so-called Dutch system, a number of long, narrow, individual pens are built around an existing stream. If no stream is available, you can use instead a series of low troughs

Temporary shelters for free-range ducks.

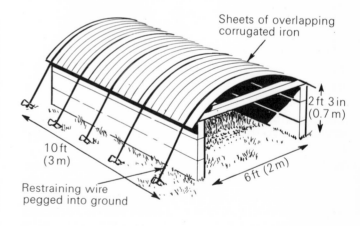

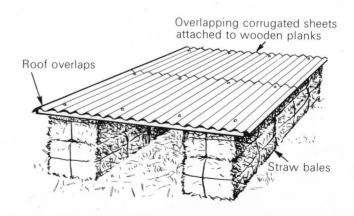

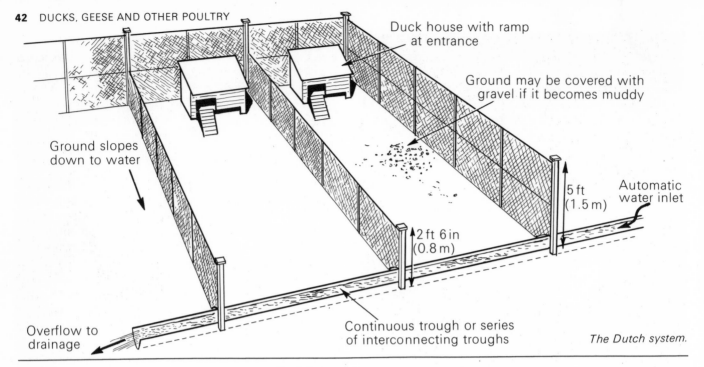

Duck house with ramp at entrance

Ground may be covered with gravel if it becomes muddy

Ground slopes down to water

5 ft (1.5 m)

Automatic water inlet

2 ft 6 in (0.8 m)

Overflow to drainage

Continuous trough or series of interconnecting troughs

The Dutch system.

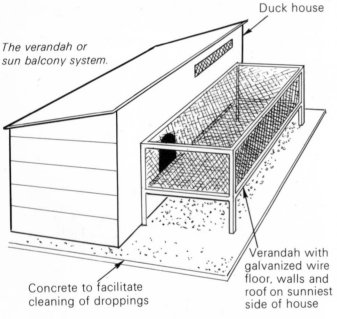

The verandah or sun balcony system.

Duck house

Verandah with galvanized wire floor, walls and roof on sunniest side of house

Concrete to facilitate cleaning of droppings

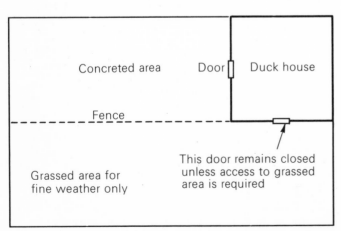

Concreted area

Door

Duck house

Fence

This door remains closed unless access to grassed area is required

Grassed area for fine weather only

Suggested lay-out for a small pen in continuous occupation.

with an automatic watering system. The ground needs to be light and well drained, and should slope down to the water. A house for about ten layers or breeders is installed at the top of each pen, and the whole area surrounded by high wire netting to deter predators. The netting separating the individual pens does not need to be more than 2 ft 6 in (0.75 m) high. This system has been widely used by duck keepers, and an added advantage is that a variety of breeds can be kept without the possibility of inter-breeding.

The grass in the pens will naturally become exhausted very quickly, and if drainage is inadequate the whole area could become a mudbath. Smooth gravel has been widely used in this situation, to provide a thick, permeable layer. As grazing is impossible, all green feed has to be provided by the keeper.

Straw yards

This system has much to recommend it and, again, is widely used – particularly for fattening table ducks which do not need access to water for swimming. A house opens onto an area which is fenced off, and this is kept covered with a layer of clean, dry straw. It is a good idea to have planks along the bottom of the netting, to stop the straw spilling out beyond the run.

The system naturally relies on a plentiful and cheap source of straw, which is continually added as the existing straw becomes spoiled. The whole lot is periodically removed and it provides an excellent source of compost. If the straw becomes at all damp it must be removed very frequently as decaying matter could be the source of botulism.

The verandah system

The verandah, or sun balcony system as it was sometimes called, is basically a galvanized wire mesh floor stretched over a frame and with a wire netting roof and walls. The whole structure is placed on the side of the poultry house where it receives the maximum amount of sun and warmth. Ideally, there should be concrete underneath the mesh floor so that the droppings which fall through can easily be disposed of. An alternative is to use a droppings board which slides out for cleaning.

The verandah system is particularly suitable for young stock, whether it be ducks, chickens or rabbits, and has been widely used for all these. It is not, however, my choice, for I feel that any livestock should be kept in conditions as close as possible to their natural ones.

The small house and run

Where space is limited, as in a garden, it may be necessary to confine ducks to a small run. The grass area will quickly become paddled down into a marsh, and alternative flooring will be needed. Some duck owners have found that concreting half of the area is one solution – particularly if the housing can be arranged in such a way that, on wet days, the grass area is closed off. If it is a very tiny area, then it may be more appropriate to concrete the whole space. Another possibility is to keep the run covered with straw, as a small-scale adaptation of the farmer's straw yard I have already mentioned.

Ducks do well on concrete, provided that it is smoothly finished, for any abrasive surface is bad for their feet. I remember one lady who kept Khaki Campbells in a walled yard measuring about 20 ft × 12 ft (6 m × 3.5 m). The duck house had been built of packing cases with a tarred felt roof, and it stood on a brick foundation to keep it clear of the ground. The ducks entered by means of a wide ramp. The yard was regularly hosed down to clear away debris, and the water drained into a gully and drain at one side of the yard.

Housing

It may come as a surprise to some who think of ducks as being continuous dabblers that they need housing at all. But of course any livestock that is being kept for its production will only give of its best if it is adequately sheltered. The duck is no exception. Another myth is that ducks are happy to sleep in damp places because they are water fowl. It is true that the duck is beautifully adapted to walk in muddy places, to swim and to repel water from its oiled feathers, but when it is resting it needs a dry, sheltered nesting place like any other bird. Waterfowl do not like wind, nor can they cope with too much hot sun – a sheltered, shady place is essential. The basic housing requirements are therefore protection against wind, rain and sun; adequate ventilation and a dry bed.

Floors

The floor area of the duck house should provide a minimum of $1\frac{1}{2}$ sq ft (0.15 sq m) for each bird. This should be kept covered with a thick layer of straw, peat, wood shavings, sawdust, dry leaves or any material which is readily avail-

able, clean, dry and hygienic. Rammed earth floors are suitable, but it may be necessary to sink wire netting under the earth to keep out rats which would otherwise burrow in. Other floorings include wire mesh, concrete or boards. If mesh or boards are used, they will need to be raised off the ground to deter rising damp. I dislike mesh floors, not only because they are cold, but also because in my experience they are difficult to clean. Theoretically, the droppings should fall straight through, but the nature of duck droppings is such that they form a crust which is difficult to remove. Concrete is also cold, but it does provide an excellent deterrent to rats. A thick layer of flooring material provides insulation.

Doors and windows

The door into the duck house should ideally be at least 2 ft (0.6 m) wide to allow for the traffic jam that ensues when the ducks are let out in the mornings. There is a real possibility that they may injure each other when they all rush forward to get out, and as ducks are particularly vulnerable to leg injuries, everything should be done to help them to get out in an ordered way. There should also be a wide ramp for them to reach the ground. For the same reason it is not a good idea to have a window which the ducks can see out of, for sudden movements or disturbances outside can cause them to panic and flutter and trample on each other.

Ventilation

Adequate ventilation is necessary and a small, open space covered with wire netting will ensure that fresh air circu-

Inside the duck house, showing a thick covering of clean straw on the floor and a wide ramp for ease of access.

lates in the house. This should be situated near the top of the sunniest wall so that the ducks are out of draughts, and can't see out of it.

Construction

The dimensions of the house itself will depend on how many ducks are to be housed, but as you do not need to provide perches, it will not have to be more than 2 ft 6 in (0.75 m) high. With this type of low house, the roof will need either to lift off completely, or to be hinged at one side for opening. Without this, it would be impossible to clean the house or to collect the eggs conveniently.

Wood is undoubtedly the best and warmest material for constructing the house. Weather boarding, or $\frac{3}{4}$-in (19 mm) tongue-and-groove boarding are both suitable. The latter may need extra proofing in the form of tarred roofing felt on the back and walls, in exposed situations. The roof itself will need to slope backwards away from the front, and to have an overhang of at least 2 ft (0.6 m). Tarred felt should also provide an extra protection for the roof.

Feeding

Ducks that are free ranging in grass and have access to a pond will naturally find some of their own food in the form of grass, insects, slugs and water fleas. They would probably survive on such a diet as wild ducks do, but they would not lay well or fatten effectively for the table.

I am constantly surprised at the number of people who complain that their ducks are not laying regularly and then, when pressed, admit to giving them 'the occasional handful of grain'. It cannot be overstated that a duck will only lay its full egg potential if it is fed regularly on food that contains the right balance of nutrients.

Ducks eat more than hens and, depending on its size, an adult duck will consume 6–7 oz (170–200 g) of food a day. It is better to feed twice a day than to give this all at the same time in one feed. I find that the most convenient times are when they are let out of the house in the morning, and when they are put to bed at night. A rough way of estimating whether the ducks are getting enough is to see whether there is any food left after twenty minutes of feeding. If there is then you are probably providing too much.

Some people favour demand feeding, particularly if the aim is to fatten the ducks quickly or to gain the maximum number of eggs. Where the ducks are allowed to help themselves when they feel like it, the only feasible system is to feed proprietary pellets in a hopper. It goes without saying that this type of feeding should take place in the duck house otherwise the wild bird population will bless you for your beneficence.

Ducks have a tendency to transfer new food to their waterers, which may help them to swallow in the case of mash but is wasteful with other foods. To prevent this, and to provide them with useful exercise, it is a good idea to place their feeders and waterers some distance apart.

It is up to the individual to decide what the basis of his feeding practice will be. If eggs are the priority, layer's mash or pellets will produce the largest number, but fatteners of course will grow more quickly on rearer's or broiler's pellets. Against these advantages must be set the cost of the commercial feedstuffs.

Ducklings

From day-old to a month, the best and most convenient start is provided by proprietary chick or broiler starter crumbs, which can be left out for the ducklings to eat at will. Specialized duck mixtures do exist, but as these are often only distributed in places where there happen to be large commercial duck farms, you may find it difficult to obtain them. However, if you can get them, you may prefer them because some chick feeds contain medicaments such as coccidiostats and antibiotics. These are unsuitable for ducks and duck mixtures generally do not include them.

The crumbs should be put in a shallow dish with fresh, clean water next to it. The ducklings will then waddle from one to the other – taking a few crumbs, a drink and then more crumbs. The water however should not be in an open container, otherwise they will use it as a paddling pool and get soaked. A hen's drinker is ideal, as long as it enables them to dip their bills and faces in it. A chick drinker is normally too small for this.

If you give them wheat, it should be coarsely ground into small particles for the ducklings. It should be emphasized however that chick or broiler starter crumbs, although relatively expensive, give the birds a first class start in life, and I would not recommend an alternative.

Four weeks old and onwards

From the age of four weeks the ducklings are growing rapidly and proprietary grower's pellets or broiler finisher pellets will provide all the table bird's nutritional needs, apart from water. Layer rearing pellets will do the same for future layers or breeders. Again the feed can be made freely available, which means that you must fill up the hoppers once a day and let the ducklings help themselves when they want to.

If you prefer wet feeding, the pellets or mash can be mixed with a little water and put in a wide, shallow dish. It is essential to have drinking water close by, so that the ducks can clean their beaks, as they tend to burrow their bills into soft food, and there is a danger that their nostrils will become clogged if the particles dry and harden before being washed off. If kitchen scraps are fed, these can be mixed into a mash and fed in the morning, while wheat or mixed grain will provide their evening meal.

Laying ducks

From the age of sixteen weeks onwards the ducks can be treated like adults and fed accordingly. They will come into lay around this time, and layer's pellets or mash can be given either dry or wet. In my experience pellets are fine given dry, but the powdery mash is really better dampened until it resembles crumbs.

If you give your ducks wheat, it should be fed in the ratio or 2–3 oz (60–90g) wheat in the evening and 4 oz (120g) layer's pellets in the morning. If you are not giving any proprietary foods at all, limestone grit should be made available otherwise egg shell formation and quality may suffer. Other suitable foods are finely chopped fruit such as windfall apples and, of course, kitchen scraps.

Fattening ducks for the table

Table ducks, such as Pekins or Aylesburies or surplus drakes, should be separated from the flock and confined in a run or outbuilding. If proprietary food is given it should be fed ad lib in hoppers and the ducks should be killed when they are nine to ten weeks old. Alternatively they may be allowed to free range and given supplementary food in the form of cooked potatoes or stale bread. In this way they can be taken up to fifteen or twenty weeks. This latter method is, of course, quite unsuitable for commercial purposes. Commercially, the age at which to kill is related to detailed economic factors and a duckling is killed as soon as it reaches acceptable table weight, which may be as early as seven or eight weeks old.

Meat and down

Killing

A duck can be killed either by decapitation, by stunning and then cutting the jugular vein in the throat, or by the traditional method already described for fowl (see page 24). Killing should never be undertaken until you have first been shown how to do it by an expert. It should be immediate and should cause no suffering to the bird.

Plucking

Plucking is more difficult with a duck than with a hen, and should be done immediately after killing before the carcass cools. Hot wax is frequently used – after rough plucking, the carcass is dipped in wax at a temperature of 140°F (60°C), and then immediately plunged in cold water to set the wax. After about ten minutes the wax and attached down feathers are removed. With this method of course, the down feathers, which represent a valuable commodity with the increasing interest in continental quilts, are wasted. Some people pour boiling water over the carcass immediately before plucking it. (It can be helpful to add a little non-perfumed detergent to the scald water – ducks' feathers are oily and the detergent helps the water to penetrate to the skin.) This is a satisfactory method, but is unacceptable commercially because it damages the skin.

I have found that the best method is to dry pluck the wings and tail, then hold the carcass above hot water so that the steam permeates through the down feathers. A small fire outside with an old saucepan full of water will provide the steam, but you should wear rubber gloves to avoid scalding your fingers.

Using the down

The down feathers of both ducks and geese are much in demand for quilts of all kinds as they provide an ideal light, warm bedcover. The feathers should be stripped from the carcass while it is still warm, or by the 'steam' method described above which is equally satisfactory. As feathers fly everywhere it is advisable to wear an overall, cover your head and wear a face mask – if you use the steaming method you will find they do not fly about as much as they do with dry plucking. Any larger feathers from the wings or tail should be discarded.

Place the soft down feathers in muslin bags and, if the weather is warm and dry, hang them up on a line outside. If it is wet, a warm kitchen will do, particularly if you are lucky enough to have beams from which to suspend them. Shake the bags frequently to allow air to get to all the feathers, and leave them suspended for at least four days. Finally, place the bags in a cool oven for half an hour.

Duck feathers do tend to have a 'ducky' smell, and this may last for a few weeks, but it will eventually disappear, particularly if the feathers have been aired for long enough.

Egg production

There are many myths about duck eggs, the most common being that 'they taste strong', 'you have to boil them for fifteen minutes' and 'you get salmonella food poisoning from them'. I have never tasted the eggs of wild or semi-wild ducks, and so am not in a position to comment on their taste or quality. I can, however, state from considerable experience that eggs from domestic ducks which are properly fed and housed in hygienic conditions taste fresh, sweet and delicious, just as new-laid hen's eggs do.

In the past there have been a number of salmonella scares involving duck eggs as one of several sources. Immediately there was a reaction against them, and some commercial duck egg producers went out of business virtually overnight. In some countries, including Australia, commercial egg production has never prospered since. It is true that salmonella can be transmitted via duck eggs that are laid on damp, dirty straw or at the edge of a muddy pond. However hen's eggs would be just as likely to be affected if they were laid in such conditions. (It is also worth knowing that the organisms only multiply at temperatures above 41°F (5°C) so storage under refrigeration is best.) Statistics indicate that the prime sources of salmonella are processed and cooked meats, which account for more than half of all outbreaks of the disease.

The shells of duck eggs are more porous than those of hen's eggs and it is obviously important to ensure that they are not laid in damp or dirty places. If clean, fresh straw, woodshavings or any other bedding is put in the duck house regularly, and the eggs are removed as soon as possible after laying, there is no reason why they should be regarded with suspicion.

If you want to clean the eggs you can remove isolated spots of dirt with sandpaper. Excessively dirty eggs can be immersed in warm water – about 115°F (46°C) – for not more than two minutes. If you add sanitizing agents to the water the eggs need not be rinsed afterwards which is an advantage as rinsing prevents the formation of a protective film.

Ducks lay during the night or early in the morning and so they should be confined until about 9.30–10.00 a.m., which will ensure that when they are let out they will already have produced their eggs. I put my ducks to bed in the early evening, as soon as they have had their second meal of the day. When they are released in the morning they are then given their first feed. They quickly adapt to this kind of routine, and will even congregate outside their house when they think it is bedtime.

The onset of lay

Many people will say that the onset of lay is marked by a dropping of the duck's abdomen and that this is the only clear indication. In my experience this is not necessarily true, and first-year layers may show no such sign. What is true is that when the ducks reach sexual maturity, they develop a particular pattern of behaviour – the characteristic head bobbing or nodding between the duck and the drake. Once you notice this you can be fairly certain that egg laying will start within a week or so. It is not necessary to have a drake in order to encourage egg laying as some people believe, but it may be useful to keep one in view of the distinctive head bobbing. For breeding purposes you will need one drake to every four to six ducks, depending on whether they are light or heavy breeds.

Ducks which hatch in the winter and reach the age of sixteen weeks in the spring can come into lay at this time. If they mature in the summer, however, they will probably start laying at twenty weeks. Any late hatched ones which mature in winter may not come into lay until they are six or even seven months old. The only way of getting these late ones to lay during the winter is to give them artificial light, just as one would for hens. Artificial light will of course increase overall egg production anyway.

Moulting

Moulting, or replacement of old feathers with new ones, normally takes place at the beginning of the autumn and may last from six to eight weeks. Egg production will drop, and may even disappear completely during this period, unless your ducks are of a particularly good laying strain. Ducks that were hatched in the winter may not necessarily moult in their first autumn, but go on until the next year. It is not a bad idea to give them extra protein during the moulting and a small amount of fish-meal added to their normal food, will help to keep them in good condition during a time which inevitably makes extra demands on their energies.

Using duck eggs

There is no mystique about duck eggs, and they can be used for all the things that hen's eggs would normally be used for. As I have already said, the taste does not differ from that of hen's eggs, but the albumen or white does tend to be more rubbery. For this reason, many people prefer to keep duck eggs for cooking rather than to eat them as they are.

Breeding and rearing

Breeding

Ducks will continue laying reasonably well for about three years before there is a radical decline in their yield. There will come a time, however, when you will probably wish to raise replacement ducklings, or possibly to raise some fatteners for the table. Not all ducks make good mothers, and in many cases may make downright bad ones. It is not unusual for a duck to lay a clutch of eggs, start sitting on them and then desert them after a few days. Many people have preferred to use broody hens for hatching duck eggs for this reason.

Khaki Campbells, Aylesburies and Welsh Harlequins are all reputed to be poor mothers, although it must be said that I have known them to hatch and rear their young successfully and protectively. The Cayuga and Black East Indie ducks are good sitters, and have a reputation for being good mothers. The mother supreme of the duck world is the Muscovy which will sit and hatch any eggs you give her once she is broody. It is not uncommon for a Muscovy to lay a clutch of infertile eggs in a hedge and continue sitting on them even when they are rotten and when many weeks have gone by since they would have hatched, had they been fertile.

Ducks will mate on land and in water, usually after the head bobbing or nodding that I referred to earlier. According to some authorities the fertility of the eggs is increased if mating occurs in water, but there appears to be no substantial evidence to support this other than hearsay.

The broody duck

A broody duck behaves in a similar way to a broody hen, in that she fluffs up her feathers on your approach and will peck if any attempt is made to touch the eggs. She will normally leave the nest once a day and, if she has access to a pond, will go for a short swim before returning to the eggs. This is important, because in this way the humidity of the eggs is maintained. If she has no access to water for dipping her whole body, it will be necessary to sprinkle the eggs regularly to stop the membranes drying. The time when the broody leaves the nest is a good opportunity to feed her, and if it is done regularly she will emerge at the same time each day.

Incubation

A duck's egg takes twenty-eight days to hatch, and the process is the same as that for a hen's egg. The duck eggs can be candled after a week if they are being incubated artificially, in order to establish whether they are fertile or not. The main thing to remember when using an incubator is that duck eggs do need a higher level of humidity than hen's eggs, and it is important to keep the water level topped up in the reservoir.

Care of the young ducklings

Ducklings are not difficult to rear if they are properly looked after and fed, and it is not unusual for all the members of a clutch to survive. It can happen though, that there is a weak one, and it will soon manifest itself by throwing its head back along the back of its neck. If this happens there is nothing you can do, for this is part of the behaviour of a duckling that is about to die.

Unless they are being reared by a mother or foster mother, the ducklings will need to be brooded or kept in artificial warmth for the first few weeks. For the first few days they should ideally be in a temperature range of 85°F–90°F (21°C–32°C). After four days the temperature can gradually be reduced, by about 2°F (1°C) a day until at about two weeks they can be hardened off. If the weather is particularly cold, they will need supplementary heat for a longer period.

Chick or broiler starter crumbs are the best food for ducklings, and can be left out for them in a shallow container. Water must be available at all times, but for the first week this should be in an enclosed drinker, rather than in a dish which they can use as a paddling pool. They should be able to drink easily without getting wet. I have known ducklings die of cold because they have got wet too early, and if they have no mother to preen them they have no defence against water. The ducklings should be at least eight weeks old before they are allowed to swim, but before this they will have been put in a small pen, with water deep enough for them just to dip their heads in it. From eight weeks onwards they can be treated as adults, although if any are to be fattened, they should be separated and confined at this age.

Sexing ducklings

Day-old ducks can be sexed by the so-called Japanese or vent method. This is comparatively easy, but care should be taken to avoid causing any pain or damage. Hold the duckling gently but firmly, so as to expose the vent. With the finger and thumb of the other hand push back gently on either side, so that the vent is extended and opened. In a male bird, the penis will be seen as a small organ attached to the inside top of the vent.

Wing clipping

With some breeds of duck it may be necessary to clip their wings to stop them escaping, or flying over fences. In my experience, this is particularly true of the Muscovy – other breeds such as Aylesburies are less likely to want to fly.

The primary feathers, on one wing only, are clipped with sharp scissors so that when the bird tries to fly it loses its balance. There is no harm done as long as the feathers are not cut where there is a blood vessel. If the wing is held up to the light the blood vessels can be seen, and the cut made accordingly. This causes the duck no pain and is roughly equivalent to cutting hair in humans. The secondary, or rounded-tip, feathers of the wing should not be cut as these help to keep the bird warm. The process will need to be repeated every autumn when the new primary feathers have grown.

Wing clipping should not be confused with pinioning,

Clipping the primary feathers on one wing only.

where newly hatched waterfowl have the tip of the wing (known as the bastard wing) cut off. The latter causes permanent disfiguration to the wing structure, and is not recommended.

Geese

Geese are not really suitable for anyone with limited space, but where grass is available they are first class foragers and grazers. It must be admitted that they are generally noisy, and if you or your neighbours object to cheerful and voluble honking, then geese are not for you. However, this characteristic does mean that they are excellent watchdogs and they will let you know of the approach of anything that moves, be it stealthy burglar or just a visiting cat. A company in Scotland actually got rid of their guard dogs in favour of geese as an economic measure, thinking that it would reduce their feeding and lawn maintenance costs. Unfortunately they did not foresee that the Christmas season would bring out determined poachers in search of free Christmas dinners, and one dark night the whole patrol disappeared.

Background and breeds

It is thought that the Greylag goose was the ancestor of most domestic geese. No one knows exactly when domestication

Embden x Roman

Chinese

Roman

The Roman goose is medium-sized and white, and was introduced into Britain from Italy at the beginning of this century. Some people claim that this is the original goose kept by the Romans, but there seems to be no evidence to substantiate this. It is particularly suitable for killing at a young age and is at its best when it has reached its first full feathering. The Crested Roman has been bred from the Roman in the USA, but in Britain crested specimens usually only appear as sports.

Pilgrim

These geese were so-called because they were taken to America by the Pilgrim Fathers. The breed is an interesting one because its plumage is sex-linked, the gander being white and the goose light grey. Before its transportation by the Pilgrim Fathers it appears to have been fairly common in the west of England as a farmyard goose with this feature of sex-linked plumage, but, as far as can be established, it did not have a specific name. Now it is referred to as the Pilgrim on both sides of the Atlantic.

Buff

Buff geese originally appeared as sports from the common grey goose but there is no evidence that they were selectively bred until 1928 when Sir Rhys Llewellyn obtained some from a Breconshire hill farm and called them Brecon Buff geese. They are medium-sized with buff-coloured plumage and markings similar to those of the Toulouse. In the USA they are called American Buff geese and strains have been developed for uniform quality and colour by specialist breeders. Neither the Buff nor the Pilgrim breeds are kept in Australia.

Chinese

These are thought to have originated from the wild goose of China rather than from the Greylag. They are graceful and swan-like in appearance and are either white or brown. They are excellent foragers and lay more eggs than any other breed. The flesh is excellent for the table – it is darker than that of other breeds, but less greasy. The birds are ready for killing from eight weeks of age onwards. Chinese geese are not as hardy as some of the other breeds and adequate winter shelter is particularly important for them in more exposed areas.

African

This breed is more common in North America than it is in Britain. It is similar in appearance to the Chinese, but much bigger, and it is likely that they have common ancestry. The African goose has a large dewlap or fold in the skin of the throat, similar to that of the Toulouse. It is a good market goose and lays moderately well.

Sebastopol

This breed of goose has distinctive white frizzled feathers, giving it a rather untidy appearance. It is kept mainly as an exhibition bird but does have good qualities as a medium-

began, but we do know that the ancient Egyptians, Chinese, Greeks and Romans kept them. Geese, like ducks, are classified as either ornamental or domesticated. For the purposes of this book I shall confine myself to a description of the main domestic breeds.

Embden

This is one of the most important breeds as it is the largest of the utility birds and produces a fine carcass for the table. It is a white bird which originated in Germany. It reached Britain from Holland and was subsequently introduced to other parts of the world. In Australia it is now the most popular breed. It has often been crossed with what is euphemistically called the British goose – a mongrel of uncertain background which was the common farmyard type for many hundreds of years in Britain.

Toulouse

The grey Toulouse from France has also been frequently crossed with the smaller common British goose. The Toulouse, which is recognized by the large 'dewlap' under its throat, was originally bred for the table. When introduced into Britain it was also crossed with the Embden which resulted in increased body weight. In the USA there is also a Buff Toulouse and a Giant Dewlap Toulouse.

Geese do not need a pond but must have enough depth of water to immerse their heads.

weight table bird which lays moderately well. I would not recommend the Sebastopol goose as a utility bird however, as its long trailing feathers are difficult to keep clean in muddy conditions.

Housing

Do geese need water?

Geese are waterfowl and will spend a certain proportion of their time in water if they have access to it. However, they spend far more time on land than ducks do as, unlike ducks, they derive most of their food from the land. My own geese were originally allowed to free range around the pond and lawn area, but they spent most of their time searching for grass to eat and only occasionally went into the pond. I subsequently decided to transfer them to a field where there was more grazing, and provided them with a large tank which could be filled up periodically from a hosepipe. The geese need to have sufficient depth of water to immerse their heads completely but, provided they have this, do not need access to a pond.

Shelter

Like ducks, geese need protection from wind, excessive sun and weather extremes. Having said that, their needs are simple. A basic three-sided shelter is quite adequate for

them, the important thing being that there should be enough room for the 'sets' or family breeding units (see *Breeding* section page 52). Geese are more particular about their breeding partners than many other birds and in some cases have been known to ignore a particular goose entirely, or even to chase her away.

Housing can be made out of straw bales as described in the section on *Ducks* (see page 43), or in the three-sided

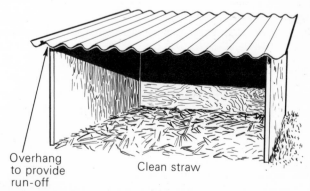

Overhang to provide run-off

Clean straw

A simple three-sided shelter for geese. Where foxes are a problem it may be necessary to incorporate wire mesh and a door across the open side so that the geese can be locked up at night.

Grass is the best food for geese. Note the aggressive pose of the gander.

shelters I have just mentioned. Chicken houses and arks are also suitable but you should take out any nestboxes and perches. The important thing is that there should be enough grazing for the number of birds. One acre (0.5 ha) will support nine heavy-weight geese on a permanent basis, as long as it is good, well-drained pasture. Lighter breeds can be in the ratio of twelve to 1 acre (0.5 ha). Some people keep geese in individual pens, but this causes a good deal of work both in supplementary feeding and maintaining their pens. Unless you are interested in specialized breeding, I do not think that it is worth keeping more geese than the natural grazing system can support.

Feeding

Traditionally, the production of table geese was intimately linked with the growth and decline of grass, and this is still the most economic way to keep them. The young goslings begin to graze just as the new grass appears in the spring and this grazing, together with water, provides all their nutritional requirements until the autumn when the grass begins to decline and the geese are killed. In Britain this means that they can be harvested just before 29 September, in time for the feast of St Michaelmas. If the geese are kept until Christmas, they can be given additional grain to compensate for the progressively diminishing food value of grass as the winter sets in. In the USA geese are fattened for Thanksgiving Day at the end of November, rather than for Michaelmas.

Goslings
Goslings do very well on a diet of proprietary chick crumbs and water for the first few weeks of life, but they do need to have something green to graze as well. They are born with the instinct to pick at grass stalks and, right from the first couple of days, need something to graze, as otherwise they

may pick at each other's down feathers or at bits of straw. This characteristic obviously only occurs in goslings which are being artificially brooded in protected conditions. I find that chopped chives or onion tops are welcomed, and particularly successful is a section of clean, ungrazed turf, out of which they can pick the grass. This is far better than cutting grass for them because they find difficulty in picking it up if it is already cut. They should not be put outside to graze until they are three weeks old, and even then they should be brought back inside for night-time protection until they are feathered enough to withstand the drop in temperature. This will depend very much of course on the climate in your area.

Growing geese
Young geese that have access to enough good pasture do not need supplementary feeding. Where grazing is only available in limited quantities they will need other rations, but keeping geese without pasture is not a very good idea as it is messy, time-consuming and costly in terms of bought feed. In this situation it is more expensive to produce your own goose for the table than it would be to buy an oven-ready one. However, if you do keep your geese confined the best all-round feed is the following mixture (fed twice a day):

> 1 part oats
> 3 parts chopped greens
> 2 parts wheat

The greens can be any leafy vegetable, such as lettuce, cabbage or celery. Dandelion leaves are also good. Proprietary grower's pellets such as those used for chickens are suitable, provided they do not contain medications which may result in digestive upsets. This is the most expensive way of feeding.

As grass declines in the autumn any geese kept for the winter will need supplementary food. The ration of oats, chopped greens and wheat described above is suitable, but if you want an extra fattening ration you can give them the following:

> 1 part wheat
> 1 part barley
> 1 part oats

fed either dry or soaked. Greens, as available, will be needed along with this ration, and, in the USA, corn is often included too. If skimmed milk is available from your own cow or goat you can use this to mix up the dry feed to crumb consistency and this mixture will provide a good 'finish' to a fattening bird. You should feed the geese twice a day, morning and afternoon.

Over-wintering geese
During the over-wintering period, no grazing will be available and although the birds will not need a fattening ration, they will need enough nutrients to keep them in

The correct way to catch a goose.   Lifting the goose up . . .   and carrying it.

good health and to combat the cold. Wheat and oats is a good combination in the proportion of 3 parts wheat to 1 part oats. In addition, the outer leaves of winter vegetables such as brassicas are useful, as are apple peelings, and, in the USA, alfalfa hay is often used. Feeding should take place twice a day. In winter you will need to keep a close eye on the water supply, and to break the ice which will form in zero temperatures.

Geese as weeders and foragers

Geese, like ducks, make efficient weeders and are sometimes turned out into commercial strawberry fields after the crop. They are also frequently kept in orchards where they forage for windfall apples. This is fine for breeding stock which is being over-wintered, but is not a good idea for Christmas fattening geese. In my experience they tend to eat too many apples at the expense of other foods and therefore do not fatten effectively. I have also known them to damage young apple trees by pulling down and sometimes breaking the lower branches in their efforts to reach the apples.

Geese for the table

If you are giving your geese supplementary feed in any case, it is a good idea to confine them in a colony run for three weeks before killing them. They should be kept all together as they do not adapt well to individual separation. Straw in the run will prevent a mud-bath developing, but you will need to replenish it at frequent intervals, depending on how many geese there are. A fattening ration (see *Feeding* section above) is given twice a day and ample water should be available at all times.

Geese are more difficult to kill than other poultry and it is a good practice to have two people – one to do the actual killing and one to stand by and help. The second person will, for example, be able to help control the sudden flapping of the goose's wings if the first person loses his grip. The wings are extremely powerful and can break a man's arm.

The goose is placed on the ground, with the wings held firmly, one in each hand. As it is lowered, the goose's neck will automatically stretch out. A broom handle is then placed over the goose's neck by one of the two people and the other puts his feet on the handle, one on either side of the bird's head. As he presses his feet down on the handle the goose's body is jerked upwards simultaneously, thus breaking the bird's neck and killing it. It can also be done by suspending the goose by its feet and cutting the jugular vein with a sharp knife.

Plucking and gutting are done in the same way as previously described for other poultry, and the down may be used in the same way as ducks' feathers.

Method of killing a goose.

Breeding

Generally speaking the mating ratio for geese is smaller for the heavier breeds than for the light ones. For heavy breeds such as Toulouse or Embden one gander will accept three geese in his 'set' or family. For lighter breeds such as Roman, Chinese or Buff the ratio is one gander to four or five geese. Ideally they should be allowed to form their sets before mid-winter if you are raising table geese the following season. The individual sets can all be in the same field without being penned off from each other, but it is a good idea to provide separate housing for each family. For the initial setting, however, the group will need to have been separately confined until the families are sorted out and then transferred to the field.

Egg laying

Geese have not been selectively bred for increased or intensive egg production in the way that chickens and, to a lesser degree, ducks have been. They therefore produce comparatively few eggs and only during their natural breeding season. Some breeds are more productive than others and, generally speaking, the lighter varieties lay more eggs – the Chinese is the best layer, producing forty to fifty eggs a year.

The onset of lay may be characterized by a dropping of the abdomen, but as older geese already have dropped abdomens this is not necessarily a reliable sign. A far more obvious indication is the behaviour of the gander who will become much more aggressive than before. A gander at this time can be dangerous, particularly to small children, and I must emphasize that the geese should be restricted to a field or orchard where they are securely confined. I have known a small child to receive a broken leg after a blow from a gander's wing. The bill can also inflict injury, by giving not so much a peck as a painful biting pinch which causes severe bruising. Anyone who has been attacked by a gander will know what I mean. Ganders will not usually attack their owners, although this is by no means certain and I know of one lady who had to feed her geese armed with a dustbin lid as a shield during the breeding season.

Egg laying begins in the late winter and, as each egg is laid, the goose will cover it over with straw or whatever material she can find. She will lay one a day until there is a clutch of about twenty and then proceed to sit on them. Geese can be extremely frustrating in their choice of nests. I have many times provided large nesting boxes with straw, only to have them spurned in favour of seemingly quite unsuitable places where the only bedding material is a few broken twigs. There is no point in trying to move them, and I have found that the best policy is for 'the mountain to go to Mohammed'. Once the geese are settled in their chosen spot I construct a shelter around them using straw bales, roofed with old wooden doors.

Most domestic geese are good sitters and will sit patiently for the thirty days of the incubation period. During this time the gander is extremely protective and a call from the sitting goose will bring him honking and flapping furiously in the direction of the intruder. Geese occasionally have the irritating habit of wanting to share the same nest and the clutches of eggs become hopelessly mixed. This can be a nuisance if the eggs have begun incubation at different periods, as both geese may end up with newly hatched as well as later eggs, and some of the goslings may die in the shell as a result. Again there is little that can be done to dissuade the geese from nest sharing, and the best bet is to mark each egg clutch before they get a chance to become mixed up, and return them to the appropriate goose.

Using a broody hen

Broody hens have been successfully used for hatching goose eggs, but as the eggs are too heavy for them to turn, you will need to do this for them several times a day. An 'X' marked on one side will help you to ensure that alternate sides face upwards. The eggs will also need to be sprinkled occasionally to make sure the internal membranes do not dry up. Two eggs are usually enough for one hen, unless she is particularly big in which case she can manage three.

Artificial incubation

Eggs incubated in an incubator are treated in the same way as chicken's eggs (see page 33). The only added factor is that a slightly higher humidity is required for goose eggs. Normally incubators have a built-in humidifier and precise instructions will be provided for specific eggs. With a general all-purpose incubator I have achieved satisfactory results by giving the eggs a fine spray from a small garden sprayer every day.

Brooding

The mother will normally brood her goslings, but is nowhere near as protective as most broody hens. Although the ganders are attentive to the brooding geese they seem quite ineffective against rats. I once lost four goslings in the space of twenty minutes and could not understand where they had gone until I lifted up a piece of corrugated iron and found two rats busily gorging themselves on the corpses. It may be better, if you are afraid of losses from rats, to take the goslings away from the mother and rear them in a safe warm place until they are big enough to defend themselves.

Details on feeding goslings are given in the *Feeding* section above. On no account should they be put outside to graze too early, and three weeks old is soon enough. Access to water for swimming should also be denied until they are fully feathered and have lost their baby down. When they do go out to graze they should ideally go out onto fresh pasture in order to minimize the possibility of parasitic infection. The gizzard worm which infests heavily grazed pasture in some areas is responsible for killing many goslings. Adults can tolerate a certain level of parasites but the young ones cannot.

Using goose eggs

If you do not wish to incubate all the eggs there is no reason why you should not use them. They are, of course, much larger than hen's eggs and anyone who eats a boiled goose egg for breakfast will probably not want anything else until quite late in the day. The shell is very strong and difficult to break, but the eggs make superb omelettes and are also excellent for use in cooking. Many people use the shells for decorative purposes, painting them with beautiful and elaborate designs.

Turkeys

Most people enjoy eating turkeys, and as a Christmas or other festival dish they have become firmly established. There is no doubt that they put on weight rapidly and for this reason they have been commercially exploited to extremes. It is now possible to buy maxi, midi or mini turkeys, depending upon the size of your oven. They have also been selectively bred to have shorter legs and enormous breasts, to the extent that natural mating is impossible for some of the heavy breeds. Commercially, all breeding is by artificial insemination – which increases the overall rate of fertility.

When it comes to keeping turkeys, people seem to be either all for them, or dead against them. It must be admitted that they are not the most attractive of birds; the combination of small head, bulky body and flapping hysteria puts many people off. They are certainly not ideal for people with limited space but, in the right conditions and properly managed, they are full of interest and provide a useful source of income.

It should be emphasized however, that the small turkey rearer will not be able to compete with the huge broiler turkey industry. The cheapest turkey is the frozen one which anyone can buy in the supermarket, but there is a growing minority of people who wish to buy freshly killed birds that have been raised under natural conditions, and it is here that the smallholder has his potential market – although it will be a limited, localized one. I do not keep turkeys commercially, and rear them only for my own family's use, but a neighbour regularly rears about a hundred for the Christmas market, and starts taking orders for freshly killed birds from October onwards. He can never produce enough for all the orders he receives and has a good return.

Breeds

When the early colonists arrived in the Americas they found the woods teeming with large bronze birds, which they discovered were turkeys. There were at least five different varieties recorded from Central Mexico up to Florida and New England.

The wild turkeys were hunted almost to extermination in many areas, but meanwhile they were also being domesticated and selective breeding was beginning. Original stock taken from America to Europe was developed and then reintroduced. Gradually a heavier, meatier bird began to appear and in the mid-nineteenth century the famous

American Mammoth Bronze cock

Bronze turkey was developed. From this, newer strains such as the Broad-Breasted Bronze and Broad-Breasted White have been evolved and these types are still some of the most popular.

In North America there was a larger choice of breeds than in Europe, and breeds such as White Holland, Beltsville, Narragansett and Bourbon Red were frequently kept. In Britain, until the commercial turkey industry evolved, the choice for the farmer was usually between Mammoth Bronze, White Austrian or Black Norfolk, although Cambridge Bronze (a small version of the American variety) was also popular. In Australia both bronze and white varieties were kept.

When the turkey industry developed, the emphasis was on hybridization to produce quick-growing hybrids, and these commercial strains, such as Arnewood International's Double AA or Treble CCC or the Nicholas in North America, are now more readily available than some of the older varieties. From the smallholder's point of view, these commercial hybrid strains are the best proposition, unless he has a particular interest in raising older varieties.

Housing

Turkeys can of course be kept on free range as they are reasonably hardy birds. This method is not very common, although it can still be seen in parts of North America, in Essex in Britain and in areas of France such as the Dordogne. There is a greater risk of disease with free ranging, and turkeys should only be kept on ground which has not been previously used by poultry. Damp, badly drained soil is

A good example of small turkey housing which provides weather protection as well as ventilation.

Automatic watering laid on outside the turkey house.

A commercial version of the traditional pole barn.

A stag displaying.

the worst type of terrain for them, for it is in these conditions that the disease Blackhead is commonly found. Dry, chalky soil is best, particularly where trees can provide natural shade. Turkeys are tree perchers by nature, but now that they have been developed as heavy breeds, they are not as nimble as their wild forbears, and flying up into trees can be hazardous. In fact, a friend of mine lost one of his birds when it tried to fly up and promptly fell and broke its neck.

The fold method, as used for chickens, has also been applied to turkeys. It confines the birds to one area and the fold unit is moved on at regular intervals to a fresh piece of ground.

The verandah or sun-porch system I have described in the *Ducks* section on page 42 has been widely used in North America, and this is suitable for a small number of

birds, particularly where weather conditions are not extreme.

Most commercial turkeys are raised nowadays in the same conditions as broiler chickens with breeding stock housed in purpose-built battery cages. In the past, however, most turkeys were housed and reared in a variation of the pole-barn system. This is still favoured by many of the smaller turkey rearers and provides a hardy, healthy and relatively natural environment for the birds. Basically the housing provides a roof, a dry floor with a thick layer of clean litter such as woodshavings, and relatively open walls. These are usually poles or wooden slats covered with wire mesh which confines the stock, while providing good ventilation.

For those who wish to raise only a small number of birds, any sound, dry shed or building will do, provided there is adequate ventilation and you do not overstock. Too many birds in too small a space produces a high risk of infection.

A newly hatched turkey poult.

Newly hatched turkey poults in a brooder. Note the feeder, small drinkers and overhead lamp.

Young poults, four weeks old, transferred from the brooder to a larger house, but still needing artificial heat.

Young turkeys transferred to an open house. At this stage the sexes are separated.

Feeding

The best feeders and drinkers are the ones which can be suspended so that litter is not scratched into the food and water. Automatic watering is of course the ideal.

The general pattern for feeding turkeys is similar to that for broiler chickens. Like chickens they require a starter crumb feed for the first four weeks, followed by rearer's ration until two weeks before killing, when a finisher ration is given. If they are not fed on proprietary feeds they will put on weight more slowly. The most convenient method is to leave the feed permanently available for the turkeys to eat at will.

Traditionally, turkeys kept in small numbers on general farms were given a home-mixed feed of bran, maize meal and fish meal, together with mixed grain, twice a day.

Breeding

Generally speaking, it is not worth taking the trouble to breed your own turkey chicks or poults, as they are called, for they are difficult to rear in the early stages and quickly succumb to chills if there are any temperature fluctuations. It is much better to buy in young poults and rear them to table weight. If you get them from a reputable source they will have been treated against Blackhead, and will stand a good chance of doing well. If you are determined to breed your own, it is better to incubate the eggs artificially or under a broody hen. The hen turkey lays an average of about 100 eggs in one breeding season, which lasts for about five months, although she can be brought into lay at any time by the provision of artificial light.

The mating ratio is one male, or stag, to every ten hens,

and incubation takes twenty-eight days. After hatching, the young poults should be treated in the same way as young chicks, and kept in a brooder with artificial heat until they are hardy enough to cope – usually at four to five weeks old.

Turkeys for the table

Your decision on when to kill will depend entirely on the type and weight of the bird. The heavy strains are ready at about 32 lb (14.5 kg) which is reached in approximately twenty-four weeks. The medium weights achieve about 15 lb (7 kg) by their killing time of sixteen weeks, and the small strains average 9 lb (4 kg) at twelve weeks. It is important to know precisely which strains you have, and how long you should keep them before slaughtering, because your feeding costs will need to be calculated accordingly. Another factor is that hen turkeys have a slower growth rate than the stags, and for this reason they are usually segregated at about seven weeks old. The males are then given a higher protein ration to cater for their more rapid growth. As this special feed is expensive, it would obviously be uneconomic to feed it to the slower-growing hens.

Turkeys are killed by neck dislocation and it is the custom to bleed them immediately, so that the flesh is whiter than it would otherwise be. This can be achieved by cutting the jugular vein in the throat and catching the blood in a container.

Plucking is the same for turkeys as for other poultry and can be done wet, dry or by the use of wax. The legs of turkeys, however, are endowed with particularly strong tendons or sinews, and it is the practice to remove these. This involves cutting along the shank bone above the foot, and then drawing the stiff white tendons out one at a time. A large nail or a hook is the best tool for the job.

Guinea fowl

Guinea fowl belong to the pheasant family and derive their name from the Guinea Coast of Africa which is where they are thought to have originated. The Greeks and Romans are known to have kept them for the table, and throughout Europe they were prized as a rare delicacy, masquerading under the name of Tudor Turkeys.

The flesh is white and delicate with a distinctive game flavour, and carcasses weigh $1\frac{1}{2}$–$2\frac{1}{2}$ lb (0.5–1.1 kg) at nine to ten weeks which makes them suitable table birds for the small family.

It is only fair to warn those who have never kept guinea fowl before, that they are flighty to the extreme and, if allowed to free range, may wander long distances and fall prey to foxes. They will also perch in trees and are capable of flying over high fences unless the flight feathers of one wing are kept trimmed (see *Wing clipping* on page 47).

They are nervous, easily agitated birds and the females in particular will utter loud calls at anyone's approach. This makes them good watchdogs, but is extremely trying if you

happen to dislike their particular warbling shrieks. They are not suitable for anyone with limited space or with neighbours living close by. Where space is available, however, a stocking rate of 500 birds to the acre (0.5 ha) is the average for good pasture. They are particularly good for clearing land of insect pests, and are frequently used in this way after a crop has been harvested.

Breeds

The three most common varieties of guinea fowl are the Pearl, the White and the Lavender. The Pearl, sometimes referred to as the Grey, has purplish grey feathers with white markings. The White is altogether lighter in its markings, although it may still have some grey in its feathers. The Lavender has light grey feathers marked with white – and if this leads you to think that it might be difficult to tell some of the varieties apart, you are quite correct.

These varieties are suitable for the smallholder who wishes to keep a few around the place, either as watchdogs or to provide the occasional meal. For anyone who wishes to keep guinea fowl on a commercial basis however, it is essential to acquire one of the hybrid strains that have been selectively bred for their quick-growing capacity. Usually it is only specialist breeders who can supply these strains, and they will often refuse to supply breeding stock, preferring to sell only day-old keets for you to rear for meat. In this way you are forced to keep going back to him for new stock. Generally speaking hybridization of guinea fowl for meat strains has been far more widespread in Britain and France than elsewhere.

Housing

Commercially guinea fowl are kept confined in suitably adapted buildings on a 3 in (75 mm) litter of woodshavings. Adequate ventilation is important because the droppings are much drier than those of other poultry, and this leads to a dustier atmosphere and therefore to an increased risk of Aspergillosis or Farmer's Lung disease. For the same reason you should avoid overstocking.

Plan of a guinea fowl house.

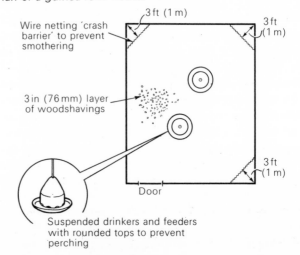

3 ft (1 m)
3 ft (1 m)
Wire netting 'crash barrier' to prevent smothering
3 in (76 mm) layer of woodshavings
3 ft (1 m)
Door
Suspended drinkers and feeders with rounded tops to prevent perching

Guinea fowl keets being started off with artificial heat in a brooder.

Young guinea fowl just taken off heat.

If you are keeping large numbers it is worth putting internal wire netting across the corners of the building, about 3 ft (1 m) away from the wall. This provides a crash barrier and, in the event of a sudden crush, will prevent the birds from getting smothered. Guinea fowl are easily panicked and will stampede away from the source of disturbance, which could be something as trivial as the opening of the door.

Feeding

Feeders and drinkers should be provided in the proportion of one to every twenty birds. Those inside should be of the suspended, rounded type which discourage perching.

Guinea fowl have relatively small crops and therefore need feeding more frequently than do other poultry.

Commercially it is therefore best to make the feed permanently available as labour costs would otherwise be very high.

Proprietary chick feed is suitable for the young guinea fowl (or keets, as they are called), followed by a normal grower's ration. The best feed conversion ratio is gained from a broiler-type ration, similar to one that would be used for broiler chickens. You must, however, take care over your choice of broiler ration, as there are some coccidiostat additives which are toxic to guinea fowl if they exceed certain levels. You must check this point with your feed supplier and local poultry officer before giving your birds such a ration.

Traditionally, keets were given a mixture of hard-boiled eggs, bread crumbs and biscuit meal moistened with skim milk for the first week. This was followed by a mixture of barley, wheat and meat meal, ground oats and cooked rice, fed as crumbs. From four weeks onwards dry, mixed grain was added to their diet.

Breeding

Guinea fowl are not particularly good mothers, and the eggs are best hatched under broody hens or in incubators. The incubation period is twenty-eight days.

Artificial heat in brooders will be needed for the young keets and should be adjusted as follows:

Day-olds	95°F (35°C)
After 3 days	88°F (31°C)
After 10 days	84°F (29°C)

Thereafter reduce the temperature gradually until by the twenty-eighth day, it is down to 70°F (21°C). After this period, a temperature of about 65°F (18°C) is the ideal, but they can obviously withstand fluctuations once they are old enough. One of the important things to remember is that guinea fowl keets are particularly vulnerable to damp, so you should make sure the brooder is kept dry. By contrast, the adults are extremely hardy and can withstand quite severe weather conditions without any ill effects.

Guinea fowl for the table

If they are housed indoors and fed on proprietary rations, the birds will reach their killing weight of 1½–2½ lb (0.5–1.1 kg) by the age of nine or ten weeks. If they are kept outside, on a non-intensive basis, this will take nearer sixteen weeks in the case of a commercial strain, or, with the older slow-growing types it may take as long as six months.

Killing, plucking and dressing is the same as for chickens, and the birds freeze extremely well.

The guinea fowl is attracting increasing attention as a supermarket and restaurant delicacy, which makes it of special interest to anyone wanting to keep poultry as a small-scale commercial venture. Before embarking on such a project, however, you would do well to seek advice on the marketing conditions in your area.

External view of a pigeon loft, showing alighting platform. Note the ventilators and the brick supports to deter rising damp.

Inside a pigeon loft, showing locker-type box perches, a nestbox, feeder and drinker. (The birds shown here are not meat breeds.)

Pigeons for squabs

Pigeons are usually kept as show birds or for sport as racing birds and their well-developed homing instinct, which means that they will fly back to their own loft from wherever they are released, is well known. This feature has been used in the past as a means of communication – the message being borne in the pigeon's leg ring. Even Reuters, the international news agency, has in the past transmitted its news via the humble pigeon.

The other aspect of pigeon keeping (which has now largely fallen into abeyance) is that of raising squabs for meat. Squabs are young pigeons, and at one time they were a common source of fresh meat for the winter when not much else was available. The Romans kept them and in medieval times in Britain it was the custom for every manor to have its dovecote. It was in fact against the law for anyone other than the lord of the manor to own one.

The best breeds for meat squabs are the Mondain, the King and the Carneau. The Mondain is not available in Australia where the most popular breeds are the White King and Red Carneau.

Housing

The best form of housing is a pigeon loft and there are several types available on the market. A house should provide an external landing place for the pigeon when it alights, shelter from the elements and good ventilation. Inside it should be equipped with a feeder, drinker, perches and nesting boxes.

The perches are normally square, locker-type openings

measuring approximately 11 in × 11 in × 3 in deep (280 mm × 280 mm × 75 mm). Nestboxes are placed lower down than the perches to dissuade the birds from sleeping in them.

Pigeons are a favourite target for many predators and pests and it is therefore sometimes desirable to confine them rather than to allow them to fly free. A wire fly pen can be attached to the front of the pigeon house or dovecote. If it is large enough, the nests as well as the feeders and waterers can be placed in the protected area and the birds can exercise and sun themselves in the fly pen. In more severe climates it may even be best to have a totally enclosed building with a fly pen attached to its front.

Feeding

A normal pigeon mixture available from feed suppliers is adequate, or you can feed them a grower ration like that used for chickens. Traditionally, pigeons were given a mixture of home-ground maize, maple peas, dari and wheat. This is a fattening diet particularly suitable for meat production. Pigeons can be selective in their eating habits and often prefer whole grains which have not been ground. The most convenient method of feeding is to keep hopper feeders permanently filled with whatever mixture you are using and leave the pigeons to eat from them at will.

Breeding

Pigeons lay two eggs at a time, but to make up for this small number they are prolific breeders and will produce up to eight pairs of young a year. Good working pigeons will start building their new nest about ten days after hatching their last eggs, and while they are still feeding the squabs.

Incubation starts immediately after the second egg is laid, and lasts for sixteen days. The male and female take it in turns to sit on the eggs, and for the first six days after they hatch will feed the young on 'pigeon's milk'. This is a food which forms in the crop during incubation and is regularly regurgitated by the adults.

The young are kept covered for two to three weeks, by when they will have enough feathers to resist the cold. By this time they will begin to feed themselves, and when they reach twenty-eight days of age they will be entirely weaned from their parents.

At this stage they should be culled for the table, while they are at their most tender and before they start to become flighty and lose condition. They are killed by neck dislocation and plucked in the same way as other fowl. However, their skins are very tender so you must take extra care not to damage them during the dressing process. They can be roasted, stewed, barbecued or used in the traditional delicacy of pigeon pie.

Quail

Most people, it seems, have never seen quail, and often the only thing they know about these birds is that they produce the delicacy of 'quail's eggs in aspic'. When pressed, they will usually admit that they have never tasted that either. It sometimes surprises these people to learn that the quail is not a chicken-sized domestic bird, but a small, shy little creature with an adult weight of only 5–8 oz (140–225 g).

Breeds

The Japanese breed, Coturnix japonica, is the best one to keep. Under the right management conditions this breed is capable of producing nearly 300 eggs a year. For the table the Bob White Colinus Virginianus may be more suitable. It is heavier than the other breeds, and is the one most frequently kept in North America.

In Japan and China the quail industry is far bigger than it is in the West, where it is only in the USA that there has been any significant development. In Australia it is attracting increasing interest although it is still as yet a small industry.

Housing

Quail are hardy, and for those who are only interested in keeping small numbers a small chicken house and run is adequate. Being small birds they are more vulnerable to rats than other poultry and you should take this into consideration if you are using outdoor housing.

Commercially, quail are kept inside, away from extreme changes in temperature, and they are separated into fattening and egg-laying stock. Cages are frequently used, but I personally favour the litter rearing method. With litter rearing however, you must take care to avoid overstocking which can lead to cannibalism. Woodshavings form a suitable litter and they should cover the floor to a depth of about 2 in (50 mm).

One method of housing is to put each pair in a small wire-floored run about 1 ft 6 in (0.5 m) wide and 4 ft 6 in (1.5 m) long. These can be arranged in continuous rows with nesting compartments at one end and protected feeders at the other. Where quail are kept in flocks it is a good idea to put up solid partitions between the compartments as quail are much less nervous if they can't see their neighbours or pen mates all the time.

Feeding

The only possible method for feeding such small birds is to leave their feed out for them all the time. A turkey or starter ration is suitable, followed by a grower-type feed. Water should, of course, be available at all times. Generally speaking, the birds need 30 per cent protein for the first week of life, reducing to 26 per cent for the next two weeks, and finally 20 per cent from four weeks onwards.

Breeding

Laying begins when the birds are ten to twelve weeks old and the usual mating ratio is one male to three females. The incubation period for quail's eggs is eighteen days and they are best hatched artificially. Japanese quail are wholly incapable of going broody and hatching their own eggs. Brooding will be similar to that of other poultry, but you will need to take extra care to ensure that they have no means of escape. They are extremely small and will get out through the tiniest crevice. It is important too to place pebbles in the chick drinkers to make the water shallower, otherwise there is a risk that the tiny chicks will drown.

Quail for meat and eggs

Killing for meat
When newly hatched the young quail weigh not more than $\frac{1}{4}$ oz (7 g), but by the time they are six weeks old they have reached a killing weight of approximately $\frac{1}{2}$ lb (227 g). The feed consumption during this period is in the region of $10\frac{1}{2}$–11 lb (4.7–4.9 kg).

Killing, plucking and dressing is done in the same way as for chickens but with extreme care. Quail are normally marketed in pairs or in fours, arranged in packs.

Eggs
If eggs are your main priority, the laying stock is best kept indoors on litter in the way described earlier.

The birds come into lay at ten to twelve weeks old and are capable of producing up to 300 eggs a year. In order to achieve this level of production, they do need to be properly fed and a laying quail will require just under $\frac{1}{2}$ oz (15 g) of feed per day plus oystershell, ground small.

The eggs are small and are usually served in one of the following ways:

1. Hard-boiled for decorating salads
2. Hard-boiled then pickled as hors d'oeuvres or as a cocktail accompaniment
3. In aspic jelly – make the aspic by mixing $\frac{1}{2}$ oz (15 g) gelatine in $1\frac{1}{3}$ pt (0.9 l) of stock. Season and heat to boiling point. Stir well, strain and, when it is beginning to cool, pour it over the hard-boiled eggs in a mould. Leave until the aspic jelly has set and serve with green salad.

Health and disease

It is of course important to make every effort to keep your poultry free from disease, and the preventive measures described in the *Chickens* chapter on page 37 hold good for all fowl. However, disease may still occur even under the most carefully controlled conditions, and it is useful to know how to recognize and deal with the ailments you are most likely to come across.

Signs of ill health

A healthy bird is normally alert, curious, interested in its food, bright-eyed and with an upright stance. Any fowl standing in a dejected, drooping position or huddled up

Stance of the sick bird. Its closed eyes, hunched position and trailing wing feathers all indicate that it is far from well.

into its feathers (see photograph) should be isolated from the others without delay. I have a small pen which is called 'the hospital' where any suspects are put for a few days until they recover. Symptoms to look out for are diarrhoea, lack of appetite, coughing or wheezing, runny nostrils, eye discharge, lameness, trembling, sudden pecking of the feathers, or indeed any other unusual behaviour. Always attend to the sick birds after the healthy ones to avoid transfering the disease.

Diseases

Aspergillosis
This disease, which is also called Brooder Pneumonia or Farmer's Lung, is caused by a fungus inhaled by the birds, usually from mouldy litter or food. The symptoms are rapid breathing, gasping and possibly inflamed eyes. There is no known cure so it is best to cull affected birds and replace any musty litter or feed. The condition can affect all birds, and also human beings. Hasty harvesting results in mouldy hay and straw that throw off dust containing the dangerous fungus spores. Only clean hay and straw should be used as litter.

Avian tuberculosis
Older birds are the most vulnerable to this disease, and for this reason it is unwise to keep poultry too long, or to allow young and old birds to mix. Symptoms are loss of weight, lameness and progressive swelling of the joints. It is usually found in chickens and turkeys but ducks and geese may also be affected by avian tuberculosis, so it is vitally important not to let poultry free-range all over the farm. The disease is also carried by wild birds. Diagnosis is a job for the veterinary surgeon and, as there is no cure, it may be best to cull the entire flock on his advice. In Australia this disease must be notified to a Government Veterinary Officer.

Blackhead (Infectious Enterohepatitis)

Turkeys are more susceptible to Blackhead than other birds, although chickens may also be affected. It is caused by a protozoan transmitted either via the droppings of an infected bird, or in the eggs of the cecal worm passed out in the same way. It can survive in the soil for several months. Symptoms are weight loss, general droopiness and yellow diarrhoea, together with the discolouration of the head, which gives the disease its name. It is usually fatal in young turkey poults up to the age of twelve weeks and the only possible action is to inject all young stock as a preventive measure. Turkeys should always be kept separate from chickens and should not be allowed on ground where chickens have been grazing.

Botulism (or Limber Neck)

This condition, where the birds lose control of their neck muscles, is caused by bacteria found in decaying organic matter. It is more common in waterfowl than in other poultry because of their tendency to dabble in mud, although losses in domestic waterfowl are much lower than those sustained in the wild. If diagnosed in time, half a teaspoonful of Epsom salts in a dessertspoonful of water may improve the condition and eventually bring about a cure. It is most important to remove decaying vegetation from pools and runs, and to maintain the highest standards of cleanliness.

Bumblefoot

The first time I encountered Bumblefoot was in a two-year-old Rhode × Light Sussex hen which was limping and generally bumbling along in the wake of everything else. An examination revealed a hard abscess on the ball of the foot.

The condition is caused by a small wound which heals on the outside, but remains infected inside and gradually forms a hard core of pus. In chickens it is often the result of having a perch too high so that the bird lands awkwardly on the hard floor. The height of perches should be no more than 30 in (0.8 m). The treatment is to cut open the abscess with a sharp sterilized scalpel, squeeze out all the pus, including the hard core, and then apply an antiseptic. Keep the bird by itself in a clean straw pen for a few days, until the wound has healed. All birds can be affected but it is most common among chickens and ducks. Abscesses may occur on any part of the body as a result of undetected injury and the treatment procedure is the same, but if there is any doubt as to the nature of the swelling, or if it is in a vulnerable place such as the neck, seek the advice of the veterinary surgeon.

Chronic Respiratory Disease

Poultry are just as susceptible as we are to the common cold, particularly if their houses are damp and draughty. Colds usually disappear of their own accord, but it is important not to ignore them, for if the symptoms persist they may be paving the way for more serious complaints. It is as well to separate the bird from the flock and keep it in a warm straw pen. The nostrils and eyes may be sponged with a 10 per cent boracic acid solution if they are blocked by thick mucous discharge.

Coccidiosis

This disease is caused by small protozoan parasites which attack the intestines and are transmitted via the droppings. Symptoms are a watery diarrhoea which may be blood-stained, and a sick, ruffled-looking bird. Sulphonamide drugs are available from veterinary sources and can be administered in the drinking water. However, whilst these are effective, you must at the same time pay attention to hygiene avoiding a build-up of droppings and periodically moving the birds to fresh ground.

Cramp

Little is known about this condition, which appears to affect ducks more than any other poultry. The cause might be fatigue, lack of exercise, faulty feeding or cold. The first indication is when a duck is found to be hobbling badly and swinging its tail over to one side in an effort to maintain its balance. The condition may become so acute that walking is impossible. I find that the best course of action is to isolate the duck in a sheltered, straw-lined pen and leave it there until it recovers. This may occur in a day, but I have known cramp to continue for as long as four days, after which the bird is back to normal and can rejoin its friends. The condition can be brought on by deficiencies in calcium or vitamin D in which case supplements of these should be included in the treatment.

Dropped wing

This condition is most frequently found in geese, although it has been known in ducks as well. It appears to be a result of muscular weakness, making the wing hang down lower than normal, in its resting position (see photograph). Slipped wing is a variation of the same condition, where the flight feathers stick out at right angles. Both conditions are harmless if somewhat unsightly, and do not appear to cause any distress.

A Muscovy drake with 'dropped wing'.

Duck virus hepatitis

This is a highly infectious and fatal condition affecting young ducklings. The first indication is when their movements are seen to be uneven and wobbly. They then quickly succumb and die. There is no cure, but a preventive vaccine is available.

Egg-laying disorders

Full details of these are given in the egg-laying section of the *Chickens* chapter (see page 29).

Feather and vent pecking

These bad habits are usually the result of boredom or overcrowded conditions and are not usually prevalent among free-range birds. If this kind of pecking does occur, the culprit should be separated immediately and left for a few days in solitary confinement to try and break the habit. Attention should be given to improving the environment and management, and only as a last resort should you de-beak the bird. De-beaking involves removing the tip of the upper beak, but as pecking for food is a natural behaviour pattern of the hen I am personally against interfering with her ability to peck unless absolutely necessary.

Fowl cholera

This disease is found in ducks, geese, chickens and turkeys. It is an infection transmitted through contaminated water, soil or feed and symptoms are acute thirst, difficult breathing, loss of weight and a bluish tinge to the combs or wattles. Infected birds should be isolated and treated with sulphonamide drugs which you can obtain from the vet. Vaccination will prevent infection. It is more common in North America than elsewhere.

Fowl pest (or Newcastle disease)

Young chicks are particularly at the mercy of this virus which strikes suddenly and apparently from nowhere. All poultry (but most commonly chickens and turkeys) can be affected. There is no treatment other than preventive vaccination in the web of the foot. In chicks the disease is usually fatal, but older birds often recover although their egg production never returns to its previous level. Symptoms are sneezing and coughing followed by nervous disorders such as twitching or spasms, and there is characteristic green diarrhoea. There are other similar diseases and diagnosis is a job for the expert. Fowl pest has not yet occurred in Australia but there is always a danger that it will be introduced and any suspected cases should be notified to a Government Veterinary Surgeon.

Fowl pox

This is a virus infection which affects chickens and turkeys. It is recognized by the wart-like swellings on the head and around the beak. There is no cure and prevention by vaccination is the only way to deal with this condition.

Fowl typhoid

Fowl typhoid is a bacterial infection which can affect all fowl and is characterized by a droopy appearance and ruffled feathers. The combs and wattles become pale and there is an overall loss of appetite and lassitude. Sulphonamide drugs in the drinking water will cure it.

Frozen combs and wattles

This is a condition which sometimes affects poultry overwintering outside in particularly cold winters and in northern areas of the world. It is usually caused by drinking which makes the comb or wattles wet and vulnerable to frost-bite. Treatment is as follows: rub the affected parts with a mixture of 5 tsp (25 ml) vaseline, 1 tsp (5 ml) glycerine and 1 tsp (5 ml) of turpentine, twice a day.

Gapeworms

If you see a bird repeatedly opening its beak and gaping without making any sound, you should suspect an infestation of gapeworms. The worms are transmitted via contaminated soil or food, and migrate to the lungs and trachea where they become embedded. A heavy infestation leads to breathing and feeding difficulties, and the bird will become dull and lose its appetite. It affects chickens more than other fowl. The traditional treatment was to take a quill feather and strip it clear except for a small tuft at the end. This was dipped in turpentine, then carefully inserted into the bird's windpipe, rotated, and withdrawn. Many of the worms were brought up in this way, but there was the danger of causing suffocation or lacerating the windpipe. A more effective treatment is to isolate the infected bird and treat it with a poultry worm medicine available from a vet. The only long-term solution, if poultry are free ranging, is to ensure regular rotation of the ground so that a build-up of parasites does not occur.

Gizzard worm

The gizzard may become infested by fine, cotton-like worms which become attached to its inner lining. The condition is primarily found in geese, as a result of overgrazing the same piece of ground. Older geese can resist a certain level of infestation, but young ones can quickly succumb. This is why it is so important to have fresh pasture for young geese when they start grazing in the spring. The first symptom is when the gosling develops a slow, staggering gait, and quickly weakens and wastes away. A worm medicine from the veterinary surgeon added to the drinking water will kill off the parasites and save the birds, if it is administered in time, but the only long-term preventive is adequate rotation of pasture. Other poultry should not be allowed to run with geese because of the possibility of picking up this parasite.

Infectious coryza

This bacterial infection affects chickens only and is often transmitted by dust in the air. For this reason adequate ventilation is essential. The symptoms are an inflammation of the eyes and nostrils, and a general swelling of the face. One or both of the eyes may be closed with conjunctivitis. It also produces a distinctive odour in the hen house.

Infected birds should be isolated immediately. Sulphonamide drugs in the drinking water will bring about a cure but the bird may be left in a permanently weakened condition, and it may be best to cull it.

Leukosis

This is an infectious form of cancer caused by a virus which is transmitted from parent to offspring through the egg. Tumours form in the adult's internal organs and cause diarrhoea, wasting and death. There is no known cure.

Lice

A bird with lice will be seen to start frequently and to peck its feathers. When the feathers are parted the lice will be seen scurrying across the skin. They are small, greyish parasites which feed on blood, skin debris or the roots of feathers, depending on the type – and there are over forty species! Louse powder is effective in controlling them and should be applied once every four days for about a fortnight. It is important to treat the houses and perches as well, and it may be quickest to remove the birds and fumigate their house. Remember to give the dust bath a periodic dust of louse powder too.

Marek's disease (or fowl paralysis virus)

This is a highly contagious virus infection which affects young chickens and results in paralysis. There is no cure, but you can vaccinate day-old chicks against it.

Pullorum (or bacillary white diarrhoea)

This infection is transmitted via the egg and it is therefore vital to breed only from stock which has been blood-tested and is known to be free of the disease. Most reputable suppliers of young stork or of day-olds should be able to give the buyer an assurance that his breeding stock has been so tested. In Australia they are required by law to carry out these tests. Symptoms in the young stock are listlessness, and huddling together, followed by rapid death. There is no cure.

Red mite

This is one of the most common mites affecting poultry and, although the parasites are grey in colour they soon become red after a good feed of blood from their host. Measuring little more than $\frac{1}{50}$ in (0.5 mm), they spend most of their time hiding in crevices in poultry houses, coming out (particularly at night) to feed on the roosting birds. The first step in eradicating them is to give the house, perch and nestboxes a thorough clean. Creosote painted inside and outside the house is effective in getting rid of them, but I generally find that it is more satisfactory to do this in the winter when the hens are being over-wintered elsewhere for birds cannot be re-introduced for several weeks after creosoting or they may suffer 'burns' to the breast feathers and legs when perching. The use of a blow-lamp is an effective alternative to creosoting.

There are other mites which spend the entire time on the bird, such as the black mite in the USA and the northern

Scaly leg mite infestation has pushed up the leg scales of this Silkie.

fowl mite in Australia. These should be treated with dust or spray insecticides.

Roundworms

Most birds can tolerate a certain level of roundworms, but these parasites may build up to an intolerable level under unhygienic conditions. Here again the importance of making sure that the droppings do not build up and of moving the birds onto fresh, uncontaminated ground from time to time cannot be overemphasized. Many drugs are available from veterinary sources and are easy to administer, either in the food or water.

Scaly leg

A small mite is the cause of this irritating complaint. It finds its way under the scales of the leg, bores its way through the skin, and forms crusty deposits which force the scales outwards. Chickens are most commonly affected, although turkeys, guinea fowl, quail and pigeons may also get it. Discomfort and lameness result and the bird should be isolated and treated without delay. The crusts can be easily removed with warm soapy water and an old toothbrush, but a preparation such as benzyl benzoate is needed to kill off the mites. The traditional cure was to apply an ointment of 1 part caraway to 5 parts vaseline, but paraffin is equally effective, as long as the treatment is continued until all the mites are killed off. The proprietary medication available from the veterinary surgeon is much quicker and more effective than the traditional treatment. Before you begin the treatment the housing should be thoroughly cleaned and the litter burned to prevent re-infection.

Wry neck

This condition differs from limber neck or botulism in that the neck is bent tautly over the body. It is usually the result of brain or nerve damage in the young bird, and it is best to put down any bird which is suffering from it.

RABBITS

Buying rabbits

Why keep rabbits?

If you are looking for livestock that is inexpensive to buy, easy to look after and does not make a noise, then you could do worse than to get rabbits.

No one is quite certain where they originated, or for how long they have been kept as domestic animals, but the Romans were enterprising enough to keep them in special enclosures called 'leporaria'. From these they were regularly culled for the table and were sufficiently highly thought of to have had their image engraved on the coins of Emperor Hadrian.

Rabbits provide an inexpensive source of meat that by comparison with beef, lamb and pork, is low in cholesterol and yet high in protein value. They can be kept in a relatively small area and are particularly suitable for people whose access to land is limited. Even urban dwellers have been known to produce their own meat from hutch rabbits in back yards or on balconies. For those who wish to make an income from their livestock there is a steady demand for meat rabbits, not only from local butchers but also (in Britain, although not in the USA) from the meat packers who are supplying the supermarkets.

Which to choose

All breeds will provide meat for the table including wild rabbits. The latter, however, are a pest in large numbers and in Australia it is illegal to keep them. Restrictions on keeping rabbits in Australia are laid down by the individual states. In some it may be necessary to apply for a permit to keep rabbits so you should check up on the current legislation in your area.

There are only a few breeds which are specially bred as commercial strains. The main ones are the New Zealand White and the California. Both of these have a greater meat to bone ratio than other breeds and have been selectively bred for rapid and efficient growth. For anyone starting with meat rabbits these are the two breeds that I would recommend. Other breeds which are kept for meat, but to a far lesser extent, are Champagne d'Argent, New Zealand Reds and Blacks, Satins and Chinchillas.

Angoras are frequently kept by spinners for their fur, which is akin to wool and can be spun into quality fibres. The skins of all breeds can be tanned, and there is often a demand for top grade pelts. California and New Zealand are the ones most frequently requested.

How to buy

The best place to buy stock is undoubtedly from a reputable commercial breeder. He will have good quality stock that has been specially bred for the meat trade, and he should provide a guarantee that the rabbits are healthy and disease-free. Check that they have been vaccinated against myxomatosis.

If you buy rabbits from an amateur keeper who is otherwise unknown to you, you stand a good chance of acquiring weakly stock whose ancestry is unclear, and which may prove to be more trouble than they are worth. Breeders of show rabbits too may not be a good source, for although they may be reputable and experienced, their rabbits are not necessarily good meat strains.

Housing and equipment

There are basically three ways of keeping rabbits, not counting the traditional method of semi-artificial warrens where the wild rabbits were fed through the winter, then regularly trapped. This method was wasteful in that the rabbits were frequently caught by predators, or else they escaped and raided the farmer's crops, and diseases such as myxomatosis could wipe out the whole colony. The more efficient and most commonly used ways of keeping rabbits are either in hutches, cages or grazing arks.

California

New Zealand

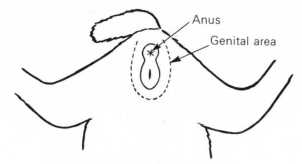

A single outdoor hutch in a three-sided building.

(a) In a female the organ is slit-like.

(b) In a male it is rounded.

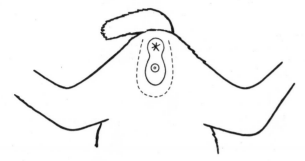

Sexing a rabbit.

Hutches

The best hutches are made of wood which is warm and draught-proof. They can either be used indoors or outdoors. The indoor hutches can conveniently be situated in a disused shed or barn, provided that ventilation is adequate. They should be free of draughts and damp but, unless winter temperatures are particularly low, will not need supplementary heating. The rabbit is a hardy animal and can withstand the cold well, but a combination of damp and draughts can quickly be fatal. Electric lighting in the shed will make the task of feeding on winter evenings a more pleasant one for the rabbit keeper, and will also improve the rabbits' productivity.

Outdoor hutches obviously need to be proofed against the weather. Tongue-and-groove board, or indeed any other wooden construction, will need to have a roof cover of bitumous roofing felt. In particularly exposed situations the back of the hutch should also be proofed in this way. The roof should slope from front to back, with an overhang of at least 2 in (50 mm) all round to allow rain to run off.

The minimum dimensions for a one-rabbit hutch, as recommended by the Royal Society for the Prevention of Cruelty to Animals, are as follows: 48 in × 24 in × 24 in (1.2 m × 0.6 m × 0.6 m). For two females the overall length should be increased to 60 in (1.5 m). On no account should males and females share the same compartment. The buck should have his own quarters to which the female is introduced only at mating time. Bedding material, which should be either straw or sawdust, will need to be cleared away frequently to avoid a build-up of droppings.

A tiered hutch.

A small Morant-type ark for controlled grazing.

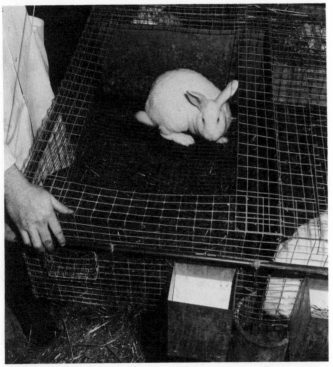

A cage fitted with feed hopper and automatic watering.

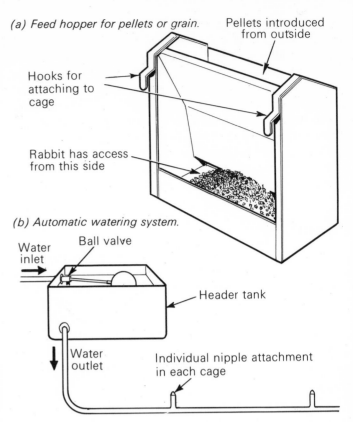

(a) Feed hopper for pellets or grain.

Pellets introduced from outside

Hooks for attaching to cage

Rabbit has access from this side

(b) Automatic watering system.

Water inlet

Ball valve

Header tank

Water outlet

Individual nipple attachment in each cage

Equipment for the rabbitry

Cages

Commercially, rabbits are housed in wire cages in a purpose-built building where the environment (lighting, temperature etc.) can be controlled. Cages are undoubtedly hygienic as the droppings and urine fall through the wire floor onto concrete below. From here the droppings are removed and the concrete washed down regularly. However, many people who are raising rabbits on a non-commercial scale do not wish to keep their stock exclusively in such highly artificial conditions. As an alternative small rabbit keepers are finding that a compromise between cages and natural grazing works well.

Ark grazing

The so-called Morant system of grazing rabbits is named after a Major Morant who, in the latter part of the last century, devised a system of movable arks (see illustration) which allowed the rabbits to graze on fresh grass every day.

My own system of rabbit keeping is based on this method. My New Zealand Whites spend the night in cages which I keep in an outhouse. During the day they are put in home-made arks on the grass. In this way they have the benefit of natural grazing conditions, while I benefit from not having to clean out hutches. In the USA, however, this system is rarely used and cages are recommended for fanciers as well as for commercial breeders.

Equipment

Feeding and drinking equipment needs to be easily cleaned and effective in use. Loose containers are not as satisfactory as fixed ones because of the possibility that the food will be contaminated by droppings. If you do use them, make sure they are heavy-based and therefore difficult to overturn. The best type of feeder for pellets is the hopper, which is fitted onto the hutch or cage mesh and can be filled from the outside while the rabbits have access from the inside.

The best type of drinker is the automatically filled one, but this system is often beyond the scope of the small rabbit keeper. Bottle drinkers with specially made tube outlets from which the rabbits can suck at will are the best and most hygienic of the non-automatic types. If you want to feed your rabbits hay, a useful piece of equipment is a small hayrack so that the hay is kept clear of the floor and is not contaminated by urine and droppings.

A triangular scraper, rather like a paint scraper, is a good tool for removing compacted droppings from the hutch. A small stiff brush is also useful. Water bottles will need to be cleaned with a baby bottle brush, and people who want to show their rabbits will need a grooming brush. Where fluff has accumulated on netting or on cage floors a stiff wire brush will remove it or, in extreme cases, a blow-lamp.

Feeding

A rabbit's teeth are adapted for gnawing and, as they keep on growing throughout life, need foods which will wear them down. The rabbit is a herbivore, or plant eater, and its digestive system is therefore adapted to coping with fibrous plant matter such as grass or hay. Like all animals, rabbits need protection for growth and body repair, carbohydrates and oils for energy and warmth, vitamins and minerals for the maintenance of good health, and water at all times.

Proprietary feeds

Rabbit pellets are formulated to provide the rabbit with all its nutritional needs, apart from water, throughout its life. They normally have a protein content of 16 to 18 per cent, a fibre content of 12 to 15 per cent and an oil content of 3 per cent. Also included are minerals, vitamins and coccidiostats or antibiotics for the prevention of the disease coccidiosis. Commercially, pellets are fed exclusively and a typical diet pattern for the does would be as follows (although amounts vary according to the size of the rabbit):

After mating	– 4 oz (113 g) subsistence allowance of pellets per day
Three weeks after mating	– 8 oz (226 g) pellets per day
After kindling	– as much feed as they want until weaning

The disadvantages of pellet feeding are two-fold. It is expensive by comparison with more traditional feeds and it includes antibiotics which may, to many people, make it unacceptable on health grounds. Preliminary research findings also indicate that rabbits fed exclusively on pellets produce meat with a higher cholesterol value than that from rabbits with a more varied diet.

Traditional feeds

The traditional diet for meat rabbits was made up of oats, bran mash, greens, roots and hay. A typical diet was as follows:

Morning feed	– 2 oz (56 g) of either bran mash, oats, barley, dried bread or cooked potatoes
Evening feed	– 2 oz (56 g) mixed greens, roots and hay

If you are following a traditional diet it is not a good idea to give pregnant does bread or potatoes as they are too fattening – oats are better. On the other hand if you want a fattening ration a mixture of 2 oz (56 g) cooked potato mixed with 1 oz (28 g) barley can be given in addition to hay and greens.

Combined diet

Small rabbit keepers, who are daunted by the high cost of pellets and have a plentiful supply of garden greens, often prefer to feed a combined diet of proprietary pellets and traditional feed. This is the pattern that I follow:

Morning feed	– 2 oz (56 g) pellets 2 oz (56 g) oats or barley
Evening feed	– mixed vegetable and herb greens and chopped up roots (in season)
At all times	– hay (in a rack)

Rabbits will eat most garden vegetables, but it is important not to give too much of any one kind. The ideal is a mixture of whatever happens to be in season, but beware of changing from dry to green feed too often as this can cause diarrhoea.

Breeding

Breeding stock should be vigorous, hardy and healthy. The does should have eight visible teats, and both does and bucks should have come from good strains. At twenty weeks the doe is ready for mating, and the buck is ready at twenty-five weeks. Although they will mate earlier than this it is better to delay it until they are fully developed and mature.

There is no period of 'heat' as in other animals, and the doe will ovulate in response to the buck. When mating is required, you should put the doe in the buck's quarters – never the other way round otherwise there will be a fight. The doe will usually stand still for the buck to mount her, but she may occasionally run around. Commercially it is common to practise 'assisted' mating where the doe is held while the buck mounts her. Once mating has taken place the doe should be removed and returned to her own quarters.

Pregnancy

The pregnant doe should have adequate food and water to keep her in good health, but fattening foods should be avoided. A traditional addition to her normal diet is a few raspberry leaves as the effect of these is to ease the muscular action during birth.

Pregnancy lasts thirty-one days, and three weeks after mating the doe should be given a nestbox. This can be a simple wooden box measuring 18 in × 12 in × 6 in (460 mm × 300 mm × 150 mm) and with a height of 10 in (250 mm). Put a layer of hay at the bottom, and a day or two before kindling the mother will pull out some of her fur to make a soft lining.

Buck and doe mating.

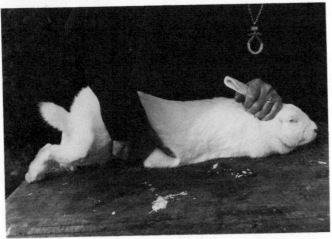

In assisted mating the doe is held with the hindquarters raised to allow easier access for the buck.

Doe with young in nestbox.

Kindling

During the birth of the young, the doe is best left alone with as little disturbance as possible. Once they are all born they can be quickly examined for defects or still-birth, and dead or damaged ones can be removed. Here, the good relationship already established between the doe and her owner is all important.

The young will quickly begin to suckle and the mother will carefully cover them up with fur each time she leaves the nest. It is often a good idea to have mated two does at the same time so that they have their litters simultaneously. Then, if one has a large litter and the other a small one, the young can be shared out equally between the does.

Lactation and weaning

For the first three weeks the young will feed exclusively on the mother's milk. After this period they will begin to nibble solids, although they may go on suckling occasionally throughout the total lactation period of seven weeks. Commercially the young are weaned much earlier than this, usually at three weeks, and the mothers are remated three to four weeks after kindling. For the smallholder who may be producing carcasses for his own use, this intensive production is probably unnecessary, and he will know how many litters per year will cater for his needs.

Keeping records

It is obviously vital to keep adequate records when breeding and generally managing a rabbitry. Breeding does do need some form of identification and most people use either ear tags or ear tattoos. As there is a possibility with a tag that

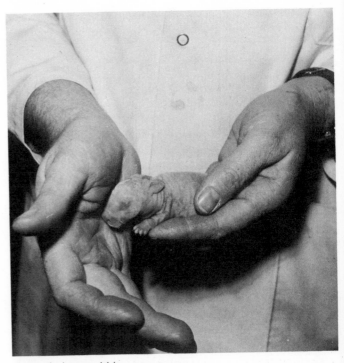

A newly born rabbit.

it may get caught and damage the ear, many people prefer tattooing. Record cards can be used to record details of mating, kindling, litter production and weight gain and these cards, along with other accessories, are usually obtainable from commercial breeders. In the USA commercial feed companies will supply their customers with record cards.

The frequency of mating depends on how many carcasses you are aiming for. Commercially does are mated from five to seven times a year, and the aim is to produce an average of nine young per litter. For this level of production artificial lighting is necessary in the rabbit house and it is important to plan the year ahead and decide what breeding programme will meet your needs.

Rabbits for meat and pelts

A commercial producer would aim to rear table rabbits of $4\frac{1}{2}$–5 lb (2 kg) in nine to ten weeks. To do this, he would need to feed an unlimited supply of proprietary pellets to ensure that adequate sources of protein were available during this period of rapid growth. The non-commercial producer, who is rearing and feeding less intensively, can afford to keep his stock longer and have a more prolonged period of growth. In both cases a set of weighing scales is a useful piece of equipment to keep a check on the progress of the stock.

Killing

The rabbit should not be fed for twenty-four hours prior to killing. For commercial rabbits dislocation is the better method as it does not cause damage or discolouration of the

Killing by neck dislocation.

Young rabbits, three weeks old.

Killing by striking the back of the neck.

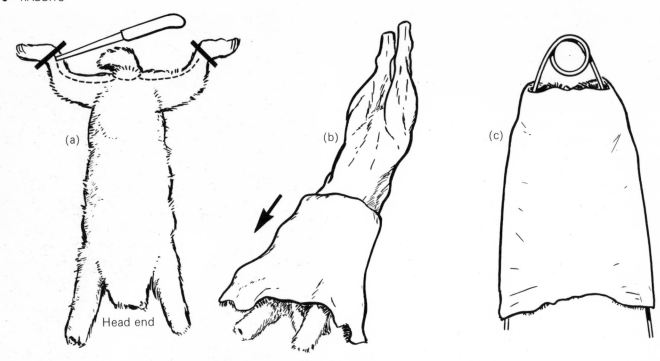

Skinning: *(a) Cut from one leg to another, across vent. Then cut around the tail and loosen the skin.* *(b) Pull the skin downwards, inside out.* *(c) Skin stretched, fur-side downwards, on 9-gauge spring wire shaped like a hairpin.*

meat. Hold the back legs with your left hand and grasp the neck with your right. Place your thumb on the back of the neck just behind the ears, with your fingers grasping under-neath the neck. Press down on your thumb while quickly pulling the rabbit upwards. This is a quick and humane method of killing, but never attempt it until you have watched an expert demonstrate it first.

An alternative method is to place the rabbit on its paws on a flat surface and to strike it sharply on the back of the head, just behind the ears, with the side of your hand or with a heavy piece of wood. After killing there is a certain amount of reflex kicking; this is normal and will soon cease.

Skinning

Cut off the head and feet, and then make a cut through the skin on the inside of the back legs. Continue each cut from one leg to another, right to left, and cut off the tail so that the skin can be rolled down like removing a glove. Continue pulling the skin down until it reaches the front legs, and loosen it in the same way as before, by cutting around the legs.

Once it has been removed the pelt should be dried, either by being stretched over wire or by being slit up the middle and tacked, fur down, on a board. Scrape off as much of the fat and flesh as possible, using a blunt knife which will not damage the skin. Then wash the skins in warm soapy water, rinse and spread them out to dry, pulling and stretching them as they do so. Rub salt well into the flesh side and then leave them to dry out completely. This is a 'green' hide and will keep about five months before real tanning takes place.

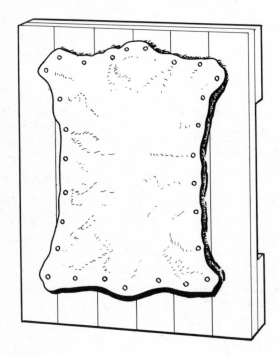

Skin stretched out and pinned, fur-side down, onto a board for scraping off fat and tissue.

Paunching

Make a cut in the abdomen, taking care not to puncture the entrails. Remove the innards carefully and save the kidneys, liver and heart if these are wanted. If the carcass is being kept whole, bend it inwards with the legs towards the belly so that it is then ready for roasting. If preferred, the carcass can be jointed into individual portions.

Tanning

There are many tanning methods, but a quick and easy way is to make up the following mixture:

1 part saltpetre
2 parts alum
5 parts bran

Mix with warm water to a paste and cover the flesh side of the skins with it. Roll them up and leave for a week. Then scrape off the mixture and rub the flesh side of the skins on a surface such as a table edge to make them soft and supple.

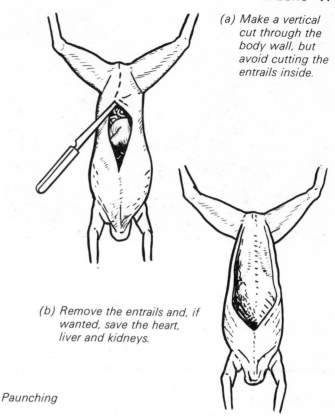

(a) Make a vertical cut through the body wall, but avoid cutting the entrails inside.

(b) Remove the entrails and, if wanted, save the heart, liver and kidneys.

Paunching

Health and disease

Preventive measures

Rabbits are generally hardy and free from disease, provided they are fed adequately and housed in a dry, draught-free environment with adequate ventilation, and provided you take the proper precautions over cleanliness and hygiene. Ideally rabbits like a constant temperature of about 60°F (16°C), but unless you go to the huge expense of constructing and maintaining an environmentally controlled building, the temperature will obviously vary throughout the year. Adequate insulation of buildings, hutches and nest-boxes will generally provide enough protection against the cold for the rabbit's own body warmth to supply the rest, but in areas where the winter temperatures fall dramatically there may be a need for artificial heating.

Apart from generally making sure that there are no sickly rabbits and that housing, food and water are satisfactory, there are several other things which are important. Rabbits with colds will need to have their nostrils and eyes bathed if matter has collected there, and any ailing stock will need attention. If the rabbits' fur has become matted for any reason it will need to be groomed with a stiff brush, particularly in the case of show animals.

Claw trimming

Unless the rabbits are able to scratch, their claws will gradually get longer and will periodically need a trim. To do this you should hold the animal firmly and clip the claws with sharp scissors or with purpose-made veterinary clippers, taking care not to cut the claw to the quick. You can make sure of avoiding this by holding the foot up to the light so that you can see where the blood vessels are in the first section of the claw. (See photograph next page.)

When held, a rabbit should always be supported from below.

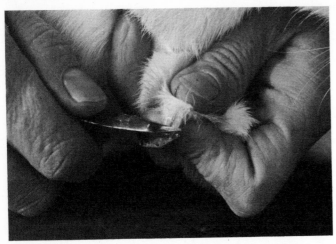

A rabbit's claws will need regular trimming about once a month.

Monitoring deaths

You are almost sure to have some deaths in any breeding project, particularly with new-born rabbits during their first few hours of life. However any unusual deaths, or deaths among adult stock, should be investigated. This can be organized through your veterinary surgeon. It will involve you in some expense but authoritative information on what is killing your rabbits may help you to cure, or better still prevent, the disease in future.

Diseases

Blows

This is a form of constipation which arises when gas forms in the caecum. The rabbit adopts a huddled position and does not move. Avoid giving green food, particularly brassicas, and just give a few pellets and hay until it recovers.

Canker

Ear canker is caused by a mite which sets up an irritation and contagious discharge. Any rabbit shaking its head and trying to scratch its ear should be investigated for canker. A veterinary preparation of parasiticide drops applied daily will kill off the mites.

Coccidiosis

This is a serious disease and has been responsible for wiping out rabbitries in the past. Parasites of the liver cause loss of appetite, an inflamed rectum, listlessness and diarrhoea, and lead ultimately to death. It is particularly serious in young rabbits up to two months old. Prevention is the best approach and every effort should be made to avoid any build-up of droppings on hutch floors. The parasites are transferred via the droppings, and for this reason wire floors are much more hygienic than solid floors. Rabbits fed on proprietary feeds are protected against the disease by a coccidiostat included in the pellets.

Colds (Rhinitis)

Affected rabbits wash their faces frequently, sneeze and may have a discharge from the nostrils and eyes. If the discharge is severe, wash the area around the nostrils and eyes gently, with a 10 per cent solution of boracic acid. Keep the animal warm and give a bran mash with a little chopped onion and mint in it.

Mange

Mites on the skin set up an irritation which leads to raw, red patches as the rabbit scratches. A veterinary preparation as for ear canker will kill the mites.

Mastitis

This occurs when there is an infection and inflammation of the teats, which become hard and sore. Antibiotics will clear up the condition but, as it has a tendency to recur, it may be unwise to continue breeding from that doe.

Myxomatosis

In the wild, this virus infection has drastically reduced the rabbit population in many areas. A rabbit with the disease will have watering eyes which swell and bulge and a swollen nose, mouth and anus. It will develop a high fever and become emaciated before it finally dies. There is no cure and all stock should be vaccinated to prevent infection.

Pasteurellosis

This often fatal and air-borne condition is most likely to develop in unhygienic and badly ventilated houses. The symptoms are rapid breathing, diarrhoea and sudden death. If discovered in time, it can be treated with antibiotics, either in the food or in water.

Snuffles (Chronic Rhinitis)

This starts like an ordinary cold but quickly develops into an infection which often leads to Pasteurellosis. In severe cases it is better to cull the rabbits.

Sore eyes

A rabbit may physically damage an eye by accident or it may have runny eyes as a result of a cold. Frequently, however, the eyes are irritated by ammonia from urine which has not been cleared out, and the need for hygienic conditions cannot be overemphasized.

Sore hocks

Occasionally rabbits may develop sore hocks as a result of being on wire flooring. It is essential, if you are using cages, to acquire rabbits from a commercial breeder whose stock has been bred to these conditions. My own New Zealand White meat strain, for example, has thick bushy pads on its hocks and is quite adapted to wire flooring. It is sometimes a good idea to provide resting boards if you are keeping large breeds on wire. If you put small wooden boards about 8 in × 12 in (200 mm × 300 mm) and 1 in (25 mm) thick into the cage for the rabbits to rest on this will help prevent or alleviate sore hocks.

GOATS

Background

Although goats are perhaps agriculturally less significant in the so-called advanced countries, on a world scale they form a substantial proportion (around one in seven) of the domesticated grazing animals. The greatest concentrations of goats are to be found in regions with a dry, tropical climate and in such areas they are often the most important livestock, both numerically and economically, that are kept. It seems likely that the goat originated in western Asia and was one of the first animals, and certainly the first ruminant, to be domesticated.

Although the goat first evolved in the arid tropics, it was in the cooler, temperate regions of the world, with their narrower daily temperature ranges and lusher grazings due to higher rainfall, that its milk-producing qualities could be identified and bred for selectively. Indeed, the modern dairy goat is a triumph of the breeder's art, a fairly average specimen (if such a thing exists) being capable of producing its own body weight in milk in about fourteen days. Goat's milk is particularly good for specialized diets for people who have difficulty in digesting cow's milk. Although the goat is valued as a dairy animal in temperate regions, it is perhaps more important in world-wide terms as a meat producer. Other important goat products include skins for leather manufacture, hair (mohair and cashmere) and, on a local scale, manure.

Goats, then, are more important in those countries with a subsistence, or at least a peasant, agriculture. To put it another way, when the object is to provide the family or small human unit with milk or meat, and labour is cheap or free, the goat comes into its own. In countries such as Britain, North America and Australia, goats remain economically of less importance than other farmstock such as cattle and sheep, and official census figures for goat populations are not available. However, looking at the number of goats registered each year by the registration societies provides a good indication of whether the goat populations of these countries are changing. What do these figures show? In all cases there appears to be a phenomenal increase in the number of goats registered. In the USA, the number of goats registered increased ten-fold in the thirteen years from 1964, with over 32,000 registrations in 1976. Registrations in Britain tell a similar story, with over 6000 recorded in 1977, nearly six times as many as in 1960.

It is interesting to speculate on the possible causes of this dramatic increase in interest. What human social factors are at work? Is it due to increased leisure time, a creative urge to produce more of one's own food, a result of the back to the land, self-sufficient movement? We will perhaps never know. One thing which is certain is that goats tend to be kept in relatively small numbers (the average herd size is about five or six) by small farmers and hobbyists with limited amounts of land.

What then is the role of the goat in the economy of the small farm or smallholding? In many cases goats are the central livestock enterprise, not because they are necessarily kept in the greatest numbers but rather because of the major effects that they have on other activities. Where family labour is available, and the burden of regular twice-daily milking is shouldered, two or more milking goats will accelerate other farming activities. Goats will produce milk from which the family can manufacture all its dairy produce needs: yoghurt, cream, butter and hard and soft cheese. By-products (such as skimmed milk, whey and buttermilk) together with any surplus whole milk can be fed to other stock and used, for example, to rear, or help to fatten, lambs, calves, pigs and poultry. Manure from the goat pens will be an excellent fertilizer for the arable side of the farm, whether used direct after it has rotted, or as an accelerator for compost, and can thus help to grow more food – vegetables for the house and crops for livestock. Remember that nothing should ever be wasted in the benign and everlasting circle of fertility. Although they must be well fed, of course, if they are to produce copious amounts of milk, goats will utilize marginal land that is too steep for the plough or otherwise unsuitable for other livestock. Being browsers, rather than grazers by nature, they will do well on scrubby land. Indeed we have used them as pioneers to help clear land overgrown with blackberry and other bushes, as a first step to reclamation. In southern New South Wales and other parts of Australia goats are recognized as valuable controllers of certain pernicious weeds. In rainfall areas where ryegrass, white clover and other pasture species have been contaminated by weeds such as star thistle and nodding thistle, herds of goats have often been successfully used to save the expense of weedicide sprays. (By the same token, though, do not turn goats loose in your orchard if you value the trees.) Surplus male kids make good eating for the family and their skins can be tanned for leather or cured and used to make a variety of craft items.

We said earlier that the average herd size was fairly small. Growth in goat populations has been principally due to an increased number of goatkeepers with small herds rather than to the development of large commercial herds. Although such enterprises do exist, they are rather few and far between, if one counts as commercial only those that provide the sole or main source of income for their owners. Commercial enterprises of this nature, particularly those dependent on paid labour, are unlikely to succeed unless some, or all, of the following conditions are met: the cost of feeding is minimal (e.g. the goats are kept on marginal land unsuitable or uneconomic for other livestock and the bulk of the food required is grown on the farm); milk and other dairy products are produced in sufficient quantity to justify the use of at least some labour-saving equipment (such as milking machines and jacketed vats); milk is processed into cheese, yoghurt or other products that represent a better return to the farmer than does liquid milk, and this processing is done on the holding; the milk and milk products command premium prices on the market, when compared with cow's milk and its products; the enterprise is close to the market so that the time and cost of transport is minimized and, finally, that the market exists. If it does not, it will have to be created by the hard-worked goatkeeper; official agencies interested in buying goat's milk and milk products can occasionally be found but are rare, and in many countries, including Britain, do not exist at all.

Breeds

Many different varieties of goats exist throughout the world, the majority of them of purely local significance. Since goats have primarily been important in peasant agricultural economies, the differences between varieties have arisen due to geographical isolation and natural selection rather than to planned breeding programmes. The establishment of distinctive breeds of improved goats is relatively recent (the last 100 to 150 years or so) and took place, for the most part, in Europe.

Saanen

The Saanen goat is probably the breed which most resembles the non-goatkeeper's image of a typical goat. The Saanen is white in colour, medium to large in size – females around 150 lb (70 kg), males 180 lb (80 kg) – with prick ears. With its dished face, slender neck and freedom from coarseness, the female Saanen is the epitome of femininity and 'milkiness' sought by breeders. The coat is normally short and fine although some individuals may have longer fringes of hair. Conformation is generally good and milk yields are high. The Saanen has an international reputation and has had the greatest effect of all the goat breeds. In Britain and the USA the Saanen is maintained as a closed breed i.e. both the sire and the dam of the goat to be registered in this section of the Herd Book must be of the Saanen breed. All Saanens, therefore, trace their ancestry back to goats that originated in the Saane and Simmental valleys of Switzerland although importations of goats may not have been directly from Switzerland. For example, the most recent importation of Saanens and Toggenburgs into Britain (in 1965) was from Switzerland, but the previous Saanen one (1922) came from Holland.

Although populations of pure-bred Saanens are relatively small, most countries have in addition, versions of the breed that have been created by upgrading indigenous goats by the repeated use of pure-bred sires. To be eligible for registration as a British Saanen or an American Saanen, for example, female goats must be $\frac{7}{8}$ pure-bred. These breeds, therefore, are open, which means that goats can be registered therein as a result of upgrading. In general, the resulting animals are excellent. The admixture of different bloodlines has resulted in larger goats with similar or improved milking characteristics when compared with the pure-bred parent breed, but which have yet retained the placidity and ideal udder shape of the pure-bred. The derived Saanen is undoubtedly an ideal goat for the small farmer whose object is the production of liquid milk for sale. In Britain the British Saanen is, according to the number of annual registrations, the most popular of all breeds.

Anglo-Nubian

The Anglo-Nubian breed can justly claim to be not only the most distinctive of the British breeds but also Britain's greatest contribution to world goat breeding. (Although we are trying to be objective we must declare an interest since this is the breed we favour.) The Anglo-Nubian was developed in Britain by crossing goats from the East with indigenous stock. It is generally accepted that two breeds were principally involved in the formation of the Anglo-Nubian: the Zariby from the Sudan and the Jumna Pari from India. Of these, the Jumna Pari is probably the most important, and all Anglo-Nubians are descended from four male goats imported into Britain at the end of the nineteenth and beginning of the twentieth centuries. The characteristics of the Anglo-Nubian breed are obviously strongly dominant over those of other breeds. Breed type has been maintained since those original imports (and without importations of fresh blood) in spite of the fact that the Anglo-Nubian has always been an open breed and stock have been upgraded into it.

Anglo-Nubians have been exported from Britain to all parts of the world. They have been in particular demand in recent years for tropical countries (such as Trinidad, Nigeria, Brazil, the Yemen, Malaysia etc.) where they have been used to improve indigenous goats for both milk and meat production. Earlier exports to Australia and the USA founded the Nubian breed in those countries. In the USA, the Nubian is the most popular of all breeds with around 28 per cent of all registrations. In Britain the rate of growth in numbers of Anglo-Nubians registered has climbed faster than that for any other breed.

The appearance of the Anglo-Nubian is totally unlike that of any other breed. With its high head carriage, pendulous ears and Roman nose it looks almost aristocratic. Its general body conformation differs slightly from that of the prick-eared Swiss-type goats in that, due to its long legs and upright stance, some specimens show a slight dip in the back behind the withers. This is not generally objected to provided, of course, that it is not pronounced or indicative of any weakness in the back. Anglo-Nubians come in a variety of colours – black, red tan, grey, white, cream and mixtures giving roans and mottled or dappled coats.

Some specimens show the so-called Swiss markings (white facial stripes similar to those of the Toggenburg breed) which are often viewed unfavourably by show-ring judges. This is somewhat surprising since at least one of the breeds on which the Anglo-Nubian was founded shows this characteristic in its native country. Breeders in Britain during the last twenty years have made considerable pro-

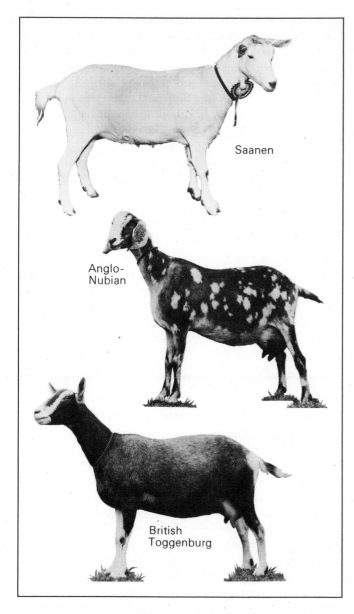

Saanen

Anglo-Nubian

British Toggenburg

gress in improving the shape of udders and teats. Milk yields tend to be lower than those of other breeds, but butterfat and solids not fat contents are higher making the Anglo-Nubian a good choice for the farmer who requires a source of high quality milk for the house or for processing into butter and cheese. The breed readily puts on flesh and surplus male kids can be castrated and successfully fattened for meat.

Toggenburg

Like the Saanen, the Toggenburg is Swiss in origin and from its native land has spread to many parts of the world. Indeed, in the USA it is more popular than the Saanen and produces comparable milk yields. In Britain the Toggenburg is numerically the weakest of all breeds (with the exception of the more recently recognized Golden Guernsey). Just over 1 per cent of all goats registered in 1977 were of this breed. Although in Switzerland and the USA it is described as being of medium or average size, Toggenburgs

in Britain can only be described as small. Mature females weigh around 120 lb (54 kg).

Coat colour varies from fawn to light brown with the distinctive white Swiss markings on the face, lower leg and around the tail. The hair is usually soft, fine and of medium length although some individuals show longer, silky hair along the spine and on the flanks, and some males have this longer hair over the greater part of the body.

Milk yields and butterfat levels from Toggenburgs in Britain are not high although in other countries they are sometimes better. The principal recommendations for keeping the breed are that it forms a useful genetic pool and that it generally does well on extensive, free-range systems of management.

Like the Saanen, the Toggenburg is a closed breed and, also like the Saanen, it given rise to localized breeds, such as the British Toggenburg, which are open. The derived breeds tend to be larger than the pure-bred parents and capable of higher milk yields. The British Toggenburg, like the British Saanen, is much more popular than its pure-bred parent breed.

French Alpine

Although it originated, as its name suggests, in the French Alps, this breed is firmly established in the USA and is next in popularity after the Nubian. It is a large and rangy animal with a variety of coat colours and patterns; some of these are so well fixed that they have retained their French names such as Cou Noir, Cou Clair, Cou Blanc (Black, Tan and White Neck respectively). Other markings include Sundgau (which is black and white, often similar to the British Alpine) and Chamoisee (which is like the wild chamois in colour).

British Alpine

The British Alpine was developed in Britain, primarily in the eastern counties. It is not related to the French Alpine breed although its colour and markings are similar to some of the Sundgau French Alpines. In appearance it is a large black goat with the white Swiss markings typical of the Toggenburg and British Toggenburg breeds. The typical British Alpine is a large, rangy animal with a good reputation for performance under extensive systems of management. Its relative popularity (as a percentage of all goats registered in Britain) has declined over recent years.

La Mancha

This breed was developed in the USA fairly recently. The first goats were registered in the late 1950s. The breed was founded by crossing goats of Spanish origin with pure-bred animals of the Swiss breeds. The principal characteristic of the La Mancha breed is the external ears which may be either absent ('gopher' ears) or vestigial ('cookie' ears). The character governing ear size is genetically dominant.

British Alpine

Golden Guernsey

Golden Guernsey

As its name suggests, this breed evolved in the Channel Islands. It is surprising not only that the geographic isolation of its island home resulted in both a distinctive breed of goats and a distinctive breed of cattle, but also that the coat colours of the two animals should be so similar. A Herd Book was opened on the island in 1922, and in 1971 the British Goat Society recognized the breed on the mainland by opening the Golden Guernsey register for imported pure-bred animals and their descendants. In 1975 further registers were opened for the progeny of Golden Guernsey females and Saanen or British Saanen males, thus laying the foundations for the formation of an open English Guernsey breed.

The Golden Guernsey is a small goat, with golden coloured hair and skin. The hair may be long on the quarters and along the spine. Although milk yields might be considered low when compared with those of other breeds, they are reasonable when related to the size of the animal and to the fact that, on its native island, concentrate feeding is rare.

Angora

The Angora originated in Asia Minor and has been prized for centuries for its fleece, which is a mixture of hair and a fine, soft-handling fibre called cashmere. The breed has risen to some prominence in Australia where it is particularly favoured by hobby farmers. It is regarded as quite a

good milker and yields a fleece of high-quality fibre which can command a very good price. However, shearing is sometimes more difficult with the Angora as its fleece yields no yolk to lubricate the cutter against the comb and the shearer may need to grease it artificially.

In Australia, where this breed is growing so rapidly in popularity, you can go to one of the major sheep shows or Royal shows on Angora judging day and see what the best Angoras in the land look like, besides receiving a wealth of information and advice from the Angora Society of Australia (address: Royal Showgrounds, Epsom Road, Ascot Vale, Victoria).

Cross-bred and grading-up goats

Most registering bodies have sections in their Herd Books where crosses between recognized breeds and the progeny of unregistered females by pedigree males can be registered. Cross-bred goats are often vigorous and capable of high milk yields. The purchase of grade stock (i.e. stock from lines in the process of being upgraded to an open breed) can be a cheap way of starting a herd provided that care is taken in the selection of the foundation stock and in the choice of males used thereafter.

Buying goats

Before you choose your goats, you should have some idea of what you are letting yourself in for. So, if you are embarking on it for the first time, do gather as much information as you can. Read what we have written here and any other books and leaflets you can obtain. Join your local or national goatkeepers' organization and get to some of their meetings so that you can make contact with experienced goatkeepers and pick their brains. Visit goat shows, in particular the larger ones held in conjunction with the major agricultural shows, so that you can see the various breeds of goats and understand why one animal is considered to be a better specimen than another.

Whilst you are learning what might be termed the theory of the subject, do not forget the practicalities. You will have housing to prepare, feeding stuffs to organize, cropping programmes to plan and equipment to obtain. Can you milk, trim hooves, administer worming medicines, take a goat's temperature? If not, spend a little time with experienced goatkeepers learning how these things are done. Then you can set about choosing the goats to buy.

In choosing prospective purchases, you will have to take the following factors into account.

How many to buy

Goats are naturally gregarious. The minimum number you should have at any one time should be two. A goat kept on its own will often show signs of disturbance and become noisy and unmanageable. It may show such an intense desire for company that it will jump out of the best-fenced field to get to its owners or other stock. Through no fault of its own, such an animal earns its species a bad name and is condemned to spend its life on a tether. Two females, on the other hand, will generally be much more settled and easier to control. (Note that we said females. There is no real need to keep your own male goat, particularly if you intend to keep a limited number of milkers to supply the house. But more of this later.)

The maximum number of goats to keep or buy is a matter for you to decide in the light of the resources you have available. Your resources will be: money (for the purchase of stock), time, housing and availability of feed. As advice to the novice we would only say that whatever your ultimate plans, start small and let your herd expand in keeping with your experience. You will probably do better to spend more time on fewer goats than to spread your time thinly over a greater number.

What age to buy

Again, it is difficult to offer hard and fast rules. Different ages of stock have advantages and disadvantages. The stock best adapted to your methods of feeding and management and most resistant to disease will be that which you have reared and, preferably, bred on your own holding although, obviously, such animals are out of the question if you have no goats already!

The next choice would be to buy in young kids (at least four days old, so that they have received colostrum and hence have obtained some resistance to disease). However, kids such as these will require milk feeding for a considerable period (say four to six months) which will necessitate the purchase of goat's milk or milk substitute, both of which can be expensive.

Weaned kids or goatlings (maiden stock between one and two years of age) will, perhaps, be more expensive to buy but should be cheaper to feed than younger animals, whilst still having a reasonable time to adapt to your methods before they kid. A potential disadvantage is that first kidders can sometimes be nervous and take time to settle to being milked. If both you and the goat are novices, the whole exercise could start to seem more trouble than it is worth!

Your final choice is whether or not to buy a milker. Initial costs will be higher, but you get an immediate return on your investment, assuming of course that you are sufficiently skilled to keep her milking well. However, all animals are creatures of habit and none is more conservative than an elderly goat. Even if your management is of the highest, the change of situation will probably cause a reduction in milk yield.

Male or female?

We have assumed throughout this section of the book that the prime objective of the goat-keeping smallholder is to produce milk for the house. If that is the case, and the eventual number of milkers to be kept is small, there is no great need to keep a male goat unless you opt for a breed that is numerically weak in your area. Indeed, in that situation, you might consider keeping a male goat for stud and making his services available to other goatkeepers. (This is unlikely to be a success unless you choose a male from a very productive line.) In most circumstances it will be more convenient and probably cheaper, to keep female goats only and use whatever males are standing at stud in your locality.

Which to choose

Breed

The short answer to the question of which breed to buy is 'whatever you fancy'. Essentially, it is a matter of personal preference. Choose the breed or type that you find most appealing since you will have to spend a considerable amount of time in company with your stock. Animals that are easy on the eye make the work involved more satisfying. Avoid the mistake of keeping too many different breeds. Your job of building a level herd of good productivity and conformation will be easier if you restrict yourself to one, or at the most, two breeds.

Conformation

The structure of a working animal (as against the points of a fancy one) is important because, as a general rule, goat breeders over the years have been successful in linking aspects of structure with productivity. The factors that are of prime practical importance with a dairy animal are, naturally enough, those that relate to milk production and longevity.

The ideal animal is wedge-shaped in appearance when viewed from above and the side. The back should be strong and straight with no tendency to weakness. It is the support from which the weight of internal organs and developing kids is slung. As a converter of bulk food, the goat requires adequate capacity, so look for hollows in the flanks, depth through the body and a wide spread to the ribs. These factors combine to give the wedge shape referred to above.

Strong legs and feet are a prime requirement for a grazing animal to enable it to get about and obtain its food. The weight that the goat will carry in late pregnancy means that weak or dropped pasterns must be avoided.

For any milking animal, the udder is of first importance. As a result of successive lactations, there will be a tendency for the udder to elongate. A near-spherical shape and attachment to the goat's body over as wide an area as possible will ensure that the udder wears well. Strong forward attachment of the udder (avoiding the formation of a 'neck') will be particularly valuable. To make milking easier (particularly if you are considering machine milking) the teats should be distinct from the udder and not merely a continuation of it. They should be of adequate size for ease of milking, but not so large as to be difficult for the kids to suck. Abnormalities of the teats, such as supernumerary teats or teat orifices, are often inherited. At best they make milking awkward and, at worst, they predispose the animal to infections of the udder. Size of the udder is not, of itself, indicative of milk yield. Be guided more by the goat's own milk records or, in the case of young stock, by those of its dam, its sister, its sire's dam and his immediate female relations. If such records are official so much the better.

Avoid buying a goat with horns; they can cause damage to owner and other goats alike. Check that any polled (naturally hornless) female you contemplate buying is not a hermaphrodite. The commonest sign of this condition is an enlarged clitoris and animals so affected will be sterile.

Condition

Condition in all classes of stock is governed by two principal factors: good feeding (the old adage that 'half the pedigree goes down the throat' should serve to remind you that, no matter how well an animal is bred, it will not perform its function unless its level of nutrition is sufficient to allow it to realize its genetic potential) and freedom from disease.

The signs of good condition you should look for include a bright eye and an alert, interested appearance, healthy pink colour of the mucous membranes (such as, for example, the inner surfaces of the eyelids) and a smooth and glossy coat. Note the appearance of the dung. The pellets should be round, firm and separate. Clumps of droppings are often indicative of worm infestation and pale, puffy droppings of coccidiosis.

Housing

We have not sufficient space here to go into detailed plans and descriptions of all the possible varieties of housing for goats. Most of us do not anyway have the time or resources to construct new buildings and have to adapt those that we inherited when we moved onto our holdings. All we can do here is to indicate some of the general principles.

Functions of housing

There are two principal objectives in housing goats. First to protect them from extreme weather conditions, and second to exercise a measure of control over them (particularly while they are eating concentrates). To these we could add a third factor, namely, that while satisfying these two objectives the housing must also be convenient for the stockman.

Walls and roofs

Goats are not unduly affected by cold as such, but they do need to be dry and protected from draughts. Obviously therefore, the walls and roof of the house need to be waterproof and free of holes. The walls and roof should be as well insulated as possible, not so much for the purposes of heat retention in winter as to avoid convection down draughts on the animals in cold weather, and to reduce excessive heat gain in summer.

Thin materials such as corrugated asbestos or steel sheet can be insulated with polystyrene or fibreglass, although an impervious vapour barrier must be placed between the insulation and the warm, moisture-laden air in the goat house. Needless to say, insulating materials must be kept out of reach of the animals or separated from them by being lined on the inside. Where this is not practicable, one of the proprietary plastic-faced insulation boards might be the best answer.

With all buildings it may be an advantage to arrange for straw and hay to be stored over the goats' quarters in order to prevent convection down draughts.

Floors

The two principal factors to be considered are the nature of the floor surface (which affects the ease, or otherwise, with which the pens can be cleaned out) and the insulating properties of the material chosen.

For ease of cleaning, a hard, relatively smooth surface is best. If you are converting old farm buildings such as stables or cow stalls you may find that the floor is made of bricks laid on their edges in sand. This forms an acceptable floor and is certainly not worth digging up or replacing. It is easily cleaned, and the dry joints permit some drainage into the underlying soil, thereby preventing the bedding from becoming sodden.

If you are laying a new floor from scratch, your only reasonable choice is concrete. For pens, and passageways that are not subject to traffic from heavy machinery such as tractors, a 3 in (75 mm) slab or a mixture of 1 part cement to 4½ parts all-in aggregate will be more than adequate. The concrete should be well-tamped and finished with a wood float to give a smooth, non-slip surface. Ideally, floors should be laid with a slight fall (say 1 in, 25 mm) from the back of the pen to the front. Goats do not produce large volumes of liquid manure and such an arrangement will make it easier to dispose of washing-down water. The concrete should be laid on a layer of well rammed hardcore and blinded with fine material such as sand or sieved ashes (not wood ash – keep that for the garden).

Concrete is cold. If you are laying a new floor, you can improve its insulating properties in two ways. First, you can include a damp-proof membrane (e.g. heavy-gauge polythene – chemical fertilizer bags will do, as long as you overlap them by at least 9 in, 225 mm) under the concrete to prevent the dampness of the soil being drawn up into it – damp concrete is colder than dry concrete. Avoid puncturing the membrane as you lay the concrete. Second, you can put an insulating layer between the damp-proof membrane and the concrete slab. The insulation is intended to trap a layer of still air underneath it, so ensure that the materials you use will not compress as the concrete is laid. A 2 in (50 mm) layer of polystyrene, insulated board, or hollow material such as clay land-drain tiles or glass bottles and jars can all be used to good effect. Sawdust concrete, in which the sand content is replaced by coarse sawdust, can also make a good flooring material.

General view of the interior of a large goat house, showing individual penning.

Because they attract rats and suffer from the difficulties of drainage common to all soils except sands and gravels, earth floors have little to recommend them.

Ventilation

Housing for goats must have ventilation but no draughts. In most buildings this does not present great difficulties. In larger goat houses some form of outlet on the ridge of the roof may be necessary. Hopper windows that open inwards will direct incoming air upwards and away from the animals and would be worth considering if you were undertaking a major reconstruction or erecting a new building. Remember that you may have to protect any glass with wire netting if the goats have easy access to it.

Lay-out

The general internal arrangement of the goat shed or barn depends on a number of factors including the size and shape of the building, the number of goats you intend to keep and the degree of control that you wish to exercise over your animals, particularly whilst they are being fed their concentrate ration. At this stage some control will be essential to ensure that individual animals receive their allotted amounts of concentrates.

Even if goats are to be housed communally, separate penning will be needed for stock of different ages, and individual pens will be required for goats that are about to kid or are newly kidded. With this form of housing, concentrates will either be fed whilst the goats are being milked or whilst they are restrained in some way. If you watch a goat suckling a kid you will see that it is relaxed and placidly chewing the cud. When we are milking we are attempting to imitate what the kid does, so in order to aid let-down it would seem sensible not to feed at this time. This will make for a peaceful milking time and avoid forcing the goatkeeper to wait while the goat finishes her ration. It also minimizes the amount of dust raised in the goat house which will inevitably find its way into the newly drawn milk. Restraint in communal housing could take many forms of which the simplest, but perhaps the most time-consuming, is to chain the animals around the walls of the shed or barn, out of reach of each other. A considerable improvement on this is to provide a feeding rack. Essentially, this consists of a fence or partition constructed of wooden slats or steel tubes, which provide slots or holes through which the goats can put their heads to reach feeding buckets on the other side. The goats are restrained by lowering a bar, and not released until all have finished feeding.

As we remarked earlier, goats are usually kept in fairly small numbers and normal practice is to provide each

A suspended feed trough.

'Keyhole' access to buckets.

Feed and water buckets supported outside the pen by metal rings.

milker with a separate pen (although goatlings and kids are often housed in groups). In this case the minimum area for each milker should be around 25 sq ft (2.3 sq m) although 30 sq ft (2.8 sq m) would be better. Ideally, the pen should be rectangular, say 6 ft (1.8 m) by 4 ft (1.2 m), rather than square. Pen divisions can be of whatever materials are easily available e.g. concrete blocks, bricks, timber, plywood, chain-link etc. as long as the material chosen is resistant to prolonged contact with dung and urine. Solid walls of bricks, blocks or boards will give some protection from cross-draughts. They must be substantial enough to take the weight of a 160 lb (74 kg) goat pressing repeatedly against it. All partition walls should be 4 ft 3 in to 4 ft 6 in (1.4 m) in height.

Pen doorways should be at least 2 ft 6 in (0.76 m) and preferably 3 ft (0.9 m) wide. Doors must open outwards. Try to arrange the housing so that pens open off a central passageway or, at least, that there is separate access to each pen. Interconnecting pens should be avoided at all costs. The passageway itself should be a minimum of 3 ft (0.9 m) in width although, within reason, the wider the better. If you intend to muck the pens out into a tractor-drawn trailer, you will need a passageway of, probably, 10 ft (3 m) or 12 ft (3.7 m) minimum width.

Pen fittings

Each pen needs to be fitted out with food and water containers. Bulky foods such as hay, straw and kale are better fed from racks. Haynets can be used but are dangerous for young kids and time-consuming for the goatkeeper. To avoid the need to enter each pen to replenish racks, they should be placed on the front wall or dividing wall so that they can be reached from the passageway. The gaps between the bars or slats should be about 2 in (50 mm) and the bottom of the rack needs to be about 2 ft 6 in (0.76 m) to 3 ft (0.9 m) from the floor, to minimize the risk of foodstuffs being fouled by the goats. To reduce wastage, racks can be fitted with lids, hinged on the side nearest to the goat. A further refinement is to build what is known in Britain as an Egerton rack, which consists of a second set of slats about 6½ in (165 mm) apart some 1 ft 6 in (0.5 m) in front of the hay rack and extending from the floor to the level of the top of the rack. This arrangement allows the goat's head and neck through, but not her shoulders and therefore prevents her from trampling upon any hay that she pulls from the rack.

Water and concentrates are fed from buckets which can be hung on the outside of the door or the front wall of the pen. The goat has access to them via a hole about 12 in (300 mm) by 10 in (250 mm) in size. This arrangement will allow the goatkeeper to empty or fill the buckets easily and help to prevent the goats fouling or breaking them. Bucket holders can be made from mild steel rod, bent to shape with the ends inserted into two large screw eyes or a specially made bracket mounted on the wall or door. By this means the bucket holders will fold down when not in use, thereby taking up less room in the passageway. Bucket holders can also be made from heavy-duty Vee belting fixed with a couple of large staples. Although these do not fold down, they will push out of the way when goatkeepers pass by with barrows or stock.

A conventional barrel bolt will be adequate to secure most doors although occasionally one comes across a cunning goat who has learned to manipulate these and can let out not only herself but also her herd mates. A simple remedy is to use a pad bolt and fit a clip, like those used on dogs' leads, through the hole where the padlock would go.

Milking arrangements

Those of us producing milk for home consumption can rarely afford the luxury of a separate milking parlour. However, we should at least provide an area in the goat house to which we can take each goat to be milked and thus avoid milking animals in their pens. The requirements are simple: adequate lighting, enough space to enable the goatkeeper to sit comfortably alongside the goat, and somewhere to tie up the goat by her collar. In addition, the floor and walls alongside should be made of impervious, dust-free materials that can be washed down. If you wish to bring the goat up to your level, rather than descend to hers, a milking stand or platform could be useful. Milking stands can vary from the simple, which can be just an old table with its legs sawn down to bring the top to about 1 ft 6 in (0.5 m) from the floor, to the more sophisticated. You can make a space-saving stand that folds up against a wall when not in use by putting up a wide shelf, hinged to the wall, with several legs attached by hinges underneath the edge furthest from the wall.

Dairies

Dairies, strictly speaking, are beyond the scope of this chapter. However, if you are contemplating commercial goatkeeping a dairy will be necessary and a few words on the subject will not be out of place.

The dairy is the place where the milk is handled, processed and stored and where the milking equipment is cleaned and stored. Floors, walls and ceilings must be easy to wash down and therefore surfaces such as tiles, laminates or gloss paint are ideal. There should be a minimum of horizontal surfaces to collect dust. Hot and cold water should be provided and you will need a separate sink, hand basin and dairy wash tanks. Adequate storage cupboards and slatted drying racks for utensils are also important. Refrigerated storage for milk and milk products will probably be needed and can be provided either within the dairy or separately.

Storage buildings

Apart from housing for the animals themselves, you must also provide storage for foodstuffs and straw. Hay and straw should be stacked undercover in an airy building, such as a Dutch barn, preferably as close as possible to the consumers. Good access for vehicles is an advantage as is any step you can take to keep an air space under the stored material e.g. by stacking it on baulks of lumber.

Concentrates are stored in rodent-proof bins preferably away from the stock to reduce the likelihood of a goat escaping from its pen and gorging itself.

Control

Control is probably one of the biggest problems that a goatkeeper comes across. Goats are notorious escapers, partly through sheer curiosity and partly through the desire to search for different foods. It is difficult to keep any livestock on overgrazed paddocks when there is lush pasture in sight, and well-nigh impossible to restrain goats in these conditions.

Before spending vast sums on fencing, sit down and work out how best your land can be used. If it is too small to make successful grazing for the number of animals you plan to keep, then opt for stall feeding and spend the money on a small amount of fencing round a concrete yard. If you have access to common land or verges and lanes which are traffic-free you may decide to tether. If you have adequate grazing and wish to make full use of it, fencing is essential and there are various types of fencing from which to make your choice.

Exercise yard

The design of this will depend very much on available materials, but if it can be built alongside a building in such a position as to give shade from the sun and shelter from the strongest winds, it should be suitable for use throughout the year. Make the fencing out of sturdy materials as the goats will probably spend a fair amount of time trying to climb up it. Brick or stone walls are suitable or post and rail fencing with sturdy netting such as chain-link nailed onto it. Make the fencing high enough to prevent the goats from jumping out, 4 ft 6 in (1.5 m) should be sufficient.

Tethering

This method is frowned on by many goatkeepers, mainly because one associates the miserable, mangy, half-starved goats that are sometimes seen on the roadside with tethering. It can be a very successful way of using grazing as long as you can be there to move the animals frequently to new grass (they will not eat the grass they have trampled), and as long as you can get them to shelter in the event of rainstorms. It is a time-consuming method of control as the goats will have to be checked frequently to make sure that the chains do not get caught up, which could even strangle them.

Use swivel links to attach the chain both to the goat's

Male goat houses with exercise yards.

collar and to the stake. Hammer the stake firmly into the ground and make sure that the goats are sufficiently widely spaced to prevent them from getting tangled in each other's chains. Never be tempted to tether kids – they are far too active and there would be a high risk of them strangling themselves. It is unwise to tether any goat of an excitable or nervous disposition because when tethered they are unable to escape from dangers such as dogs or noisy children, and react by panicking. A goat which pulls on its collar can cause pressure on a nerve in the neck. As a result the goat 'faints' due to temporary paralysis of the heart and diaphragm. Although recovery is normally rapid, if the goat is unable to get up (e.g. if it is tethered on a steep bank) it could die.

Running tethers are a slightly more sophisticated method of tethering, using two stakes with a wire run between them and the swivel link attached to the wire. This gives a wider area over which the goats can graze and means that they do not need to be moved so often.

Fencing

The three forms of fencing in common use are chain-link, post and wire and electric fencing. Chain-link is prohibitively expensive, although it is effective. Post and wire, with the wires close enough together at the bottom to prevent young stock from climbing through, can be successful but the goats will tend to try to climb it. Again 4 ft 6 in (1.5 m) will be about the right height, and the wires will require straining when the fence is erected to make it strong and durable.

Electric fencing is highly successful as long as it is run along in front of another boundary such as a hedge or post and wire fence. It is sometimes possible to strip graze (i.e. to graze only one section of the land at a time so that each area has a chance to rest) with electric fencing, but goats do not have the same respect for the wires that cows do, and will frequently barge straight through if the pasture on the other side looks more interesting. Three strands of wire are recommended. Heights of 4 ft, 2 ft and 1 ft (1.2 m, 0.6 m and 0.3 m) above the ground are satisfactory, and the top two should be electrified all the time. The bottom one can be disconnected, particularly when there are no young kids grazing with the herd. You can buy either mains-powered or battery-powered units and it is advisable to have a look at the different models available before deciding on the one most suited to your needs.

The wire used can be stranded wire or synthetic strands with wire run through. The former is more durable but less easy to move if you wish to change the position of your fencing regularly, and the latter is unlikely to last as long but involves a smaller initial outlay. Electrified nettings designed for sheep can also be used, but unless the goats are grazed with sheep it is unlikely that the expense will be justified.

Feeding

The goat is naturally a browsing animal. That is, its preference is to eat bushes, trees and herbs. Goats will graze very happily, but if hedges are accessible they will make straight for these before eating the grass. Most domesticated herbivores are grazers by habit, so it often comes as a shock to new goatkeepers to realize that their animals regard hedges as a meal rather than a boundary. In the wild the goat can obtain all its nutritional requirements from browsing and grazing, but because domesticated animals are encouraged to produce more milk and meat, they need extra feeding. It is normal practice to supplement any forage that the animals may have with hay and a cereal ration often referred to as concentrates.

Forage

This includes all the browsing and grazing to which your animals may have access, plus anything you are able to cut and carry to them. If your goatkeeping project is restricted to a small amount of land, say less than half an acre (0.25 ha), it is probably not worth bothering about grazing, but would be more worthwhile to put the land down to crops which you can cut and carry to the goats in their shed or exercise yard. Although goats are extremely good at cleaning scrubland, it must be remembered that their habit of stripping bark means that they augur certain death to trees, so they should never be turned into orchards or within reach of valuable trees. Unless the scrubland you have is extensive, the goats will eat and kill the shrubs and bushes, and you will then have to clear the roots and reseed for pasture. Obviously with a larger area the damage is less severe and the scrubland will regrow of its own accord and allow continuous browsing.

Next to natural scrubland, the goat probably most enjoys pasture. Goats eat a far wider range of plants than most other domesticated herbivores, and consequently tend to find plain grassland rather boring. The best pastureland to grow is herbal ley, and although it is possible to list various mixtures, the most suitable constituents tend to vary with locality so it is probably best to consult your local seedsman. A basic mixture might contain grasses, red and white clovers, plantain, yarrow, sheep's parsley, burnet and chicory, but these are subject to much variation with soil type. The deep-rooted herbs tend to bring minerals to the surface, and the goat has a very high mineral requirement so benefits greatly from grazing on mixed leys.

Pasture management is a whole topic on its own, but the basic guidelines are:

1. Do not overstock, or the pasture will be killed off and the animals be liable to worm infestation.
2. Try to prevent the land from getting acid, by spreading lime as needed (local farmers can be helpful with advice here as to how much the soil in your area needs). Land deficient in phosphates will benefit from top dressing with basic slag which will encourage the growth of clovers. Again, seek advice from your neighbours or arrange for soil testing by local agricultural advisory services.
3. Do not let the pasture grow too fast so that the plants become rank and tasteless. The goats would then not be able to graze and the quality of the pasture would deteriorate rapidly. Ways to avoid this are to allow other people's stock to graze your land if you have insufficient stock, to top the fields (i.e. cut them with a scythe or grass-mower) or to shut a paddock up for hay and take a hay crop off it. Remember that the palatability and nutritive value of pasture declines rapidly if grasses are allowed to run up to seed.
4. Always rotate the animals round the grazing in order to rest each part of it in turn. Divide your grassland into three or four, and let them graze one section down before moving them on to the next. This has a twofold purpose. It helps to keep the parasitic worms in check, and it also means that the pasture is eaten down evenly while the other areas are regrowing. The length of time on each area will vary with its size and the speed at which the grass is growing, but try to rest each area for at least six weeks, and never leave the animals on the same area for over two months.

The other points to remember about pasturing goats are that you must always put fresh drinking water out on the paddocks and make sure that some protection from the weather, such as trees or an open shed, is available.

Crops to grow

Various crops can be grown for feeding at different times of the year. If you require foodstuffs all the year round, plan a succession of crops. This can be supplemented with wild food such as branches, and with vegetable wastes from shops and gardens.

The earliest greenstuff in the year comes from Russian comfrey which is propagated by dividing its roots. Once established, it is almost impossible to get rid of, and the leaves can be cut up to six or seven times a year. Against comfrey it has to be said that not all animals find it palatable, particularly in the cooler months as the leaves tend to become coarser.

Hungarian rye sown as a winter cover crop on bare ground in the early autumn will not only prevent soluble nutrients from being leached from the soil but will also provide early greenstuff for cutting before it becomes too coarse. The usual practice is to plough it in in the spring as a green manure but it makes a good early bite until then.

Lucerne (alfalfa) is an excellent crop to grow. It requires very clean land so preparation for sowing takes a long time. However, once the crop is established it will stand for about seven years and provide several cuts during the summer. It is palatable, highly nutritious and a great standby in times of drought since its roots penetrate to great depth.

Oats and tares is another favourite crop and can be cut as soon as it is growing well. Towards the end of the summer, fodder maize is excellent feeding, particularly in the sunnier temperate regions. Chicory also makes a very good garden crop although it tends to run up to flower in its second year and loses a lot of its leafiness. However, it will stand for several years so again requires very little attention.

Brassicas are very much enjoyed by goats, and a year-round succession of cabbages can be planned for them, using the varieties recommended by seedsmen for your area. In regions with hard winter frosts such as Britain and northern states of the USA kale and cabbages are probably the only greens that will stand the winter. Thousand Head Kale and Maris Kestrel are reasonably frost-hardy and summer sowing should give greenstuff from autumn through to the early spring.

Roots are generally liked by goats, and moderate quantities of mangolds, sugar·beet and fodder beet can be fed to females, but it is wise to avoid feeding any roots to males as they can cause urinary calculi and subsequent blockage in the urethra. Potatoes can be fed in small quantities. Some goats will take to raw ones, but any green ones should be discarded. Cooked potatoes and boiled or baked peelings tend to be preferred.

A lidded hay rack.

Hay

Goats need an enormous amount of roughage and hay should be available for them at all times, summer and winter. It is unwise to turn goats onto lush pasture without feeding some hay first as it might upset their digestive systems. Good quality hay is essential as goats are fastidious feeders and will reject mouldy or poor quality hay. Good meadow hay is ideal for the average family goat, but for a high performer or show goat, seeds, hay, clover or lucerne hays are ideal.

A milker will probably require about a bale weighing 56 lb (25 kg) per week, and dry and young stock rather less. This is very much a rough guideline as it will vary with the quality of the hay and the amount of other roughage being fed. Feed hay morning and evening after milking (if you do it before, the dust will contaminate the milk) and at midday as well if the animals are housed.

Hay should always be fed in racks if possible, with sufficient space between the bars or mesh for the goat to get its nose in and pull a mouthful of hay without wasting it. A lid on the top of the hay rack will prevent wastage, as goats tend not to touch hay that they have trampled on, and it makes expensive bedding. Slats in front of the hay rack through which the goat can get her head but not her shoulders will allow any dropped hay to be picked up and fed to other stock. Never be tempted to use hay nets. They are time-consuming to fill, dangerous to adult stock and lethal to kids. A kid can easily strangle itself in a hay net, so avoid them at all costs.

Concentrates

Concentrates are the cereal rations which enable the goats to make satisfactory growth and to produce milk over and above that required for the kids. Concentrate feeds contain much more protein than roughage foods so need only be given in relatively small quantities. A milker or working male will require a mixture with a 16 to 18 per cent protein content, but weaned kids and goatlings only require about 14 per cent.

The basic ingredients used in cereal rations include bran, rolled oats, flaked maize and a high-protein legume such as kibbled beans or seed cake (made from such things as linseed flakes, decorticated groundnut cake, or soya bean meal). Which cake you use depends on what your local miller has available. A mineral supplement such as that used for dairy cattle must be added, and various other ingredients such as molassine meal can be added when they are available, to make it more palatable.

Cereal rations may be bought ready-mixed from a local mill (either as a coarse dairy ration or mixed specially for you) and with a small number of animals this may well be the most practical method. Alternatively they can be mixed at home from bought ingredients. Very few small farmers have the space, time or facilities to grow their own cereal crops. Some suggestions for home-mixed rations are listed below (all measurements by weight):

1. 1 part flaked maize
   1 part rolled oats
   1 part broad bran
   1 part decorticated groundnut cake

2. 2 parts crushed oats
   2 parts bran
   1 part flaked maize
   2 parts kibbled beans

3. 1 part flaked maize
   1 part rolled oats
   1 part dry sugar beet pulp
   1 part linseed flakes

To all mixtures, mineral supplement should be added in the recommended quantities.

How to feed

The best containers to use for concentrates are buckets or bowls. Buckets can be supported by collapsible metal rings on the outside of a pen door, with access through a key-hole opening. Bowls can be removed immediately after feeding. Troughs, although useful for feeding large numbers of animals at a time, are less easily removed which means that they can quickly get fouled and need constant cleaning out.

How much to feed

The quantity of concentrates to feed depends on the quality and quantity of forage available and the age or stage of lactation of the goat. If you assume that a stall-fed goatling and dry adult each requires 2 lb (1 kg) per day for maintenance, and that kids should gradually be worked up to this amount at a year old, taking into account the drop in food intake at weaning, you should be able to keep your dry stock in good condition. A milker giving a gallon (4.5 l) a day will require 4 lb (2 kg) of concentrates. Beyond this stockmanship comes into play. For a further gallon of milk a day you must attempt to feed at least another 2 lb (1 kg), but the goat may either not be able to consume it, in which case high-quality forage should be used, or may show signs of needing more. Feeding a high milker is a skilled art because the aim is to allow the animal to produce its highest potential yield without upsetting its digestive system by overloading it with concentrate foods.

When to feed

The question of when to feed concentrates is one open to continual debate. It is normal to divide the feed into two portions, or possibly three for a heavy milker, and feed morning and evening. Some people like to feed while the goat is being milked, theoretically to keep the goat quiet. In fact, if you watch a goat eating, you will see that it is anything but quiet, because it is continuously moving its head to reach the food. If the food is finished before milking, the goat may well decide that the milking is also over, and walk off of her own accord or become restless.

A goat suckling its kids always stops eating and cuds and, since milking simulates the sucking of the kids, it seems only natural to us that a goat will want to cud while it is being milked. Consequently, we always feed concentrate rations after all the milking is finished.

Minerals

As we have already mentioned, goats have a very high mineral requirement. A fine balance of minerals is required as they are only needed in minute quantities and severe metabolic disorders can occur if the balance is upset. Unless you live in an area which is short of a particular mineral and it is routine practice among local farmers to give a particular mineral to the animals or put it on the soil, it is probably best to leave well alone and use only a balanced mineral supplement suitable for dairy animals. Shortages of the following minerals can cause problems.

Calcium and phosphorus

Keeping the correct balance of calcium and phosphorus in the body is extremely important to maintain the bone structure and for milk production. The calcium:phosphorus ratio will usually be about right in any animal fed a normal mixed diet, but care must be taken not to distort the ratio by feeding excessive quantities of leguminous forage and hay or seaweed meal.

Salt

Salt (sodium chloride) should be available to goats at all times. They will not take an excess of it, so a salt lick can be hung in the pens. A goat secretes in the region of $\frac{1}{4}$ oz (7 g) of salt per 10 pt (5 l) of milk, so needs to replenish its supply continuously.

Cobalt

Cobalt is required to supply the material from which vitamin B12 is synthesized. Vitamin B12 is required to keep the

Note the water bucket supported in a metal ring and the mineral lick on the wall.

animal in good health, but parasitic gut worms also need it, and will rob the goat of the vitamin if they are present in sufficiently large numbers. If the land is short of cobalt, then the goats will become deficient as a consequence. The symptoms of the deficiency include poor condition, lack of appetite and general unthriftiness, but the classic symptom is 'goaty' tasting milk. The cure is simple. Prepare a cobalt salt mix by dissolving 1 oz (25 g) cobalt sulphate in $\frac{1}{2}$ pt (0.25 l) water. Wet 6 lb (2.5 kg) salt with the solution, and dry in a low oven or in the sun. Add 1 oz (25 g) of this mixture to the concentrate feed every day for a week. We always add cobalt salt to the feed for a week once the goat is back onto a normal ration after kidding. This is a time when the goat's requirements of all foods is higher than the amount she can consume, and also a time when she is extremely susceptible to worm infestation. Using cobalt salt at this period is a preventive rather than a curative measure.

Copper

The other mineral deficiency you may well come across is copper. Certain areas in Britain and the USA and large areas of Australia and New Zealand are deficient in copper. The deficiency causes a condition known as swayback in sheep and goats. If you happen to live in a swayback area, you will have to feed your stock with copper supplements. Take advice from your vet or local farmers as to how much should be fed, but copper salts are highly toxic, and overfeeding can be more dangerous than underfeeding. Copper poisoning can result when goats have access to commercial pig rations and various mineral preparations for horses. Symptoms of the deficiency are scouring, staring, coarse coat and subsequent low fertility. Swayback itself makes the young animals unable to stand, and they consequently starve to death.

Water

Goats require clean fresh water, and a milker in full production may well require 6 gal (30l) a day. Either leave water with the animals, changing it at least twice a day, or offer them water after the morning feed, again at midday and after the evening feed. Lightly salting the water is a good way to provide salt for the goats, and molasses can be added after kidding to encourage the goat to drink plenty. Many goats like warm water, but be warned – once they are in the habit of drinking it warm they will not take kindly to a sudden change back to cold. If possible, put the water buckets outside the pens supported by a metal ring, as described for concentrates earlier. This will prevent the water being fouled.

A good animal can only be good if it is well fed. Good feeding is neither overfeeding nor underfeeding, and is the basis of stockmanship. If you watch your animals carefully day by day, you will see if they look a little dull, a little thin or a bit too fat, and this can be rectified by prompt action before you have got a major problem on your hands. Remember that the superb, sleek highly productive animals that you may see on another goatkeeper's land, or at a show are in that condition because they are fed with care and attention all the year round.

Milk

We will assume that most small farmers are keeping goats for milk production, although meat production can be a profitable sideline with the right markets and, in the case of Angora goats, the main product of course is hair. Through domestication and selective breeding, the length of a goat's lactation has been extended so that instead of only producing milk for the three or four months it suckles its kids as it would do in the wild, it continues to do so for a year or even several years without kidding again.

Hygiene

Having looked after and fed the animal to produce milk, it is only sensible to maintain the highest possible standards of cleanliness. Milk is a marvellous food for people, livestock and also for bacteria. The bacteria can make the milk unpleasant, if not downright dangerous, to drink.

First, all equipment should be clean and sterilized. Always rinse equipment in cold water immediately after use, wash thoroughly in hot water and detergent, rinse, and use a dairy sterilizer (usually of the hypochlorite type) according to the manufacturer's instructions. Boiling and steam sterilizing are suitable alternatives, but generally impractical for equipment as large as milking buckets, and unsuitable for plastic equipment.

There is a point about plastics which should be noted. Only ever use plastics which are of food grade. Ordinary household buckets and various tubs and pieces of drainpipe

Milking equipment (from left to right): seamless bucket, churn, strainer, hand can.

that are available are not suitable for storing milk or for cheese and yoghurt making. The lead levels are unacceptably high and will leach out slowly into the food and can cause chronic lead poisoning. Food grade plastics are used in ordinary kitchen ware and are also readily available from catering suppliers as well as the usual dairy suppliers.

The milking bucket should be stainless steel or tinned, and seamless, with a capacity of about 1 gal (5l). Modern tinning frequently does not stand up to hypochlorite sterilization, so if you are using a tin bucket it may well be advisable to rinse it out with boiling water rather than to use a chemical sterilizer. Never use any sort of towel for drying equipment – leave it to drain.

Personal hygiene is essential. Wash your hands thoroughly, wear a clean overall, and whenever you have an infectious disease, give the job to someone else.

Milking

Now to the goat. If at all possible, do the milking outside the pens, in a milking bay which can be hosed down after use. If you find it easier put the goat on a milking stand (see opposite). Many goats stand quietly to be milked but with a first kidder or a restless goat it is probably safer to tie her up. Do this firmly, but in such a way that she can stand in a relaxed manner and cud. Always milk from the same side and squat or sit on a milking stool by the goat's side, facing her udder. There are two schools of thought on udder washing. If a goat is used to it, it will stimulate let-down, and certainly if the udder is at all dirty it will have to be done. Washing cloths should be kept scrupulously clean, and preferably should be disposable, as the biggest argument against udder washing is the risk of spreading infection.

A goat being milked on an improvised milking stand.

Massage the udder gently to stimulate the goat to let down her milk, then take one teat in each hand, close your thumb and first finger together at the top of the teat to prevent the milk in the teat flowing back into the udder, then gently but firmly close the remaining fingers towards the palm of your hand to eject the milk. Never pull down on the teat as this can damage the udder. Do this alternately on each teat. The first few squirts should be milked into a flat-bottomed cup, checked for any clots or other signs of mastitis, and discarded. Now put the bucket under the goat and milk her until the milk ceases to flow steadily. Rub the udder a little and massage it to bring down any more milk into the teats, then milk again. Repeat a couple of times until no more milk appears to flow, then return the animal to her pen.

Straining and storing

If possible weigh the milk, as daily records are one of the best indicators of the goat's health, and then take it to your dairy or, if the milk is not for sale, to a clean part of the kitchen, and strain it. Milk strainers in which a disposable milk filter is inserted are available, but a pair of nylon sieves with a milk filter between them is suitable for small amounts of milk. Strain into a lidded jug or churn and cool immediately. Small amounts can be cooled by standing the vessel in the sink and running cold water over it for about twenty minutes. The milk is then removed and put in a refrigerator for storage. A cooling device can be improvised from hosepipe attached at one end to a tap and arranged round the neck of the churn, and perforated to allow cold water to run down the sides of the churn.

A big advantage of goat's milk over cow's milk is that it can be frozen in its natural state, straight after milking.

Milk for freezing should be strained, cooled rapidly and cartoned or bagged and put straight into the freezer. Problems are rarely experienced in freezing small quantities i.e. single pints, but freezing gallon quantities (over 3 litres) in a single block can create problems. The milk may separate on thawing and, although the reason for this is not known at the moment, it seems likely that it is because the milk freezes too slowly in a domestic freezer.

Uses

Goat's milk can obviously be used as straightforward liquid milk for drinking and cooking and can also be processed into yoghurt and cheeses, hard and soft. The cream does not rise as quickly as does that in cow's milk, so cannot easily be skimmed off. However, high butterfat milk set in a shallow bowl can be skimmed quite successfully, particularly early in the lactation. Normally, the only way to remove the cream is to use a separator. Hand or electric models are available, although they are extremely expensive. The cream can be used as fresh cream, or it can be clotted or made into butter or ice cream. The resulting skim can be used to make some cheeses or can be fed to stock, as can the whey from cheesemaking.

Goat's milk is highly suitable for rearing young animals, as the size of the fat globules is smaller than that of cow's milk, and the milk is consequently more easily digested. Puppies and kittens thrive on goat's milk, as do lambs, calves, poultry and piglets. Pigs and poultry are also useful stock if you intend to do much processing of milk, as they will happily devour unsold produce and failures such as 'yeasty' yoghurt.

Composition of goat's milk

As this book is of a practical nature we will just mention the points about the composition of goat's milk which have practical applications. Basically the chemical composition is very similar to that of cow's milk, the fat and protein levels being of the same order. The vitamin A level in goat's milk is nearly double that of cow's milk because the goat is a more efficient converter of carotene into vitamin A. Cows excrete carotene in their milk (hence the yellow colour) while goats convert it all, resulting in pure white milk.

The biggest difference between the two milks is the size of the fat globule. As we have already noted the globules in goat's milk are considerably smaller than those in cow's milk, making it more digestible and suitable for infants and people with digestive troubles.

A number of children develop allergic reactions to cow's milk which most commonly appear as infantile eczema and asthma. The allergy is to bovine protein, and goat's milk and milk products can be extremely valuable in replacing bovine products in the child's diet. The allergy frequently disappears as the child gets older, but many adult asthma and migraine sufferers find their conditions alleviated if they use goat's milk and its products rather than cow's milk, butter and cheese.

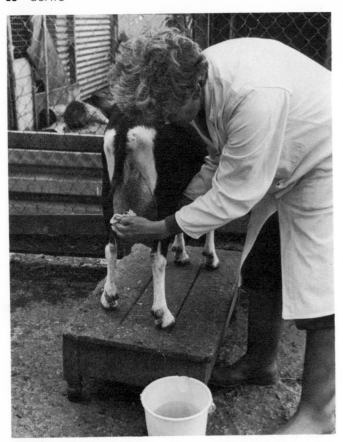

Milking: 1. *First wash the udder.*

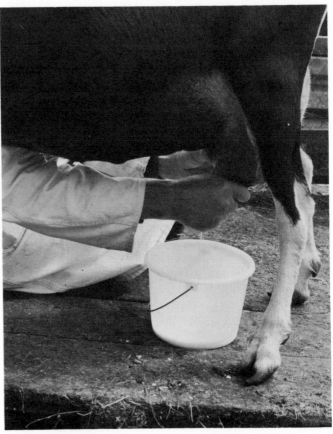

2. *The milker closes his right hand round the goat's teat to eject the milk.*

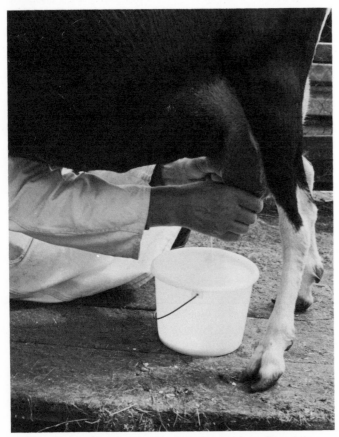

3. *The right hand relaxes, allowing the teat to refill, while the opposite teat is milked.*

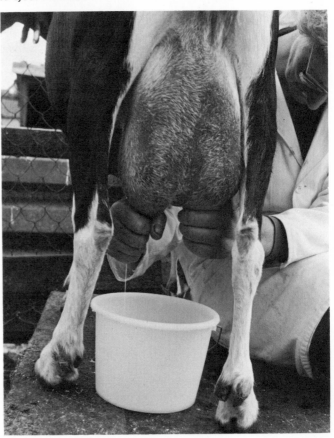

4. *Milking in progress viewed from rear.*

An in-churn cooler being inserted into a churn of milk.

The cooler in operation: cold water enters through hosepipe and circulates through the tubing in the churn.

Breeding and rearing

Signs of heat

The female goat shows that she is ready for mating by persistent bleating and tail wagging, by going off her food, reducing her usual milk production and by showing a clear glutinous discharge from her vulva which is red and swollen. This state, known variously as oestrus or being on heat or being in season, may last from a few hours up to three days. It is inadvisable, particularly at the beginning and at the end of the breeding season, to assume that the goat will remain on heat for three days.

The signs of oestrus will be repeated at intervals of about three weeks from autumn until early spring unless the goat is successfully mated. From the signs described it would seem hardly possible to miss the fact that the goat was in season but some goats show only some of the signs, and those only slightly, especially at either end of the breeding season. For this reason people with their own male can often get their goats into kid earlier than those who have to transport their goats to a male.

Mating

If you keep a male goat the in-season female will make every effort to get to him. If you do not have your own, a rag smelling of male goat (rubbed over a male goat's head and kept in a stoppered jar) will help to check if your goat is ready to be mated. Males are indifferent to the females who are not in season but, once they have gained some experience, will mate readily with any female goat that shows on heat.

If she is mated and conceives satisfactorily the goat will have her kid or kids (twins are the most common, singles and triplets not unusual) in five months. Mating during the summer months results in kids being born during the colder periods of the year and special care must be taken of the dam to ensure that she has adequate food to maintain her lactation during the winter when green feeds are likely to be in short supply, and you should make sure the kids are kept warm and get some supplementary vitamin D which is essential for normal development. In the summer with the help of sunshine they can produce their own vitamin D. In the winter sprinkling a little sterilized bone meal on the food will assist with healthy growth.

If you do not have your own male goat, it is a good idea to select the one you want to use before the breeding season starts, and let the owner know that you wish to use it. When your goat comes in season, contact the stud male's owner. Never turn up unannounced – even if the person is not frantically busy, the male might have been used by someone else five minutes before you arrive and therefore not be ready for another service. As long as the goat is properly in season and stands well for the male, service is rapid and writing out a stud certificate (required for registering the kids) may well take longer than the service itself. Many stud owners offer two services 'just to make sure', although whether this is necessary is a matter of conjecture. If you have your own male it may well be worth serving the goat on two consecutive days during the period of heat.

Check the goat three weeks and six weeks after mating, and if signs of heat reappear, get the goat mated again. It is normal practice for stud male owners to offer at least one free return service. If the goat comes into season at irregular times, or persistently does not hold to service, you may well need to seek veterinary advice. In particular if the male you are using is a kid, make sure that other females have held to service to it. Pregnancy can be confirmed by a laboratory test of a milk sample taken twenty-three days after mating.

Pregnancy

A goat's pregnancy lasts approximately five months and most kids are born between 147 and 153 days after mating, although kidding a week either side of 150 days is not unusual. For the first three months the goat is fed her normal maintenance ration plus an allowance for the amount of milk she is giving. By three months she should be dried off so that during the last two months when the kids are growing rapidly the goat's resources are not unduly drained. Some goats will dry themselves off, but others will need to be dried off by the stockman. He should not strip out the udder completely at each milking, and as the yield drops should milk once a day only, then every two days until the goat is dry.

During the last two months, the goat's concentrate ration should gradually be increased, rising to 4 lb (1.8 kg) of a milker's ration per day shortly before kidding. This practice, known as steaming up, allows the goat to build up body reserves which she can call on when her lactation begins. In the early stages of lactation when the yield is rising rapidly it is impossible for the good milker to eat sufficient food to match her energy requirements, so steaming up provides reserves for the goat to draw on at this time. Continue to feed good-quality hay and any available greenstuff throughout pregnancy, as no amount of concentrate feeding will make up for poor or inadequate bulk feeds.

A pregnant goat should be handled with care. In particular she should not be rushed through narrow doorways or be knocked on her side. The kids are carried on the right side of the goat's abdomen and the rumen on the left, so during the last two or three months of the pregnancy the goat's abdomen, particularly on the right side, will become larger.

Preparation for kidding

At kidding time the goatkeeper will need to have several items of equipment available, so it is as well to be prepared in advance.

First, make sure that you have your veterinarian's telephone number to hand and also, if you are unsure of yourself, that of an experienced goatkeeper or shepherd. You should also have the following.

1. Lambing oils or soapflakes for lubricants in case internal examination is necessary.
2. Tincture of iodine or antibiotic spray (such as the one sold for treating foot rot in sheep) to dress the kid's navel cords.
3. Old towelling for drying kids.
4. A box lined with hay for putting kids in as they are born.
5. Bran and molasses for the dam's first few feeds.
6. Salt for adding to drinking water.
7. Rubber teats and small bottles with a capacity of about ½ pt (125 ml).

The other preparation to make is to ensure that you have a clean, freshly-strawed area in which the goat will be able to kid in peace and privacy.

Kidding

Changes start to occur in the goat during the last couple of weeks before kidding. Her udder begins to fill up and should be checked daily. If it becomes distended some milk can be drawn to ease it but it should not be stripped out at this stage, as the extra fluid is colostrum which is important to the new-born kids. Due to slackening of the ligaments the spinal bones between the hips and the tail lift and become loose until it is possible to get a finger and thumb round them. During the last day or two before kidding, the uterus begins to tense up and the kids move lower in the abdomen, lining up for birth, and any movement they may previously have been making will cease. Finally a clear, slightly blood-stained discharge may appear and within hours the goat comes into labour.

To some extent goats, like sheep, have the ability to delay kidding until weather and other conditions are right i.e. the humidity is high and their owners are either absent or present, depending on which the goats prefer. As with any other animals, labour lasts a variable time and only if the goat becomes apparently distressed should you interfere. It is not uncommon for goats to scream as the kids are being born which can be alarming for someone seeing a kidding for the first time. The first sign of the kid's arrival is the transparent water bag (amniotic sac). If the presentation is normal, the front feet with the kid's nose resting on the legs will be presented first. Gradually the whole head appears, by which time the shoulders, which are the widest part, are through the pelvis and the kid then slides out rapidly.

When the kid is born, remove the water bag if it has not already broken and, using a finger, clear any mucus from the kid's nose and throat to ensure that it can breathe freely.

If the mother is standing to kid the cord may break as her offspring drops to the ground, but if she is lying down it will probably break as she turns to get to the kid. The cord has a natural point of weakness, so there is no need to cut it. Indeed, cutting can result in haemorrhage. Let the mother clean the kid up, but if another kid follows rapidly rub the first with towelling while the dam concentrates on producing the next. Make sure that the kids are kept out of the way of the milker's feet as she may inadvertently tread on one as she is straining.

Abnormal kiddings

It can be quite horrifying to read about abnormal kiddings and the things that may go wrong but always remember that, provided that the goat has been well looked after while pregnant and does not have a history of difficult kiddings, the chances are that she will kid without assistance. If the goat strains hard for a long time, and to some extent the stockman must judge for himself what a long time is, although an hour of hard straining will probably be quite enough if no apparent progress is being made, it will then be necessary to find out what might be wrong. At this stage call your vet or a local goatkeeper or stockman, or if you feel confident, do the investigation yourself.

Cleanliness is essential, so trim your nails, carefully wash your hands and forearms, lubricate your hand and arm with lambing oils, and insert your hand gently into the goat's vulva. When you reach a kid, make sure that only one is presenting itself. (If two are coming together, push one gently but firmly back behind the other.) Work your hand gradually over the kid to check if it is lying correctly (see diagram). Keep a mental picture of the correct presentation and line up the kid accordingly. The dam may well then be able to deliver it herself. If not, firm, gentle pulling of the front feet each time the goat strains should help her deliver it, provided that the kid's head is through the pelvic girdle.

If the presentation is breech (i.e. hind end first), there is no need to turn the kid round in the womb. You need only extend the hind legs and draw them back through the pelvis, giving extra help to the dam as the kid's shoulders pass through her pelvis, and pulling only when the goat strains. Remember that there is plenty of room for gentle manipulation in the space of the uterus, but unless the kid is presented correctly, it will not be able to get through the dam's pelvis.

Normal presentation of a kid.

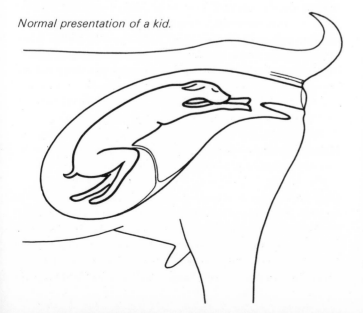

In a difficult kidding, it is the first kid that is usually the problem; others are delivered more easily. Breech presentation, although unusual for a first kid, is quite normal for second and subsequent kids. It is always wiser to call an expert than to try to struggle with a stuck kid and rapidly tiring dam for several hours. Any interference should be followed by antibiotic cover, and a vet will usually prescribe injections or pessaries to prevent infection.

When the birth of the kids is completed, the afterbirth will be left to come away. On no account should it be pulled since this could result in uterine haemorrhage. Normally the goat will give a few more contractions and expel the placenta, but this may be an hour or two after the kids are born. Sometimes the afterbirth may be retained and authorities vary on the length of time it can safely be left. However, it would be unwise to leave the goat more than twelve hours without seeking veterinary advice. The goat will frequently try to eat its afterbirth, and while this apparently causes no harm, it is best to remove it from her if you see it come away.

The kids

Strong kids will be on their feet within minutes, but weaker ones may take a few hours before they are firm on their legs. However strong or weak they are, it is essential that they get colostrum (the thick first milk from their dam) within six hours of birth. If you intend to bottle feed, a good ploy is to milk some colostrum into a sterilized, warmed bottle, put a teat on the top and offer it to the kids for their first feed. Certainly in our experience this helps the change-over to bottle feeding at four days old. The kids may need help to find the mother's teats, and gentle guidance, (never pushing on the back of their heads as this makes them back away), will ensure that they quickly learn where to suck. If the milker is old and her udder is too low for the kids to manage, or if a goat has particularly large teats, it may be necessary to bottle feed right from the start. If so, make sure that the kids are fed their own mother's milk for the first few hours to ensure that they get sufficient colostrum.

Colostrum

Colostrum is the first milk produced after the goat has given birth. It is slightly thicker than ordinary milk to which it gradually changes during the next three or four days. Colostrum is of great importance to the kid because it contains a high concentration of the mother's resistance to diseases in the form of antibodies. The kid is only able to absorb these antibodies from the colostrum for the first twenty-four hours of its life so it should be allowed and encouraged to suckle during that period. This natural preventive medicine gives the kid immediate resistance to infection which lasts for the first two months by when it has had time to develop its own antibodies. Colostrum also contains a high level of vitamin A as well as having laxative properties which help the kid to pass its first motion of black sticky meconium that has accumulated in its intestines during its time in the womb.

Recipes are sometimes given for artificial colostrum to be used if a goat dies during or after having her kid. While such mixtures are effective substitutes for some of the properties of colostrum they do not contain any antibodies. Instead of depending on substitutes it is better to deep freeze a supply of colostrum from goats kidding early. This can be used for any emergency later in the season. Some more could be preserved from goats which have their kids later to be available for the next season.

Checking the kids

Have a careful look at the kids after they are cleaned up. Check whether they are horned or polled. If they are horned, the hair should lie in whorls over tiny horn buds, and if polled, the hair lies flat and you may be able to feel ridges of bone running longitudinally along the head. Check the sex of the kids and look carefully at the teats of both sexes to ensure that there are no double teats. It is extremely unwise to keep double-teated animals for breeding purposes since a goat with this fault can be extremely difficult to milk, and furthermore the fault is inherited. Obviously such animals can be kept to be reared for meat.

The navel cord is still fleshy and is a passage through which bacteria can pass into the kid so it should always be dressed. Either dip it in tincture of iodine (an eggcup is an ideal container) or spray it with an aerosol foot rot spray. Check the kid's anogenital region to ensure that there are no signs of hermaphroditism, particularly if the parents are polled. An apparent female showing signs of intersexuality will have a swelling just inside the vulva. Make sure both testicles in a male kid are descended, although they may descend later and the male still be fertile.

Male kids (other than those carefully selected to be reared for stud purposes) can either be killed humanely or reared for meat. It is usual to castrate males that are intended to be eaten, and this should be done in the first few days of life. The rubber ring method is probably the easiest for the goatkeeper to use. There is however a good argument for not castrating meat males if they are to be killed before puberty i.e. before the meat is tainted with the secretions from the musk glands. In this case, of course, it is essential to keep the kids away from females, since they are capable of effective services from about twelve weeks of age.

Disposal of unwanted kids

The humane killing of kids is a difficult problem. The easiest and safest way is to ask your vet to do it, although the disadvantage of this is that it can rarely be done immediately after birth. Your vet may well provide you with a suitable drug if you discuss the problem with him beforehand.

Other than your veterinary surgeon, a local slaughterman or huntsman might also humanely destroy kids for you, but they should be approached before their services are required. If you are fortunate enough to live near a hunt kennels, and you have no moral objections to hunting, then the hunt will provide a useful way of disposing of all classes of stock. The service is usually free, in return for the carcass, and it has the added advantage that it takes place on your own property so that you are able to satisfy yourself that the animal has been killed.

Chloroform is probably the easiest anaesthetic agent to use. A pad of cotton wool is dampened with a little chloroform and put inside a large glass jar. The kid's head is held into the mouth of the jar for a few minutes so that it is breathing a mixture of chloroform and air, until its body goes limp. The head is then pushed into the jar and the kid left there for approximately half an hour.

Another satisfactory method is to use a captive bolt pistol. A firearms certificate is required to own one of these. To shoot the kid, stand directly behind it, hold the left ear firmly in your left hand, and with the pistol in your right hand place the muzzle on top of its head in the middle of an imaginary line drawn between the backs of its ears. The pistol is held at an angle, pointing towards the kid's mouth.

Disbudding

This is a job you could choose to ask your veterinary surgeon to do for you, although it may be more practical to be able to do it yourself. Do not, however, attempt it unless you have been properly trained, either by your vet or an experienced goatkeeper.

A naturally polled kid will have either a flat head or two ridges of bone running longitudinally along the head in front of the ears, and no obvious curls in the hair. A kid that is going to be horned will be born with horn buds, which feel like small raised pyramids on its head and are covered with a curl of hair.

If horn buds are allowed to grow for more than a few weeks, they become horns, and any operation to remove them will be major and will expose the goat to the risk of infections. Disbudding, i.e. removing of the horn buds before two weeks of age, and preferably before they are four days old, is quick and simple, and the kid's recovery is immediate.

Two methods of disbudding are used. One is to use a hot iron either electrically heated or heated in an open flame, and the other is to use a caustic solution. As the caustic method involves a slow persistent pain lasting for several days, and can cause problems when the solution runs into eyes and gets rubbed onto other kids, it is not widely used now. Hot iron disbudding is rapid – the hair over the horn area is clipped away, and the iron (heated to cherry red) is applied to the area for six seconds. Then the flat side of it is rubbed over the area to ensure that the entire bud has been removed. The horn base of males is much larger than that of females so always cover a larger area, particularly to the front of the horn, when disbudding male kids. This should prevent the unsightly scurs, which sometimes occur.

The law in Britain insists that an anaesthetic must be used but is unclear as to whether one can disbud one's own animals. Suffice it to say that it is fairly general practice for goatkeepers to do their own kids, but it is strictly illegal to disbud other people's kids. Whatever the law, it is only humane to use some form of anaesthetic as the pain is obviously intense, if only brief. Discuss anaesthetics with your vet and let him recommend the most suitable one for you to use.

The dam after kidding

The most important point to remember in the management of the freshly kidded goat is to avoid overstimulating milk secretion. The sudden drain on the goat's mineral resources can cause a drop in blood calcium level which results in milk fever. This can be fatal. Only feed concentrates very sparingly at first (if at all), increasing the amounts gradually as lactation is established.

Immediately after kidding, the goat will be hungry and thirsty, so offer her fresh warm salted water and a bran mash, possibly with molasses in it. Fill her hay rack with plenty of fresh sweet hay. She will now have an extremely large appetite for bulk foods. Over the first two days, three bran mashes a day, or dry bran feeds if she tires of bran mash, should be sufficient. Then gradually add oats, maize and other concentrate ingredients to the feed until by about a week after kidding she is eating a normal concentrate ration. The quantity should be according to yield (see *Feeding* section, pages 84–85).

The dam's mineral requirement will be very high at this time, so make sure a mineral supplement is included in the ration and that she has access to a mineral lick, preferably cobaltised. As she gets back to normal feeding, start offering green foods or allow her to go out to graze for short periods. This is a highly critical time and overfeeding during the first few days can result in a sick goat and a ruined lactation. At this time the goat is also very susceptible to worm infestation so always administer a worming agent during the first week after kidding, and preferably whilst the goat is producing colostrum. Otherwise, the milk must be discarded in accordance with the instructions supplied with the particular worming agents.

During the first couple of weeks after kidding, the goat will pass a bloody discharge from her vulva. Unless this becomes foul-smelling, it is quite normal and will eventually clear up.

It is usual to leave the kids on the dam for about four days, but if for any reason you take them away at birth, never under any circumstances strip the goat out when milking, as this can cause milk fever. There may well not be much milk at all for the first few days, so just keep the udder eased. If the kids stay on until four days, check the goat's udder since the kids may tend to drink from one side more than the other, or the udder may get distended and require easing. At four days the goat can safely be stripped out and the milk fed to the kids by bottle.

Milk-feeding kids

Clean, sterilized bottles holding about $1\frac{1}{3}$ pt (0.75 l) are used with a lamb teat. Arguments rage over whether plug-in or pull-on ones are easier or safer, but use whichever you prefer or can most easily obtain. Introduce the teat gently into the kid's mouth and persuade it to suck. It is worth

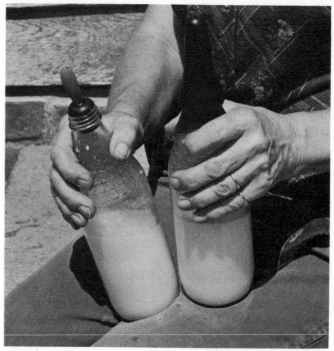

Two methods of bottle-feeding kids: The bottle on the left has a plug-in teat and that on the right a pull-on teat.

leaving the kids for several hours before their first feed when they are taken away from the dam to ensure that they are hungry. If the first bottle is unsuccessful, don't panic or get flustered. Try again a few hours later when they are even hungrier. Start with four feeds a day if you can fit them in with your routine. A $\frac{1}{4}$ pt (125 ml) per feed is quite adequate for the first day or two, increasing gradually until they are taking 4 pt (2 l) a day. At this time, probably when they are about a fortnight old, although this is very variable, reduce the number of feeds to three a day. Beware of greedy kids – it is possible for them to overdo it and give themselves digestive upsets. If this happens, miss a milk feed and give them warm water with glucose. Persistent scouring may well need veterinary attention and is often a sign of worm infestation or coccidiosis.

With more than about half a dozen kids, bottle feeding tends to become impractical and it may well be worth investing in a lamb bar and allowing the kids access to it throughout the day and night. All equipment used for milk feeding, whether it be bottles, teats or buckets, should be thoroughly washed after every feed and sterilized with hypochlorite solution or equivalent.

How long milk feeding should continue is a question open to much debate. In Britain kids are quite frequently not weaned until they are six months old. Unless you want to feed a calf weaner ration containing a high proportion of dried milk, it is probably sensible to milk feed until they are about four months old, and wean the kids gradually over a period of several weeks so that they can adjust to taking in a greater proportion of concentrates and drinking water.

It is not essential to give them fresh milk if you wish to

Kids feeding: Note the way that their necks are outstretched and their heads tilted upwards.

to establish records of improvement of progeny over their parents (which should be done by comparing their milk yields). Artificial insemination can obviously offer a wider range of males to the goatkeeper, particularly in areas where the goat population is low, and long distances may have to be travelled to reach a suitable male.

The greatest advantage of artificial insemination is for exporting, as the export of livestock is both expensive and hazardous. It is currently impossible to import goats into Australia, almost as difficult to import them into the USA and certainly not easy to get them into Canada. Import of semen is often viewed in a better light by the authorities if stringent health checks on the donor stock are carried out, and importing semen would reduce the amount of inbreeding which is unavoidably occurring within certain breeds in some countries.

Pseudopregnancy

Goats occasionally have pseudopregnancies, commonly termed as cloudbursts. The goat, either mated or unmated, will consider that she is in kid. Her abdomen will swell and after a full gestation period of five months will produce a large amount of cloudy fluid but no kids, and the goat will subsequently milk poorly, if at all. In a seasonally breeding animal this is extremely wasteful. Currently, no cause for the condition has been identified though it does appear to be more common amongst goats that have run through a two year lactation than in those that are kidded every year.

sell it or use it for other purposes. There are some excellent lamb and calf milks on the market which make good economical substitutes for fresh milk. Use them according to the manufacturer's instructions, but when the manufacturer recommends that the strength of lamb milk is doubled after a few days it is usual not to follow this but to carry on feeding the weaker strength. The strength usually works out at approximately 2 oz to a pint (50 g to 0.5 litre).

Artificial insemination

In countries where goatkeeping tends to be carried out on a fairly small scale, often as a hobby, artificial insemination has not become widely established. In France, where commercial goatkeeping on a large scale is common, artificial insemination is a practical proposition. In Britain and North America more commercial herds are becoming established, and it is possible that artificial insemination may become more widely used.

Successful techniques for collecting and preserving semen and for insemination are now well known, and the problem is only the lack of sufficient demand to establish a viable system. With cattle, commercial pressures spurred on efforts to establish an organized semen collection and insemination service. However, within the goat world commercial pressures are nothing like as great and, with a few notable exceptions, males are used too infrequently

The male goat

Deciding whether or not to keep a stud male is no easy matter. It is probably not worth attempting to keep one unless you have more than, say, four goats to be served or if the problems of getting your goats to a stud male are insurmountable. For example, if you choose to keep pure Toggenburgs or Golden Guernseys you might find that there is no male within reasonable travelling distance. In that case, having your own male at stud can be justified and may well encourage other goatkeepers in your area to keep that breed.

When deciding whether or not to keep a male, always bear in mind that it is another mouth to feed throughout the year, that it will require extremely sturdy housing, because males are large and strong, and that males smell with an all-pervading odour which makes them poor neighbours if your property is surrounded by other houses.

Selection

Having made the decision to keep a male, you then have the task of choosing a suitable one. The use of a home-bred one is somewhat limited in that it is inevitably related to some of your stock, so unless the herd is large it is common practice to buy males. Spend some time looking round at various breeding lines and families in the breed you require. It is only common sense to go for a line with good milking records, either in show results or official milk recording results.

In Britain, a milker that is officially milk recorded by the Milk Marketing Board, and has given over 2,000 lb or 1,000 kg of milk in a 365-day lactation with over 3 per cent butterfat, will carry an R prefix before her name. These are published in the British Goat Society's Monthly Journal when awarded and then again in the Herd Book. A male whose dam and whose sire's dam have both qualified will carry a section mark (§) prefix to his name which will look like this:

§ 130/145 Tooting Beck BS 14537 H

meaning that his dam had given 1300 kg and his sire's dam 1450 kg at over 3 per cent butterfat in a 365-day lactation. Double section mark (§ §) indicates higher milk records, and full details are given in the British Goat Society's Rules and Regulations. A male may also carry a dagger (†) prefix indication that his dam and his sire's dam have both qualified for star (*) or Q star (Q*) awards in twenty-four hour milking competitions at shows recognized by the British Goat Society.

Points to look for other than milk potential are conformation and hereditary faults. The conformation of a male is not terribly important as long as he is sound i.e. he has good legs and feet and a strong back, but the conformation of his progeny is important. Obviously if you are considering buying a male kid there will be no progeny to inspect, but there may well be closely related animals you can look at. Check that the goats are correctly marked for the breed, that the udder is of a good shape, that their feet are sound and their pasterns, hocks and backs are strong. Check also that no hereditary fault is prevalent in the line. Double teats are probably the biggest problem, for although a male's teats may be quite normal, if branched teats or extra teats occur frequently in his line, the male may well be able to pass the character on to his progeny.

Another problem, particularly amongst Anglo-Nubians, is wry mouths and twisted faces. Although an undershot mouth where the bottom jaw is longer than the top jaw is acceptable as long as the bottom teeth do not show, it is not acceptable to have any twisting of the jaw or nosebones and these faults tend to be inherited.

The age at which to buy should be considered carefully. A buckling (i.e. a male between one and two years old) or an adult male is likely to have progeny that you can inspect, and has also been proved to be fertile. However, you may find it preferable to book a male kid from a breeder, collect it when young and rear it, which will mean that it is well adapted to your ways of management. A male kid that is five or six months old by the beginning of the breeding season should be perfectly fit enough to serve females during the season as long as he is not used several times a day.

Stud males tend to be passed on from breeder to breeder every year or two, as their usefulness in a herd is usually at an end once their progeny reach breeding age. Consequently it may be fairly easy to buy or borrow a very well-bred male for a season or two.

Rearing

A male goat has to make enormous growth, particularly in its first year, and consequently male kids need to be extremely well reared. They should be milk fed in the same way as females, but benefit from receiving 5 pt (3 l) of milk a day, and it is normal practice to continue to give them one bottle a day after they are six months old, right through the winter if they are being used to serve females regularly.

Concentrate feed should be introduced as early as the goat will take it, building up gradually to about 1 lb (0.5 kg) per day by six months. Good quality hay should be fed at all times and ample greenstuff should be available. As with other young stock free access should be given to water and mineral licks. For preference, use rain water as hard tap water may cause urinary calculi and consequent urethral blockage.

Feeding

Once a male kid has reached six months of age, it is likely that he will be used regularly, and consequently will require rather more food both to grow and to work. Increase the concentrate ration steadily to 2 lb (1 kg) per day and continue to feed good quality hay and greens. Beware of feeding roots. Many roots, such as mangolds, contain salts which can cause urinary calculi. Various opinions exist as to which roots cause the most problems, and it is generally safest to avoid all roots (including sugar beet, fodder beet etc.) and use kale and cabbages for winter feed.

A buckling (one to two years old) and adult male goat will require approximately 2 lb (1 kg) of concentrate feed during the summer months, and can be offered up to 4 lb (2 kg) per day during the mating season. However, there is a tendency for some males to lose their appetites during the mating season, and not all of them will consume this quantity.

If at all possible, let the male out to graze. Apart from the benefits of grazing, the exercise is necessary. It is quite possible to use separate paddocks or to put the male out in the evening after the females have come in for milking. Alternatively a good-sized exercise yard should be provided.

Handling

As mentioned above, exercise is just as necessary for males as for other stock. There is a tendency for males to become aggressive during the breeding season, and allowing them adequate exercise will help them to control their aggression.

One of the big problems of handling males is their smell which is particularly pronounced during the breeding season. The smell is not only strong but it clings, particularly to clothing. Unless you wish to end up smelling the same as your male goat, always wear protective clothing when handling him, and certainly never handle him immediately before milking, as milk easily becomes tainted with the smell.

The neck of a male is very wide, and collars are often unsuccessful as a means of restraint. Head collars are a suitable alternative and you can either make them at home out of a soft rope or you can buy leather ones.

Housing

The male should be housed away from the female stock, as his smell would pervade the goathouse and, if stock which is in season can get access, you run the risk of unplanned matings.

The housing for a male must be sturdy as an adult male is an extremely strong animal and can easily break down flimsy doors and partitions. Provided that the housing provides shelter, it does not need to be solid on all sides, as males are fairly hardy. Existing buildings can be adapted for males, or you can buy and convert railway box vans, but if you are building from new concrete block or brick construction is probably the most successful. Allow at least 60 sq ft (6 sq m) floor area for an adult male. The roof should be at least 7 ft (2 m) high. The best system is probably to allow the goat free access to an exercise/service yard although if necessary a female can be successfully served within the male's pen if it is large enough.

Some breeders successfully manage to run males together, and others find that they are too aggressive for this. Probably males, if they can be reared together, find the company of another animal highly beneficial as they otherwise tend to spend a lot of their time in isolation.

The door on this male goat house can be opened from outside the yard using the handle on the left.

Using the male

It is very unlikely that males will show any interest in females that are not in season or not fully in season so it is advisable not to upset them by introducing females which are not ready to be served. Just as females are seasonal breeders so are the males, and during the summer months (particularly in hot weather) their fertility is low. For this reason, if you require out of season matings, try to let the male serve the female at least twice with an interval of several hours between services.

Take the goat in season to the male, and either let them run together in a yard, or, particularly with a goatling which may easily be frightened, hold the female and let the male run loose. He will first lick and sniff her and the female should respond by standing still and raising her tail and wagging it. The male will mount her from behind and rapidly give a thrust which indicates that he has ejaculated. It is advisable to watch to make sure that he has penetrated the female and seminal fluid is normally visible around the female's vulva after the mating. The actual mating is rapid, normally taking less than a minute, and the male and female can then be separated and a second service carried out later if desired.

An adult male is capable of serving four or five females in one day, as long as he is allowed to rest between them, but a male kid should not be expected to serve more than one goat in a day.

Health and disease

Hoof trimming

Goats are cloven-hoofed, and the horny tissue grows continuously and needs regular trimming. Goats browsing on rocky terrain or kept in a concrete exercise yard will not require too much attention, but those stall-fed or in paddocks will probably need their hooves trimmed every one or two months.

A sharp knife (very sharp penknife or Stanley knife with replacable blades) or a pair of foot rot shears is the only equipment required. Tie the goat up firmly and, standing with your back to her head, lift one leg and hold it by the pastern with one hand leaving the other hand free to use the knife. Working downwards, cut away the overgrown horny layers until they are level with the soft inner part of the goat's foot. The heel will then appear rounded and raised above the rest of the foot. Slice this off gradually until it is level with the rest of the foot. If the hoof is badly overgrown, it is better to trim it back gradually over a few weeks until you get it right, as too severe cutting at one session may cause bleeding. Repeat the procedure on each foot, then stand back and have a careful look to make sure that the goat is standing well balanced on all four feet.

The purpose of hoof trimming is to enable the goat to walk easily and to prevent it from becoming lame. A lame goat will be a poor grazer because it will spend a lot of time lying

Hoof trimming: 1. *Holding the goat.*

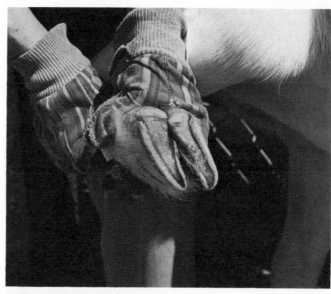

2. *The hoof requiring trimming.*

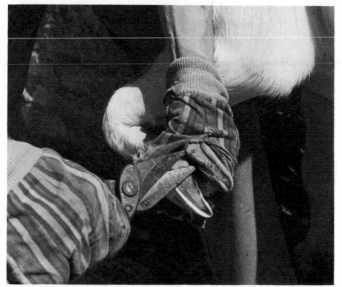

3. *The outer hoof trimmed on one claw, and the heel being cut back level.*

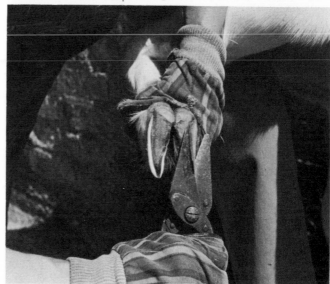

4. *The outer hoof of the second claw being cut.*

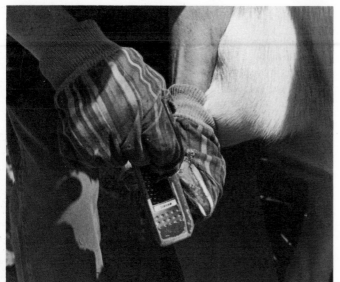

5. *The hoof being smoothed off with a rasp.*

6. *The finished hoof.*

down. A poor feeder will be a poor milker so it is in your own interests to make sure that this job is done regularly, and done well.

Preventive measures

A well fed, well housed and carefully managed goat is basically a healthy animal. Although goatkeepers should be aware of the diseases that can affect their animals, with luck most of the ailments will remain outside their experience. Goats, being closely related to sheep, share many of their diseases – particularly internal parasites, clostridial diseases and skin troubles. Ailments connected with milk production, however, and the extra stresses it puts on the goat's metabolism, have more in common with cow diseases.

Many diseases can be prevented by good management, though infectious diseases can obviously only be prevented by avoiding the source of infection, which is not always easy to identify. There are two programmes of routine prevention which are well worth carrying out on goats and these are the treatments against roundworms and/or fluke and against the clostridial diseases.

Roundworm prevention

There are many proprietary vermifuges on the market and recently more have been appearing all the time. For many years Thibenzole has been recommended, but there is now considerable evidence, although unconfirmed, that animals are developing a resistance to the drug. Many of the new drugs are related, but not identical, to thiabendazole, the active ingredient in Thibenzole. They are available from agricultural merchants or veterinary surgeons, but are usually provided in packs of a suitable size for a large sheep flock. For this reason, it may well work out cheaper to obtain the drug from your veterinary surgeon. Several formulations are available, injectable, drench and in-feed preparations. The last method is not particularly suitable for goats as goats are notorious for being able to leave the unwanted items in their food uneaten.

The frequency of worming depends on local conditions. If your goats run with sheep they will need frequent dosing, preferably with a change of drug every so often to prevent the parasites building up resistance. Unlike sheep, goats are not natural grazers and do not have the same natural resistance to the parasitic worms which spend part of their life cycle in the pasture grasses. Different worming materials will have different dosing regimes so check the instructions on the pack. However, generally, a suitable preventive programme of worming would be to treat the goat before she is put in kid, again during the first week after kidding and thereafter at four monthly intervals until the goat is mated again. Similarly, males and young stock can be dosed three times each year.

No goat is worm free. All goats carry a small worm burden, but it is only when the number of worms builds up to such a level as to cause deterioration in the goat's performance or condition that problems arise. Any signs of worm build-up

should be treated promptly by giving the animals a further dosing and moving them onto clean pasture. Kids have no natural immunity to roundworms until they are four to six months old, so consequently should be kept off any contaminated grazing. Details of signs and symptoms are given below under *Internal parasites* (see page 101).

Fluke

Flukes are internal parasites which require a species of freshwater snail in which to pass part of their life cycle. In the adult herbivore they invade the liver and heavy infestation can cause rapid deterioration in the animal's health. Any animals grazing on wet or marshy land should be routinely dosed against fluke, and several of the commercially available worming drenches also contain a drug which kills flukes.

Vaccination against clostridial diseases

The two major clostridial diseases which affect goats are enterotoxaemia and tetanus. Braxy, blackleg and pulpy kidney are others which affect sheep in particular but which are known in goats. Symptoms of enterotoxaemia are described below. Suffice it to say that the diseases are usually fatal and the vaccination relatively simple and inexpensive.

The vaccine is injected subcutaneously (under the skin) either behind the foreleg or on the neck. There is a tendency with some vaccines to leave a local reaction in the skin which appears as a hard firm nodule which may or may not become abscessed. If you find this happens, it is worthwhile trying a different brand of vaccine next time to see whether the reaction is less severe. The vaccine is usually administered to the pregnant goat two to four weeks before kidding. This confers immunity to the kids by adding a high level of immunoglobulins to the colostrum. No immunity is obtained across the placenta so a newly born kid has no immunity at all until it has received colostrum. The kids are then vaccinated at two to three months of age, and again four to six weeks later. The dam should not require revaccination until just before the next kidding, although if she is 'running through' it may well be advisable to vaccinate a year after the previous kidding. Animals being shown regularly are at risk of developing enterotoxaemia because of the stress of frequent travelling, and it is advisable to vaccinate all show animals annually. Male goats, also, should be vaccinated annually.

Vaccine is freely available from agricultural suppliers or from your veterinarian and most vaccines are effective in preventing all the various diseases caused by different species of *Clostridium*.

Care of the sick animal

Early diagnosis and treatment of illness is essential. A long illness may be severely debilitating and may stunt the growth of young or have a prolonged effect on the yield of a milker. The stockman who looks carefully at all his animals every day will easily notice a goat which is 'off colour'. The animal may not feed or cud normally, its coat may be dull and standing up, its droppings may be lumpy or loose, its milk yield may be lower than usual, it may be lame or just listless.

On finding a sick goat, check its temperature and pulse immediately.

Normal rectal temperature	102.5°–103°F (39°C)
Normal pulse rate (taken at top of front leg, inner side)	70–90 per minute
Normal respiration rate	20–24 per minute

A goat with an abnormal temperature (either high or low) should be kept warm with a rug or sack tied over her, and if she shows any sign of collapse she should be propped up between straw bales. A collapsed ruminant lying on its side will rapidly become blown (the rumen will cease to function properly and will fill up with gas causing acute discomfort). If the goat is blown when you find her, keep her on her feet and walking round. Treatment for this condition is described below under *Bloat*. Never hesitate to call veterinary assistance. Even if you live in an isolated situation and cannot easily reach your vet it is advisable to telephone him and ask for advice. Remember that it is your job as stockman to provide the first aid and nursing, and it is the veterinary surgeon's job to diagnose and treat disease.

Some diseases are notifiable by law. These are nearly always highly infectious, sometimes, as in the case of anthrax, to humans as well. Apart from your legal obligation it is also, therefore, in your own interests to notify any sudden deaths or inexplicable severe illness to your veterinary surgeon.

Keep a basic first aid box for treating accidents and first signs of disease. We will not include drugs on the list since their availability varies from one country to another as also do their names.

First aid box

Antiseptic lotion	Liquid paraffin
Antiseptic cream	Worming formula
Cotton wool	Scour mixture
Surgical gauze	1 tsp (5 ml) common salt
Gauze bandage 2 in (50 mm) wide	½ tsp (2.5 ml) sodium
Crepe bandage 2 in (50 mm) wide	bicarbonate
Scissors (rounded ends)	4 tbsp (60 ml) dextrose
Drenching bottle	or honey (*never* cane
Clinical thermometer	or beet sugar)
Udder cream	4 pt (2.3 l) water
Kaolin powder	

Diseases

The list of diseases we give here is by no means exhaustive. It deals with those diseases to which goats are most prone or which, if they do occur, are most serious. The following key may help you to identify the different diseases on the list by relating them to their typical symptoms.

Recognizing common ailments

Symptom	See under
Sudden death	Anthrax; enterotoxaemia
Collapse/coma	Acetonaemia; milk fever; poisoning; enterotoxaemia
Unsteady gait	Acetonaemia; enterotoxaemia; pregnancy toxaemia
Staring coat/poor condition	Acetonaemia; internal parasites; coccidiosis; enterotoxaemia; pregnancy toxaemia
Poor appetite	Acetonaemia; pregnancy toxaemia; internal parasites; coccidiosis
Abdominal pain	Bloat; colic; poisoning; enterotoxaemia
Scouring/diarrhoea	Diarrhoea; internal parasites; poisoning; coccidiosis
Vulval discharge	Metritis
Udder inflamed	Mastitis
Milk yield, sudden drop	Acetonaemia; enterotoxaemia; internal parasites; mastitis
Coughing	Internal parasites
Lameness	Food rot; laminitis
Joints swollen	Navel ill; rickets
Navel swollen (young kids)	Navel ill
Sores around mouth	Foot and mouth; skin diseases (orf)
Hair loss	Skin diseases
Scabbing	Skin diseases; abscesses

Abscesses

Abscesses are the result of a local invasion of bacteria, either into a wound or into an organ, causing it to erupt. Abscesses which appear in the skin are normally lanced, thoroughly washed out with antiseptic solution and then kept open for several days with frequent applications of antibiotic cream. It is advisable to ask a veterinary surgeon to undertake this procedure. In North America a condition known as caseous lymphadenitis is widespread amongst sheep and goats. A bacterium affects the superficial lymph nodes and large abscesses appear under the skin of mature goats. Deep abscesses may also occur which lead to weakness, emaciation and death. Superficial abscesses can be treated

Administering medicine by drench: *Hold the goat in the correct way. . . .*

. . . and then administer the medicine from the drenching bottle.

successfully but little can be done when the bacteria affect the deep lymph nodes.

Accidents

Cuts and wounds should be treated as soon as seen. If a foreign body, such as a thorn or sharp object, is present in the wound try to remove it carefully. If it appears to be at all stuck call veterinary assistance. After carefully inspecting the wound, bathe it thoroughly with antiseptic solution then apply an antiseptic cream. If blood is flowing very fast and you suspect a major blood vessel is punctured, apply direct pressure to the wound with a pad of gauze. Never use a tourniquet as these can be extremely dangerous to the animal. If the cut or wound is of a size which looks unlikely to heal easily it may need stitches, so call the vet. It is advisable in the case of any deep wound that the animal should receive antibiotic treatment to prevent infection. Also, if the animal is not vaccinated against tetanus, the vet should be asked to administer an anti-tetanus vaccine. In the case of a cut or wound on the udder, always call for veterinary assistance since the udder tissue can easily become infected and rapid treatment is necessary.

Fractured legs are the other serious accidents which are likely to befall goats, particularly kids. However difficult it may be, keep the animal as quiet and still as possible and call veterinary assistance immediately.

Acetonaemia

Acetonaemia or ketosis occurs when a goat cannot take in sufficient food for her body requirements, and metabolizes stored fat. This may occur just before or shortly after kidding, particularly in fat animals, when the goat's energy requirements are unusually high. Classically, the condition is characterized by the goat's breath smelling of pear drops or nail varnish, but this by no means always occurs. The animal rapidly loses condition, ceases to eat and the milk yield drops rapidly. In acute cases the animal becomes comatose within a couple of days.

Treat the goat with 6–8oz (approximately 200ml) propylene glycol orally, and repeat twice daily for two days. Alternatively dissolve one tablespoon of sodium bicarbonate in 4oz (115ml) water, administer it orally and follow it with 8oz (230ml) of dextrose or honey. The condition is serious so always call a vet and he may choose to use corticosteroid therapy. If you can get the goat to eat or drink, include molasses, black treacle or other easily digested high energy food in the diet.

Anthrax

Very often, the first symptom of anthrax is death so little can be done to save an animal. However, in most countries it is a notifiable disease so any sudden death should be reported to the vet, who will arrange a post mortem examination to ascertain the cause.

Bloat

Bloat (blown or hoven) is a symptom rather than an illness. It is caused by a build-up of gas in the rumen. The animal will first of all look abnormally distended on the left side of her abdomen, then gradually the whole of her abdomen will become distended. She will be in acute discomfort and distress. Drench the goat with 50ml of liquid (medicinal) paraffin (20ml for a kid) and add ½oz (12g) bicarbonate of soda if available. Rub the goat's sides and try to keep her moving and the condition should subside. If no improvement occurs in half or three-quarters of an hour, call the vet. The condition is usually caused through eating too much green food, particularly clover, lush or wet grass, and frosted greens, on an empty stomach. Prevention is obviously better than cure. Accustom stock gradually to lush pastures in spring by turning them onto it for limited periods only. Give them access to foods with a high fibre content such as straw and hay before turning them out in the mornings.

Brucellosis

Brucellosis is a disease, transmissible to man, which has been a problem in cattle for many years, but is now being steadily eradicated from temperate lands. *Brucella abortus* is the species which occurs in Britain but it is not transmissible to goats. *B. melitensis* (which causes Malta fever in man) occurs in North America, and although the number of cases in goats is extremely low, they may nonetheless be affected by local brucellosis eradication schemes.

Coccidiosis

Coccidiosis is caused by a microscopic protozoan and, contrary to earlier beliefs, is relatively host specific. The coccidia that affect poultry and rabbits will not affect goats, although those that affect sheep can be carried by goats.

The condition is most frequent in kids, causing scouring and poor condition. The scouring is persistent and does not clear up by missing a milk feed. Kids can rapidly infect each other, so kid quarters should be kept clean and hygienic, and any suspected coccidiosis should be treated immediately. A veterinary surgeon may well let you have sulphamethazine to treat the animals, or might prefer to treat it himself with other drugs. The condition may occur in older animals and is characterized by poor condition and scouring or constipation, which may often occur alternately.

Colic

Colic, like bloat, is a symptom of digestive upset and is characterized by severe pain. It can be caused by a variety of conditions, so administer 50ml liquid paraffin orally as a first aid measure and seek veterinary advice.

Diarrhoea (scours)

Diarrhoea is also a symptom of digestive upset and can be indicative of many conditions. It should never be ignored as severe diarrhoea can lead to dehydration and emaciation. In young milk-fed kids, miss a milk feed and give a bottle of warm water and glucose with a dessertspoon (10ml) of kaolin powder added. In adults, suspect worms if the droppings are first lumpy then loose but call the vet in any cases of persistent scouring.

Enterotoxaemia

This is one of the commonest clostridial diseases, and prevention has been discussed earlier in this section (see page 98). It is usually induced by sudden changes in circumstances such as access to fresh lush pasture, travelling (particularly if concentrates are fed before a journey) or if the animal is accidentally allowed to gorge itself on concentrates. It is frequently the thriving animals which are affected. Death is rapid, occurring within hours of the appearance of the first symptoms. These include inability to balance, drop in yield of milkers, persistent stretching of the abdomen as in colic and walking backwards a few steps. Many authorities say there is no cure, but one leading breeder suggests the following treatment when the first symptoms appear. Administer 6cc penicillin intramuscularly and a large dose of sulphamethazine (20ml of 16 per cent solution) orally, or, if the condition is advanced, intravenously (administered by a vet).

Foot and mouth disease

This is a notifiable disease, affecting all cloven-hoofed animals. It is unlikely to occur unless the disease is endemic in your area or unless there is an outbreak. The symptoms are blisters on the tongue, palate and inside of the lips, and around the top of the hoof. Any suspicion of the disease should be notified to your vet immediately.

Foot rot

This is an infection in the feet of sheep and goats, common on wet pasture, caused by an organism which invades and kills the horny part of the hoof causing severe lameness. Any infected areas should be cut away, the animal's feet dipped in a 10 per cent formalin solution and sprayed with a Terramycin foot rot spray, available from agricultural suppliers or vets. Move uninfected animals onto clean pasture and isolate infected animals until the condition has cleared.

Goat pox

This is a viral disease characterized by a mild fever and small vesicles on bare skin, particularly the udder. Affected animals should be milked last and scabs treated with boracic powder to control bacterial infection.

Internal parasites (worms and flukes)

Goats are frequently susceptible to roundworm infestation, sometimes to flukes and occasionally to tapeworms. The first symptoms of any of these are generally poor condition, with loss of yield in adult stock. Roundworms will also cause lumpy droppings or scouring and possibly cause the milk to taste 'goaty'. Lungworms similarly cause the animal to appear in poor condition and to cough persistently, but lungworms are not frequently encountered in goats.

The only way positively to identify what is causing the goat to lose condition is to ask the vet to examine faecal samples. He will arrange for a worm count to be done and advise treatment accordingly. If you suspect intestinal roundworms, dose the goat with the vermifuge you nor-

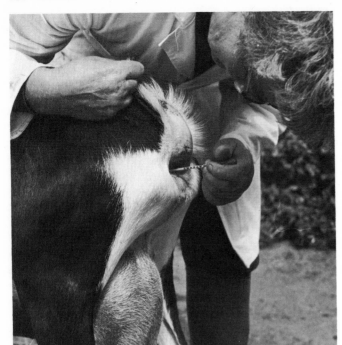

Taking a goat's temperature: The thermometer is gently introduced, bulb end first, into the rectum.

mally use, but if the condition persists consult the vet straight away. Routine preventive measures have already been discussed earlier in this chapter (see page 98).

Laminitis

Laminitis is an inflammation of the soft part of the hoof and causes severe pain to the goat, who becomes lame and possibly unwilling to walk or stand. The condition frequently occurs shortly after kidding, or as a result of having had too high a level of protein in the diet. Treat the inflammation by soaking the affected foot in a warm water bath, and then bandaging it with a hot bran poultice. If the goat has a fever or appears to get no relief from this treatment, it is advisable to call in the vet who will be able to administer drugs to reduce the swelling.

Mastitis

Mastitis is basically an infection in the udder tissue caused by bacteria, but the condition can be triggered off by a variety of causes. It can be the result of physical damage to the udder such as a bang or a cut, or it can be due to not milking the animal when it requires milking; rough handling during milking; infection carried by flies; or infection from another goat, usually carried by the person doing the milking. Any goat with suspected mastitis should always be milked last, and the person milking should wash his or her hands immediately afterwards. Needless to say, the milk from the animal should be discarded.

Acute mastitis is a severe condition and the animal will have a swollen, hot udder. She will act as if in discomfort when milked, and will run a high temperature. To save the udder and even the animal, call veterinary assistance

immediately. The goat will need to be milked frequently so that the udder does not become full and distended, and bathing the udder alternately in hot and cold water will help to reduce the swelling and discomfort. Clots and blood in the milk are generally seen in cases of sub-acute mastitis. This does not trouble the goat to the same extent, but will require antibiotic treatment which the vet can supply. Again, milk must be discarded from any animal receiving antibiotic treatment. The acute or sub-acute conditions may well damage the udder permanently, leaving lumps in the udder tissue and reducing the animal's milk yield (probably for life).

Metritis

Metritis is a uterine infection, occurring most frequently after kidding. Any manual interference at kidding should be followed by antibiotic treatment by injection or pessary, but the condition may still occur when no interference has taken place. The animal will have a high temperature, appear generally unwell and have an unpleasant smelling discharge from her vulva. Veterinary treatment is necessary.

Milk fever

This is not an accurate name for what is in fact an acute condition, usually occurring in heavy milkers shortly after kidding although it can occur at any time during the lactation. Contrary to what the name of the disease would suggest the animal's temperature is usually depressed rather than raised. She becomes unsteady on her feet and generally listless, and may collapse and become comatose. The condition is caused when calcium is removed too quickly from the blood and tissues. Calcium borogluconate (50 ml) should be injected subcutaneously and will give rapid recovery.

If the animal has collapsed call the vet to inject the borogluconate intravenously. Prompt action is essential. It can be prevented by sensible management of the dam after kidding and, in particular, by not forcing the milk yield early in the lactation.

Navel ill

Navel ill is normally seen as a swelling on the umbilicus in the first few days of life. It is painful to the touch and the kid may well run a temperature and be generally unwell. Antibiotic treatment is required, and dipping the navel cord in iodine at birth may help prevent the condition. Serious cases may cause generalized septicaemia in the animal or an arthritic condition in the joints, so treatment should be prompt.

Poisoning

The goat is an animal with an inquisitive nature and willing to eat an incredibly wide range of plants, but as long as one particular kind is not eaten to excess, particularly on an empty stomach, there are very few plants which will cause severe poisoning.

The main problem plants are *Taxus baccata* (yew); *Rhododendron sp.*; *Datura stramonium* (thorn apple);

Laburnum anagyroides (laburnum); *Conium maculatum* (hemlock); Family *Astragalus* (locoweeds) – USA, *Senecio jacobaea* (ragwort) – particularly when dead; *Rheum rhaponticum* (rhubarb leaves). In Australia the common oleander bush is particularly deadly to goats.

If a goat is seen eating any of the above, you can assume that you have got a case of poisoning on your hands. Call the vet immediately. First aid treatment for one type of poisoning may be fatal for another type so it is advisable to ask your vet for instructions.

There is an excellent book called *British Poisonous Plants* published by the Ministry of Agriculture in Britain which is an invaluable source of information on harmful plants and the appropriate first aid and veterinary measures that should be taken. In North America two useful books available on this subject are *Poisonous Plants* (W. James), published by Naturegraph and *Poisonous Plants of the Midwest and their Effects on Livestock* (R. Evers) published by the University of Illinois Press.

Pregnancy toxaemia

This condition should be suspected when an in-kid goat becomes ill during the last two months of pregnancy. It is caused when the level of nutrition is insufficient both to maintain the dam and to allow for the growth of the kids she is carrying. As a result, the foetuses begin to rob the dam of stored reserves. It frequently happens in an animal which has been fed too well in the first stages of pregnancy. Call the vet, since the condition is serious to both the goat and the unborn kids, and feed the dam as for acetonaemia.

Rickets

Rickets may occur in fast-growing kids, particularly those born during the winter or in prolonged periods of bad weather with no sunshine. Sunshine is necessary for kids to produce vitamin D which in turn is essential for normal bone growth. The first signs of the condition are swollen joints, followed by the long bones appearing bent. The vet can supply vitamin D preparations which will give rapid improvement, and sprinkling sterilized bone meal on the feed may also help.

Skin diseases

Goats are subject to a number of skin diseases, several of which they share with sheep.

1. **Orf** (contagious pustular dermatitis) is probably the commonest condition shared with sheep. It is viral in origin and causes pustules around the mouth which can be spread to the udder, particularly with suckling kids. An effective vaccine is available and if the condition is common in your area, it is advisable to vaccinate routinely. An animal which has already had the condition is likely to be immune for nine to twelve months. The sores should be treated by removing scabs and bathing the area with antiseptic solution. Antibiotic aerosols may be sprayed on the affected areas, but creams should not be used. The condition spreads rapidly from one animal to another and is transmissible to humans, so rubber gloves should be worn when handling infected animals.

2. **Mange** is a condition caused by mites which burrow in the skin setting up local irritation, causing hair loss, scabbing and often secondary infections. The scabs should be removed, and the goat treated with a BHC or Coumaphos preparation. The treatment may well fail and different treatments may be recommended by your vet, until the condition clears. Although it is infectious to other goats, some animals appear more susceptible to the condition than others.

3. **Lice** are blood sucking creatures living on the goat's skin. BHC or Coumaphos preparations are usually effective, and may be used as powders or as a bath or dip.

4. **Eczema** is a scabby condition, particularly affecting the area round the eyes and nose, for which no real cure exists. Local treatment with creams may alleviate the condition, as may changing the diet, particularly feeding less maize and more green foods. The condition is not infectious, and seems to run in families with some animals being affected and others immune. It is unsightly, and the lesions leave the animal open to secondary infections but it is not serious.

SHEEP

Background

More than ten thousand years ago man first domesticated the sheep. Modern sheep breeds are descended from the wild sheep of Europe and Asia that inhabited relatively dry, unrestricted upland areas with sweet but sparse vegetation. The transition of the sheep to modern farming with its lush grass and fenced enclosures has been achieved only by a long process of selective breeding. Many breeds still possess an inborn urge to escape from the restrictions of their allotted grazing area, and they prefer the short, sweet grasses when they are available. Many of the problems which afflict sheep, such as pneumonia and foot rot, are associated with their origins in a dry climate and their continuing struggle to adapt to wetter conditions.

Despite these problems the domestication of the sheep brought under man's control an animal that could provide him with wool and skins for his clothing, and meat and milk for his nourishment. But of equal importance in settled agricultural systems the 'golden hoof' of the sheep trod and fertilized the ground on which the farmers grew their arable crops. The sheep has continued throughout its domesticated history to provide these benefits making it in many ways perhaps the most useful of the domesticated species.

Originally the domesticated sheep was probably developed as a milch animal, being milked after it had weaned its lamb to provide milk, butter, yoghurt and cheese for human consumption. This pattern of husbandry is still practised in many parts of the world. Later the sheep was developed for meat and wool production. The selection of sheep for meat received little attention until the last two hundred years or so, but selection for higher quality wool started in the early stages of domestication. The coarse fibres in the fleece were gradually eliminated, and as long ago as three thousand years a breed of finewool sheep had been developed in Anatolia. In the Middle Ages in western Europe, between the eleventh and sixteenth centuries, wool was a major industry based on breeds such as the Spanish Merino, the finewools of Hereford in England, and the longwools of northern England.

Some modern breeds of sheep are highly specialized for the production of one commodity, while others are of a more general type and can be described as dual-purpose or even triple-purpose. On this basis most breeds of sheep can be grouped conveniently into seven categories, according to their purpose, namely milk (or milch) breeds, wool breeds, primitive breeds, hill and range breeds, crossing breeds, general-purpose breeds and meat breeds. In some countries breeds from some of these categories are crossed to combine their characteristics in a system known as stratification.

Breeds

It is not feasible to cover all the breeds of sheep within the space available, but the most important breeds are described below.

Milk breeds

Although milk is no longer a major product of the sheep industry in many developed countries, such as North America and Australasia, its importance has been maintained in other parts of the world including some European countries. The main product is cheese and sheep's milk is much richer than that of cattle or goats, having a butterfat content of 6 to 7 per cent.

Fries Melkschaap

The most important milch breed is the Fries Melkschaap which originated in Holland and is now found mainly in that country and in the north-western parts of Germany. It is a large sheep – mature ewes weigh about 150 lb (70 kg) – with a white leg and face, long ears held low and pink nostrils. It also has a long bald tail which is a distinctive feature.

The Fries Melkschaap is a specialized dairy animal and is usually kept in very small, carefully managed flocks. As a result it is a good producer with a lambing percentage well in excess of 200 per cent and an average milk yield of 140–150 gal (635–680 l) per lactation, with some individual high-performance ewes exceeding 200 gal (900 l). But other characteristics have suffered somewhat. It has a poor-quality carcass; its tooth and jaw structure is often less than satisfactory; and its feet are soft and susceptible to foot rot. These differences cause relatively little trouble in small flocks but can cause serious difficulties in larger units.

As the world's premier milch breed the Fries Melkschaap has been exported to many countries to raise the productivity of native breeds by cross-breeding. Nevertheless, it remains a minority breed with a total population of 16,000 ewes. It is an excellent one to keep if you are planning a specialist sheep dairy enterprise.

Lacaune

The most important French milch breed, the Lacaune, is much more numerous with a population of more than half a million ewes. The milk from this breed is used in the manufacture of Rocquefort cheese, and the Lacaune is found mainly in the central region of France. It is a medium-sized sheep with ewes weighing 110–140 lb (50–65 kg) when mature. The face and legs are white. The ewe is of low prolificacy averaging 100 per cent lambing while the average milk yield is 35 gal (160 l) per lactation.

Chios

The Chios, which derives its name from the Greek island on which it is mainly found, is known also as the Sakiz in Turkey. It is a fat-tailed breed which achieves a daily yield of milk of up to 5 pt (28 l) and is reasonably prolific (180 per cent lambing). The fleece is of medium quality, scoring 50–56 on the Bradford Count (see below).

Awassi

The Awassi is the dominant breed in Syria, Israel, Lebanon, Jordan and Iraq as it is well able to tolerate hot, dry climates. Its milk yield is comparable with that of the Chios – 65 gal (300 l) per lactation – but its prolificacy is lower and twins are rare. Mature ewes weigh 65–110 lb (30–50 kg) and yield a fleece of coarse wool (quality 36–46) weighing on average $3\frac{1}{2}$–4 lb (1.5–1.8 kg).

Wool and pelt breeds

Sheep have been deliberately selected to improve the quality of their wool for at least three thousand years, as shown by records of a fine-woolled breed of sheep being husbanded in Anatolia in the second millenuium BC. These records are reinforced both by legends like Jason and the Argonauts and by ancient industries such as the production of murex dye for wool by the Phoenicians. In the Middle Ages Spain and Britain competed for the premier position as producers of high-quality wool, but now the main centres are the rangelands of Australia, the Republic of South Africa, the USSR and New Zealand.

The Bradford Count: Traditionally the quality of wool has been measured by the Bradford Count, which is calculated according to the relation between the weight of the wool and its spinning ability. The number actually refers to the number of hanks which can be spun from 1 lb (450 g) of wool, and a medium-quality rating would be about 56. A simpler method of assessment now being used by some sections of the wool-buying trade is by the diameter

Chios

Australian Merino

Awassi

German Merino ram

Lincoln Longwool ram

of the wool fibre itself, measured in microns. On this calculation Merino wool measurements range from 17 microns (84s on the Bradford scale), through 21 microns (64s) to about 23 microns (60s). Cross-bred and Corriedale wool run up to a diameter of 28 and 29 microns (as strong as Bradford 36–40s).

Merino

The Merino and its many close relatives are not only the most numerous of the specialist wool-producing breeds, but they also provide by far the highest quality wool. While the wool of no other breed can achieve a quality on the Bradford Count higher than 60, the various types of Merino yield wool of a quality ranging from 56 to 80.

At times extreme types of the breed have been developed. The Vermont Merino was selected in about 1900 for excessively wrinkled skin in the belief that the increased surface area would yield a greater weight of wool, but this experiment was abandoned because of problems with blowfly. Now the small Tasmanian Merino produces the finest wool, while larger variations such as the Peppin, Delaine and South African Merinos give a higher yield of stronger wool. The Peppin strain is a large-framed animal suited to rangeland conditions. This and the Delaine varieties are recommended for hot, arid regions. The face should be a light cream colour.

The largest-framed Merino in the world is the South Australian, which is a strongwool suited to extremely low rainfall and harsh conditions. The original Merino rams were horned but in the 1920s polled Merinos were bred from sports in Australia where this is now a recognized breed. Approximately 70 per cent of Australia's sheep population are Merino and Poll Merino, and there are no less than 550 pedigree flocks in New South Wales alone.

Polwarth

The Polwarth is a breed that was created in Australia during the latter part of the nineteenth century in an attempt to produce a breed that would yield fine wool but would tolerate conditions that were too wet and cold for the Merino. The result was the Polwarth, which is approximately three-quarters Merino and a quarter Lincoln.

Longwools

The longwools are a peculiarly British variety of sheep although it is probable that they are descended from sheep introduced by the Romans. Most of the longwool breeds that continue to rely on wool as their major product are now of relatively minor importance and include the Leicester, Lincoln, Devon and Cornwall, and Cotswold. However, the Lincoln can be found in eastern Europe, South America and the USA and Australia. These are all large sheep with ewes weighing from 175–200 lb (80–90 kg) at maturity. They produce heavy fleeces of long, lustrous wool of 36–40 quality, but their prolificacy and milking ability are not good. In the current economic climate, which provides no good market for their type of wool, these breeds are actively developing other characteristics. In particular

they are exploiting their size and growth to produce heavyweight lambs.

In the last ten to fifteen years a number of specialized carpet-wool breeds have sprung up in New Zealand, most prominent of which have been the Drysdale and the Tukidale. The latter has spread with success to Australia.

Romney

This breed does not fit the typical longwool standards. Its wool is shorter and finer with less lustre, and it is a smaller animal. It has remained a very localized breed in Britain, but has achieved considerable popularity in Australia and has also become established in North America. Its wool is used in the manufacture of a range of high quality products, but in New Zealand (where it is the most popular single breed) it is used in crossing programmes to produce prime lambs. In Britain its value in this respect is limited because of its low prolificacy and milk yield.

Karakul

Some breeds of sheep have been selected for the quality of their pelt, used in the manufacture of fur coats, rather than for their wool alone. The Karakul originated in Bokhara in the USSR and belongs to the fat-tailed group of breeds. It was first imported into the USA in 1909. The highest quality pelts are obtained from lambs born prematurely or those killed at birth, but it is more usual to kill the lambs at three to ten days old, when the quality has deteriorated a little but the curl of the wool is still tight and lustrous.

Gotland

The Gotland is a breed of Swedish origin that has been developed for the production of pelts. In this case the quality of the pelts is at its best when the lamb is five months old, so the lambs slaughtered at this age can also be used for meat production. The wool is grey, lustrous and curly. The shade of grey may vary from about white to almost black, and the curls may be tight or open. Fashion dictates the particular requirements at any time. The Gotland is a fairly small, fine-boned, polled sheep, with a short tail similar to that of the primitive breeds. It is an excellent breed to keep in places where there is a high demand for pelts.

Primitive breeds

The word primitive may be misleading used in this context, for it could be interpreted as unproductive or non-commercial. But while some of the breeds in this category may be unimproved in that they have not been subjected to planned selection programmes, nevertheless they can be productive and compare favourably with the improved breeds in some circumstances. Such breeds are found in all parts of the world, generally adapted very efficiently to local conditions. Details of only one or two breeds are given here as examples.

Soay

The Soay probably approaches the wild sheep most nearly in both its appearance and characteristics. It is a small sheep, mature ewes weighing 55 lb (25 kg), with fine bones and a short tail. Its face and legs are brown and its wool is either brown or fawn. It is horned in both sexes. All existing Soay sheep are descended from the feral flock on the island of Soay in the St Kilda group beyond the Outer Hebrides in the Atlantic Ocean.

Until recently these sheep were kept for their novelty, but increasingly their remarkable commercial qualities are being realized. They are very hardy and thrifty and have been used to reclaim waste land in various places. They average 130 per cent lambing and, mated to a meat ram, they produce a heavy weight of lamb in relation to their own body weight. Likewise, their yield of milk is exceptionally high per unit body weight. But the most commercial characteristic of these sheep is their ability to produce a carcass of well-flavoured, lean meat free from undesirable layers of fat. This breed does well in 'easy care' systems, provided the fencing is good.

Hebridean and Manx Loghtan

The Hebridean and the Manx Loghtan are two unusual breeds from Britain that probably represent a transitional stage between the short-tailed primitive breeds and the popular modern breeds. They are both of particular interest because they are frequently multi-horned, usually possessing four horns, but in some cases six or even eight horns can be found. Both breeds are small (mature ewes 85–90 lb or 38–40 kg) and fine-boned, with a half-length tail that does not reach to the hocks. The Hebridean is black, the Manx Loghtan moorit or 'moor-red'. The value of both breeds lies partly in their efficiency of production (yield of lamb in relation to body weight) and partly in the demand for naturally coloured wool stimulated by the revived interest in handcrafts and self-sufficiency systems. In Australia black wool is in very high demand at the moment from the cottage industry spinning trade.

Jacob

The Jacob, and other similar breeds such as some of those herded by the Navajo Indians, are of interest because they have spotted fleeces. One of the earliest recorded examples of selective breeding is described in *Genesis*, and the Biblical breeder Jacob has given his name to the breed which is claimed to be descended from his sheep. It is likely that both the Jacob and the Navajo sheep originated in Spain where multi-horned, spotted sheep were found previously. The Jacob ewe weighs about 105 lb (68 kg) when mature, and has almost completed its transfer from the primitive to the hill breed category.

Hill or range breeds

The sheep industry, in Britain especially, is based on a system known as stratification, whereby hardy hill sheep that have become too old to withstand the rigours of a

Romney ram

Soay

Manx Loghtan

Improved Jacob

mountain environment are moved to lower ground as draft ewes. There they are mated to rams of specialist crossing breeds to produce cross-bred daughters, which are mated to meat breed rams to produce high-quality lambs for slaughter. Typical examples for Britain and Australia are shown below. The system is not widely practised in North America.

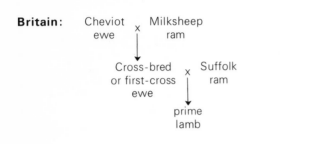

Britain: Cheviot ewe × Milksheep ram → Cross-bred or first-cross ewe × Suffolk ram → prime lamb

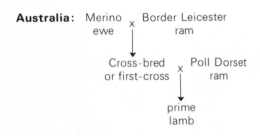

Australia: Merino ewe × Border Leicester ram → Cross-bred or first-cross × Poll Dorset ram → prime lamb

Hill and range sheep are hardy and able to survive in difficult conditions. The hill breeds of Britain are able to tolerate cold and wet conditions, while the range sheep of Australia and North America are adapted to hot, dry conditions with sparse vegetation. The latter breeds are mainly of the Merino type, which has been described already in the wool breed section. The popularity of the first-cross (Merino–Border Leicester) ewe and the Dorset Horn ram as a terminal crossing breed in Australia is well established, and it is known for combining quick maturity with good milking qualities.

Scottish Blackface and Swaledale
In northern England and much of Scotland a variety of closely related breeds can be found which are horned and blackfaced, and possess long coarse wool. They are vigorous and thrifty. The most numerous is the Scottish Blackface, but the Swaledale is currently in greatest demand. Other breeds in this group include the Dalesbred, Rough Fell and Lonk. A mature Scottish Blackface ewe weighs on average 130 lb (60 kg). It yields a fairly heavy fleece of coarse wool (Bradford Count 28–32) which commands a high price as it is greatly prized by Italian buyers for filling mattresses, apart from its value as carpet wool. The Swaledale is not so heavy as the Scottish Blackface – a mature ewe weighs 120 lb (55 kg) – but it is longer on the leg and milkier, and these two qualities have persuaded some Scottish Blackface breeders to use Swaledale rams. The Swaledale fleece is lighter than that of the Scottish Blackface, but is of higher

quality and recently a speciality Swaledale tweed has been developed. Traditionally the Swaledale has been important as the dam of both the Masham (sired by the Teeswater) and the Mule (sired by the Bluefaced Leicester), but more recently it has been mated to the British Milksheep with very promising results. The Swaledale is a good breed to keep if you want to rear hardy, productive sheep in hilly conditions.

Whitefaced Woodland and Derbyshire Gritstone
The Whitefaced Woodland is another Pennine hill breed that closely resembles the horned, blackfaced breeds, except that it has a white face and legs, with pink nostrils. It is found mainly in the southern Pennines. The same area is the home of the Derbyshire Gritstone, which is the largest of the hill breeds and does not conform to the standard type of Pennine breeds. It has a speckled face and is polled, while its wool is of high quality with a Bradford Count of 50–56. This breed has been used recently to cross with the Scottish Blackface to poll the latter breed. The advantages of polled sheep are that they are not so troubled by headfly and that the carcasses are easier to skin in compliance with the new Common Market regulations.

Cheviot
Several other hill breeds are polled or partly polled. The Cheviot is generally polled in both sexes although some rams may be horned. The majority of the Welsh Mountain breeds, the Herdwick and the Shetland have polled ewes but horned rams. The Cheviot has become separated into three separate types of which the most important now is the North Country Cheviot found mainly in northern Scotland. It is a heavy, compact sheep with mature ewes weighing 150 lb (68 kg). It has a distinctive Roman nose and large, fairly erect ears. The face and legs are white, and the wool is of high quality (50–56). Although a productive sheep, it is less hardy than the horned, blackfaced breeds, and more prone to metabolic diseases.

Welsh Mountain
The Welsh Mountain sheep are also found in several varieties. The main type is a small sheep, with mature ewes often little more than 80 lb (36 kg) body weight. They thrive in mountain areas of heavy rainfall and poor grazing, but are not suitable for improved grazings, partly because of their notorious ability to escape from enclosed areas, partly because the pasture is too rich for them, and partly because of their low productivity. The South Welsh Mountain variety grows larger with a heavier, coarser fleece, while the Black Welsh Mountain can scarcely be classed any longer as a hill breed. It has made its home in various English parks where its colour has proved an added attraction to the aesthetic appeal of large country houses, and its naturally coloured wool has been used for hand-spinning. It has become essentially a lowland breed.

Herdwick
The Herdwick, by contrast, is probably the hardiest breed

Scottish Blackface

Swaledale

Whitefaced Woodland

North Country Cheviot

Cheviot

Herdwick ram (lamb)

Shetland (lamb)

Welsh Mountain

Teeswater ram

Border Leicester ram

Blue-faced Leicester ram

British Milksheep

in Britain, living on the exposed slopes of the Lake District mountains in north-western England. Herdwick lambs at birth are black or almost black, but the colour of both the face and the wool becomes lighter with age. The wool is very coarse (quality 28–32) but nevertheless some progress has been made in popularizing speciality Herdwick tweeds.

Shetland

Another hill breed, the Shetland, produces the finest wool of any native British sheep. It has a quality of up to 60 on the Bradford Count, and is used in the Shetland Islands woollen industry to manufacture items such as the gossamer shawls and Fair Isle sweaters that are famous throughout the world. The Shetland is a small, fine-boned sheep with a short tail. Mature ewes weigh about 80 lb (36 kg) but they compare very favourably with other hill breeds with regard to efficiency of production. The Shetland is found mainly on its native islands, although in recent years an increasing number of flocks have been established in various parts of the mainland of Britain. It is a hardy breed, an efficient meat producer and has high-quality naturally coloured wool, making it in many ways an attractive choice, if you can provide strong fences to restrict its itinerant habits.

Exmoor Horn

The final hill breed in Britain is the Exmoor Horn, a localized breed found mainly in south-western England. It is a whitefaced breed and is horned in both sexes. It is a compact, medium-sized sheep, of only average performance, but yielding wool of good quality (50–56).

Crossing breeds

When the draft hill and range ewes are moved to kinder pastures they are mated with rams of breeds which have been specially developed to produce high-performance characteristics. The prolificacy, milking ability and growth rate of these breeds, combined with the hardiness and thriftiness of the hill breeds, results in a highly productive and vigorous cross-bred ewe. All the crossing ram breeds had their origins in northern England. They are polled breeds of large size.

Wensleydale and Teeswater

The Wensleydale and the Teeswater are closely related and are used to sire the Masham, previously a popular ewe although it does not at present find so much favour. The Wensleydale is now used very little and it is perhaps becoming more important for the special quality of its wool which eliminates kemp fibres in its cross-bred progeny. The Teeswater has been noted for its prolificacy for a long time. As long ago as 1802 a group of twenty-four Teeswater ewes produced seventy lambs, while one ewe produced twenty lambs in six years.

Border Leicester and Bluefaced Leicester

The Border Leicester and the Bluefaced Leicester are also closely related breeds. They have made a considerable impact on the sheep industry in both Britain and Australia, the former as the sire of the Scotch Halfbred, Welsh Half-bred, Greyface and Merino crosses, and the latter as the sire of various types of Mule. The Border Leicester is a distinguished looking sheep with its stylish carriage and high-held, Roman-nosed head but, as a result of the undue emphasis placed by the breeders on its appearance and show ring image, it has lost much of its performance standard. As a result the Bluefaced Leicester is assuming a more dominant position, although even with this breed there are signs that the obsession with points of fashion, such as the pronounced Roman nose and the 'fine skin', is having a detrimental effect on the commercial qualities of its cross-bred daughters. However, at the present time, the Bluefaced Leicester and its cross-bred progeny, the Mule, are in the forefront of popularity.

British Milksheep

A considerable amount of development work has been devoted to the creation of new breeds in this category. Genetic improvement of these breeds presents the best opportunity to improve the efficiency of the sheep industry. The British Milksheep is a new breed which has been designed to increase both prolificacy and milk yield to levels significantly higher than those achieved by traditional breeds. It achieves a lambing percentage of 275–300 per cent and yields about 140 gal (635 l) in a full lactation, which is several times more than other British breeds. The performance of its cross-bred progeny will be determined partly by the breed of the dam, but the lambing percentage will range from 195–240 per cent and the mature cross-bred ewes will yield sufficient milk to rear triplets well. The British Milksheep also yields a relatively heavy fleece – rams average 12–13 lb (5–6 kg) – of demi-lustre wool of 52–54 quality on the Bradford Count. It is crossed with a wide variety of breeds, including hill breeds to produce high-performance cross-bred daughters, and in lowland flocks to produce homebred replacements.

Pure-bred British Milksheep ewes are only kept in special flocks, but the cross-bred daughters of British Milksheep rams are well suited to anyone wanting to achieve a high level of production in lowland flocks.

General purpose breeds

Although the majority of breeding ewes in commercial lowland flocks are cross-bred, there are some breeds which bypass the first stage of the stratification process and are mated directly to rams of the meat breeds. These general purpose breeds attempt to combine the characteristics of the hill breeds and the crossing breeds, often with only limited success. There is always a danger that the attempt to select for too many characteristics can result in a lack of progress being made in any direction.

Corriedale

The Corriedale is a comparatively new breed of sheep which was developed in New Zealand in the second half of

Corriedale ram

Dorset Horn ram (lamb)

Clun Forest

the nineteenth century. It was derived from an equal admixture of the Merino and Lincoln breeds and was established in an attempt to create a general purpose breed. It combines reasonable hardiness and prolificacy with a good carcass and a 10 lb (4.5 kg) fleece of high quality wool (50–56). In some parts of Australia, such as south-western Victoria, finewool and strongwool strains of the Corriedale have been developed of which the finewool cuts the more valuable fleece. Mature ewes weigh on average 150–160 lb (68–72 kg). It now enjoys great popularity especially in the range and marginal areas and it is highly recommended. In North America several new breeds have been evolved on a pattern similar to that of the Corriedale from a cross between longwool and Merino breeds. The Columbia (Lincoln × Rambouillet), Panama (Rombouillet × Lincoln), Romedale (Romney Marsh × Rambouillet), and Targhee ($\frac{3}{4}$ Merino, $\frac{1}{4}$ longwool) fall into this category, but all are of only minor importance.

Devon Closewool and Lleyn

In Britain a cross between a longwool breed (Devon Long-wool) and a hill breed (Exmoor Horn) has led in a similar way to the creation of a new breed, the Devon Closewool. Whis is a localized breed of only modest performance characteristics. The Lleyn originated in north-western Wales from a cross between longwool breeds (Leicester and Roscommon) and a hill breed (Welsh Mountain). It is a smaller sheep than the Devon Closewool but achieves a higher standard of production. The average lambing percentage is 150–160 per cent, combined with quite good milking ability and carcass characteristics.

Dorset Horn and Poll Dorset

The Dorset Horn and its offshoot, the Poll Dorset, are general purpose breeds that have the special ability to lamb at most times of the year. Lambs forn out-of-season can be sold at a high price, although the costs incurred are frequently also much higher. The Dorset Horn ewe possesses only average prolificacy, but the milk yield is quite good. The wool is of high quality (54–58). In Australia the Dorset Horn and Poll Dorset have been developed as specialist meat breeds, and this fashion is now being reflected in some British flocks with a resulting loss of maternal performance characteristics.

Portland

The Portland is an ancestral breed of the Dorset Horn. It is a small animal with brown face and legs, and is of little importance except that it has been noted for at least two hundred years for the flavour and delicacy of its mutton. Some strains will lamb out of season. The Tunis is a breed that originated in North Africa, but is now established in North America. It will also lamb at almost any season of the year, but is of little importance.

Clun Forest

One of the most popular general purpose ewes in Britain is the Clun Forest. It is a medium-sized polled sheep – mature ewes 120 lb (55 kg) – of reasonable prolificacy (155–160 per cent), but it tends to be variable in type and performance. Some of the larger Welsh hill breeds, such as the Hill Radnor and Beulah Specklefaced are of the same type as the Clun Forest, and the Kerry Hill is also a similar breed.

Meat breeds

The meat breeds are the final link in the chain in the production of prime lamb. They specialize in good carcass quality. Meat breed rams are mated to prolific, milky, cross-bred or general purpose ewes with the intention of combining in the lambs the rapid growth rate permitted by the milk yield of the ewes with the conformation and deep fleshing inherited from the rams. The meat breeds vary in size considerably, from the massive Oxford Down to the small, squat Southdown. The growth rate of their lambs is related directly to their mature weights.

Oxford Down

Southdown

Suffolk

Texel

Oxford Down and Southdown

An Oxford Down ewe weighs on average more than 200lb (90kg). Thus the Oxford Down is used where lambs are required to grow rapidly and be ready for slaughter at a relatively heavy weight. In contrast a Southdown ewe weighs only about 130lb (60kg) and Southdown cross lambs are suitable where early maturity is more important than growth rate. The Southdown was the first of the meat breeds to be established and improved, and it was then used in the development of most of the other meat breeds.

Suffolk

The Suffolk originated from a cross between the Southdown and the Norfolk Horn. It is a blackfaced, clean-headed, polled breed which currently far exceeds all other meat breeds in popularity. It transmits good growth rate to its progeny and a relatively lean carcass. I recommend the Suffolk as the meat breed which best combines carcass quality with reasonable prolificacy. The Shropshire and the Hampshire are breeds of similar type, although both are more heavily woolled, especially on the head. The Dorset Down is intermediate in type and size between the Suffolk and the Southdown, while the Ryeland is closer in type and characteristics to the Southdown.

Texel and Wiltshire Horn

Two meat breeds vary from this standard pattern. The Texel is a breed of Dutch origin which possesses excellent carcass characteristics but only average growth rate. It enjoyed a great increase in popularity in the 1970s. The Wiltshire Horn is unusual in several respects. It is horned in both sexes; it grows a hairy coat which is shed annually; and it has a leggy conformation that is not typical of the other meat breeds.

Buying sheep

Buying your first livestock can be a daunting experience. The welter of jargon and the traditional ways of cleverly concealing faults in the animals are confusing, particularly to new buyers who often do not have a clear idea of their requirements and objectives.

Which to choose

No matter whether the flock will be large or small, pedigree or commercial, the care taken in choosing the sheep which will establish your flock will be a major factor in your future success. The four main considerations you should bear in mind when making the selection of both breed and individual animals are your own preference; your system of management and objectives; the genetic qualities of the sheep and their health.

Personal preference

This should never be underestimated as one of the essential ingredients of a successful enterprise. It is much more difficult to achieve success with sheep, or indeed any other livestock, if the breed selected does not excite the flockmaster's interest and enthusiasm. Whichever type you choose you will have to spend many hours, probably over a period of years, looking after that breed so it is better not to rush into a hasty decision, but rather to take time forming your opinion.

Environment

The type of land, housing facilities, climate and other factors including your system and quality of management will also clearly affect your choice of breed. For fertile, lowland pastures a large breed such as one of the longwools or a productive cross-bred would be suitable, whereas the Soay or Welsh Mountain would not be a sensible choice. Conversely, on exposed, poor-quality hill grazing the hill breeds, such as the hardy Herdwick or the primitive Soay, would thrive, whereas the more delicate longwool breeds would be fortunate to survive. In cold, wet conditions you might select the Swaledale and would certainly avoid the Merino – but in hot, dry conditions that choice would be reversed.

Thrifty hill sheep flocks require less intensive management than docile lowland sheep. The former range more widely and forage more keenly, whereas the latter seem reconciled to a more passive role in life – they are more reliant on the shepherd's watchful eye to help them back on their feet when they are cast or to rescue them from attacks by dogs or foxes. The independent spirit of the hill sheep may enable it to fight off an attacking fox unassisted, but also makes it a more difficult animal to control, and means that when it is moved to the lowlands the fences must be good. (Good fencing is also, and indeed even more important in the case of primitive breeds.) The basic principles of animal husbandry should be observed with all breeds, but in the case of the lowland sheep this care and attention should be intensified.

Purpose

You must be careful to choose a breed which is best able to produce the end product you want. If the purpose of your flock is to produce high quality, hand-crocheted shawls there is little point in keeping Herdwicks or Scottish Blackface. If the flock is simply a hobby, with the sole function of pleasing your eye, the choice is wide, but if maximum profit from a commercial unit is what you are after, your choice will be limited to the relatively high-performance breeds. If you are aiming to meet a speciality demand for quality carcasses, then you ought to think of including in your breeding programme breeds such as the Soay or, at the other extreme, the Texel.

Choosing particular animals

By defining your intended end product and the conditions under which the animals will be kept, and bearing in mind your preferences, you can select the breed best suited to you. However, having decided on the breed or type, the next question is which specific sheep to buy, and this should be

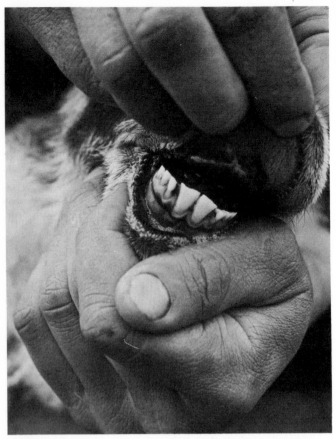

A shearling ewe with two broad (permanent incisor) and six milk (temporary) teeth. The first pair of broad teeth usually emerge at about 18 months of age.

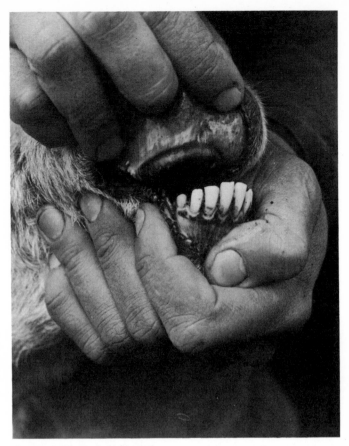

A ewe with six broad teeth.

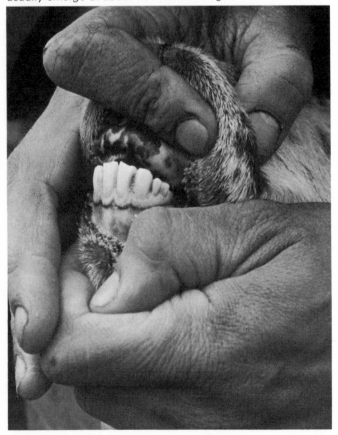

A two-year-old ewe with four broad teeth.

determined by what you know about the genetic quality and the health of the animals you are considering.

Genetic quality can only be assessed to a limited extent by simply looking at the sheep, particularly when you don't know the background history of the breed or the qualities of various bloodlines. In this case the best thing you can do is to look at any performance records that are available – as well, of course, as inspecting the sheep themselves to discard poor quality or defective animals.

Inspecting sheep to establish their health status and efficiency is often more reliable than assessing their genetic quality, although still limited in its scope. Sheep with any obvious signs of disease should be discarded, and if a problem appears widespread in a group of sheep they should all be rejected. Eye infections, for example, can spread very quickly through a bunch of sheep. Likewise foot rot is a very infectious disease. It is possible that animals may have contracted a disease but not show symptoms at the time of purchase. There is little that can be done about this except to minimize the chance by selecting animals that appear fit and healthy.

Functional defects are equally important. The teeth should always be examined. Good teeth are broad and short, fitting squarely against the dental pad of the upper jaw. Broken-mouthed sheep (those which have lost some teeth) or 'gummies' (those which have lost all their teeth) should normally be avoided although in small units and with special care they may provide a reasonable return for a small investment. The udder of the ewe should be soft and pliable and free from any hard lump which might indicate previous mastitis and maybe a blind quarter. Another fault in mature ewes is the accidental loss of one or more teats due to careless shearing, so make sure to examine the sheep for this. The ram should have two testicles of equal size and normal elasticity with a clearly defined epididymus, and the penis should be examined for any abnormality. Rams which have recently been attacked by blowfly may be infertile and should be avoided.

How to buy

Stock can be obtained from four main sources, but in all cases the inexperienced buyer must tread warily.

1. Sheep fairs
2. Agents, auctioneers or dealers
3. Flock dispersals
4. On farms

Fairs and sales
At fairs and sales sheep are usually sold in quite large lots, so that the buyer who wants only half a dozen sheep may be presented with a very limited choice, despite the large number of sheep for sale. Although the stock at such events are usually sold with a warranty, it is best to check for yourself immediately upon purchase that the animals are as described. See that the females bought for breeding do not include some wethers (castrated males) and check that wethers have been castrated correctly. Ewes should be sold with a warranty regarding both their age and their condition – 'g.t.u.' means good on tooth and udder.

The main danger of buying sheep at large fairs and sales is the risk of disease. The gathering together of sheep from many different areas in one place provides an unequalled opportunity for the diseases associated with one area to spread to sheep from another.

Agents and dealers
Most new or potential flock owners will need to obtain the services of a knowledgeable agent or dealer. The reliability of a dealer cannot be guaranteed any more than that of the sheep and it is possible to pay dearly for a bad choice of dealer or agent. On the other hand a good agent can give an invaluable start to a new flock, and although his charges may seem high they will be well worthwhile. He will be able to offer advice not only on when and how to buy, but also on the best type of animal with which to start the flock. Having given your agent a clear brief of your requirements it is preferable to give him a free hand after that.

Flock dispersals
A flock dispersal is a useful source of breeding stock because, assuming that it is a genuine dispersal, there will be a complete selection of that breeder's stock from the best to the worst. Usually a breeder sells only the animals which he does not want to keep for his own flock replacements, but

It is essential when restraining a ewe to control the animal's head by putting a hand under its chin.

in the case of a flock dispersal even the elite animals are available for sale. Clearly this is a better source of breeding stock for a pedigree unit than a sheep fair would be, as the latter is concerned mainly with commercial and cross-bred animals. When establishing a pedigree flock it is advisable to buy from as few sources as possible, and ideally from only one flock. A flock dispersal gives you a better opportunity of finding enough animals of acceptable genetic merit available for sale from the same flock.

Buying from farms
In the absence of a flock dispersal the breeder who wishes to establish a pedigree flock should try to buy direct from the farm. He should research and discover the reputable flocks within the breed that he has chosen; he should visit each of these flocks and inspect all the sheep to make a valid assessment of the overall quality and health of the stock; he should examine the records to identify the animals that meet his requirements – and he should inspect the animals individually to ensure that they are healthy and free from defects. Many buyers fail to follow this procedure and live to regret it. Some sellers are reluctant to cooperate and this usually means that they have something to hide.

If you only want one or two sheep, you can sometimes buy orphan lambs quite cheaply from local farmers. However in this instance you cannot, of course, expect them to sell you their best pedigree stock and will have to take your chance on the quality of the animals you get.

Housing and equipment

Most of the time sheep can be left to graze and need no housing beyond the natural shelter of trees and rocks. In many areas sheep do not even need to be housed in the winter. Most breeds have a thick protective fleece which enables them to withstand all but the worst storms, although some breeds such as the Merino, Finnish Landrace and Bluefaced Leicester are more vulnerable than others to bad weather, because rain can penetrate through their wool to the skin. The real reasons for housing sheep are to give the shepherd better control of his flock at lambing time and to protect the new-born lambs.

The sheep house

You can build housing for sheep quite cheaply as it is more important that it should be well designed than that it should be made from expensive materials. Perhaps the primary consideration when designing a sheep house is to ensure that there is ample ventilation. Because wild sheep inhabit dry regions, their domesticated descendants are susceptible to pneumonia in wet or humid climates and this susceptibility is accentuated in badly ventilated houses. Provided that the sheep are protected from draughts at floor level, the design of the building should permit as much air circulation as possible. Ridge ventilation and open eaves will help achieve this. In wet areas with strong winds, space boarding may be preferable. The walls should be solid to a height of 5 ft (1.5 m) and the eaves should be at least 10 ft (3 m) high.

The design of the housing should provide 12 sq ft (1 sq m) per ewe for medium-sized sheep, but extra space is required if the ewes remain in the same house after lambing. The shape of the pens is dictated mainly by the trough. Each ewe needs 18 in (0.5 m) of trough and thus each pen must be at least 8 ft (2.4 m) in depth to give the ewe the space she needs. Twenty ewes should be the maximum in a pen 30 ft × 8 ft (9 m × 2.4 m).

Feeding equipment

The design of the feeding troughs and racks should avoid wastage. Norwegian hay boxes (see illustration) or feeding barriers constructed to the same proportions are ideal. For medium-sized ewes the base board should be 14 in (0.4 m) high and the gap through which the ewes feed should be 11 in (0.3 m) wide. These proportions should be changed for ewes of different sizes. Sheep prefer running water, but this is not normally feasible unless the house can be constructed over a mountain stream. However, the water must be clean and fresh. Water bowls should be fixed 24 in (0.6 m) above the ground to prevent the water becoming soiled with dung.

Housing after lambing

Sheep are in particular need of housing immediately after lambing. Each ewe should be penned individually with her own new-born lambs until the basic tasks connected with lambing have been completed (see page 130), after which they can either be housed in groups or returned to the pasture.

Handling pens

If you have only a small number of sheep you may be able to manage without these but in general handling pens of some kind are an invaluable aid to the owner of even a small flock of sheep, and will enable you to control your animals whenever you want to carry out tasks such as trimming their feet, dipping or drenching them.

However, few things can be more frustrating than attempting to manage stubborn sheep in badly designed handling pens. The design you choose will of course depend on the size of your flock, but the items you will probably need will be:

Holding pen
Drafting race (for separating the lambs from the ewes)
Working race (for attending to the sheep)
Foot bath
Dip bath and draining pens

A typical lay-out is shown on the next page. The size of the drafting race and the working race will vary according to the breed of sheep. The floor of the races, draining pens and areas of greatest activity should be concrete. For other areas hardcore may be adequate. When designing handling pens, remember that sheep will move more readily if they are going up a slight slope, towards light or towards other sheep. The drafting race should be wide enough to permit the sheep to move along it freely but sufficiently narrow to prevent them turning round. For medium-sized sheep the race should be 16 in (0.4 m) wide at the top and should taper to 12 in (0.3 m) wide at the bottom and should be at least 10 ft (3 m) long. The working race should be wide enough for the shepherd to stand alongside a sheep in the race.

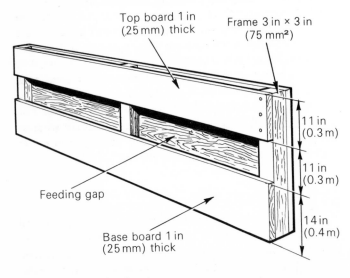

Top board 1 in (25 mm) thick

Frame 3 in × 3 in (75 mm²)

11 in (0.3 m)

11 in (0.3 m)

14 in (0.4 m)

Feeding gap

Base board 1 in (25 mm) thick

Front view of a feeding box.

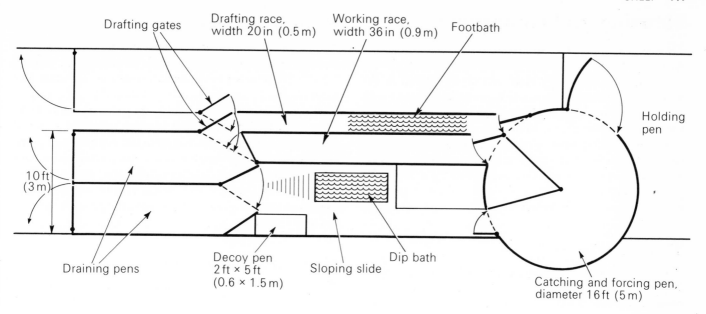

A typical lay-out for a handling pen.

Labels on diagram: Drafting gates; Drafting race, width 20 in (0.5 m); Working race, width 36 in (0.9 m); Footbath; Holding pen; 10 ft (3 m); Draining pens; Decoy pen 2 ft × 5 ft (0.6 × 1.5 m); Sloping slide; Dip bath; Catching and forcing pen, diameter 16 ft (5 m)

Dip bath

All sheep need regular dipping and a dip bath is an important part of your handling facilities. If you have only a few sheep you will probably be able to manage with any non-porous container large enough for the sheep to stand in while the solution is poured over it. For larger flocks a bath with a capacity of 160–80 gal (730–820 l) is an ideal size and suitable dimensions for a dip bath of this capacity would be as follows:

Width at top	33 in (0.8 m)
Width at bottom	12 in (0.3 m)
Depth of bath	50 in (1.3 m)
Length at top	100 in (2.5 m)
Length at bottom	44 in (1.4 m)

Details about dipping are given on page 136.

Fencing

Sheep are renowned for their ability to escape unless their grazing areas are fenced adequately. With some breeds, especially hill and mountain breeds, this ability has been developed to such a fine art that it has inspired the saying 'good fences make good neighbours'. Fences need to be particularly secure when the lambs are weaned and on many farms one or two lambs seem always to find their way back to the ewes after they have been separated from them. In some areas fields are divided by hedges or stone walls and these should be kept in good repair, but apart from these there are four basic types of fence – wire netting; high tensile; post and rail or electric.

Wire netting is usually 32–36 in (0.8–1.0 m) high with a single strand of barbed wire above. The wire is stapled to posts every 5 yards (4.5 m). High tensile fencing is cheaper than wire netting as it needs fewer posts, especially if the fences are constructed in long straight lines on flat ground. Seven-wire fences are usual, with powerful straining posts at angles in the fence.

In general, post and rail fences are too expensive, as they still need reinforcing with wire netting to make them sheep-proof. The cheapest fencing is the electric kind. For sheep a single strand of wire will often suffice, although two strands are more usual at 9 in and 18 in (0.2 m and 0.5 m) above ground level.

Feeding

Grazing

In most cases sheep can get all the nutrition they need from grazing. The growth of grass is greatest in the spring and then declines during the summer, followed by a smaller spurt of growth in the autumn. The number of ewes and lambs you keep over a certain area should be adjusted to suit the grass production of the summer and autumn. This means that there will be a surplus of grass in the spring which should be conserved either as silage or as hay. The fields which have been used for conservation will be relatively free from nematode parasites and the aftermath should be grazed by the weaned lambs.

Feed supplements

If, in the autumn and at mating time, the ewes are very thin and in poor condition it may be necessary to give them additional concentrate feed. This can be bought as pellets, pencils or cubes, or in the form of free access feeding blocks which can be left in the field for the sheep to eat at will.

Forage crops, such as rape, are sometimes used for green, succulent feed when grass is not adequate, but these are not very satisfactory at mating time as they tend to depress fertility.

Minerals should be available for the sheep at all times, in the form of general-purpose licks. In areas where specific minerals, such as copper or cobalt, are deficient you should make special provision of them, as copper deficiency will cause swayback in the lambs and cobalt deficiency will cause pine in the ewes.

Feeding at lambing time

For most of the flock the critical period of feeding will start about six weeks before lambing, but ewe hoggs and old ewes require better treatment and should be separated from the rest of the flock as long as twelve weeks before lambing. The pattern of feeding for these sheep should be designed to enable the ewe to devote more of the nutrition to her own bodily condition, and a constant ration of 12 oz (340 g) concentrate feed per day, in addition to roughage feed (hay or silage instead of grass) up to a month before lambing, should be provided. Thereafter the ewe hoggs and old ewes can receive the same ration as the other ewes.

Six weeks before lambing the young ewes and mature ewes should be given a special diet of a combination of roughage and concentrates. A daily ration for medium-sized ewes of 150–65 lb (70–5 kg) with an expected prolificacy of 160–175 per cent lambing might be as follows:

Weeks before lambing	Roughage lb	(kg)	Concentrate feed lb	(kg)
Six	4	(1.8)	$\frac{1}{2}$	(0.2)
Five	4	(1.8)	$\frac{3}{4}$	(0.3)
Four	3	(1.4)	1	(0.4)
Three	3	(1.4)	$1\frac{1}{4}$	(0.5)
Two	2	(0.9)	$1\frac{1}{2}$	(0.6)
One	2	(0.9)	$1\frac{1}{2}$	(0.6)

The level of feeding should be varied according to the size of the ewe and her expected level of performance. The concentrate should contain 14 per cent protein and may be either a proprietary compound or a home-mix of cereals plus a high protein ingredient.

If the flock remains housed for some time after lambing the level of feeding should be increased. Once the capacity of the ewe's digestive system is no longer limited by the presence of the unborn lambs she is able to cope with bulky and succulent feeds which help to stimulate milk production. Silage and mangolds are suitable, although the best feed is grass. Ewes rearing twins should be given up to 3 lb (1.3 kg) of concentrate feed per day. This will apply in the case of many pedigree flocks where the feeding levels may be increased even further in order to achieve maximum growth rates, but most commercial sheep will be turned out to grass as soon as the lambs are strong enough.

Lambs

Details on feeding lambs from birth until weaning are given in the *Rearing* section on pages 132–134.

Wool, meat and milk

As described in the section on *Breeds*, different sheep are bred for different purposes, and it will depend on the breed you keep whether you are interested primarily in its wool or its milk or in selling it for meat. However, although some breeds specialize in one aspect or another, they are all essentially dual-purpose and will all be sheared and all end up eventually being sold for meat.

Wool

Shearing

The first major task to be carried out during the grazing season is shearing. This should be done in late spring or very early summer, after the wool has 'risen'. The rise can be recognized when the yellow, greasy wool, which lies next to the skin during the winter, is lifted by the growth of new wool which is white and quite distinct. The shears cut through this new wool with ease, but if an attempt is made to shear a sheep before the rise the task is much more difficult and unpleasant for both sheep and shearer.

Sheep should be shorn when their stomachs are relatively empty and when their fleece is dry. Housing them overnight and shearing them in the morning to avoid stressing them in the heat of the day is a good policy. About three weeks after they have been shorn they should be dipped against parasites (see page 136).

Shearing is not an easy art to master and it may be better to employ contract shearers at first, meanwhile attending classes run by qualified instructors. The sheep should be brought in small groups into a catching pen, and a clean shearing area should be provided. When the fleece has been removed it should be lifted carefully in a bundle and spread, shorn side downwards, on a clean board or table. Dirt and soiled wool should be removed. The sides of the fleece are turned towards the middle so that it is 18–20 in (460–500 mm) wide and the fleece is rolled as shown in the photographs on the facing page.

Selling wool

Flockmasters in Britain who own four or more sheep must be registered with the British Wool Marketing Board and must sell their wool to the Board. Large wool sheets are provided into which approximately thirty fleeces can be packed. It is essential that the wool is dry when it is packed, otherwise it will rot. The wool sheets are collected by the Board's local agents.

In Australia local shearing contractors will take care of the whole shearing operation up to the despatch of the wool and then a local woolbroker's agent will arrange for the clip to be sold at auction.

Shearing with hand shears.

Rolling a fleece: 1. *The fleece is spread out, with the inside facing down.*

2. *The sides are turned towards the centre and it is rolled tightly, starting at the rear end.*

3. *The fleece is rolled except for the neck wool, which is twisted into a band. This is then wrapped around the fleece and tucked in to make the bundle secure.*

Shearing: 1. *The starting position. The sheep is leaning slightly backwards and to the right.*

2. *The first blows, clearing the wool from the belly.*

5. *The shearer steps over the sheep's hind legs and clears the wool from its back.*

6. *He steps round further and pulls the sheep's head between his knees to remove the wool from the neck and shoulders.*

3. *The far hind leg is shorn, making sure the shears reach over the top of the tail.*

4. *The wool is removed by short blows from the side of the sheep.*

7. *He moves backwards pulling the sheep over, so that the final side can be finished.*

8. *The task is completed. The sheep departs, leaving the fleece on the ground.*

Generally speaking in the USA wool is sold to local buyers of hide, wool and fur. In some states it is also marketed through the Midwest Wool Growers Cooperative.

Lambs for meat

When to sell
Some of your lambs will probably be sold before weaning. Lambs sold early in the season fetch higher prices and also relieve the pressure on your grazing, thereby allowing the other lambs to grow more quickly. In Britain it is usual for the price of lambs to fall rapidly in mid-summer and thus the maximum sale of lambs should be achieved before this time, even if it means the expense of providing them with extra concentrate feed. The relatively low price that can be obtained for lambs in the late summer and early autumn does not justify similar expense.

If your stocking density is high the remaining lambs should be sold as stores – that is, sold before they are ready for slaughtering to someone who will continue the fattening process. However, if you have sufficient grazing the lambs can be run on, and probably fattened on forage crops such as rape and turnips, to take advantage of the higher prices that can be obtained for lambs and hoggs in the late autumn and winter.

Lambs should be sold for slaughter when they are judged to have reached the ideal carcass quality. This is based on the conformation and proportion of fat in the carcass. As a rough guide lambs will be ready for slaughter when they are half the average weight of their sire and dam. Thus the progeny of a 180 lb (63 kg) ram and a 140 lb (82 kg) ewe will be ready for slaughter at about 80 lb (36 kg) live weight, to yield a carcass of about 37–38 lb (17 kg). The final decision on the readiness of each lamb can be made by feeling the degree of fleshing on the animal's back. There should be a deep layer of firm flesh which almost covers the backbone. The demand now is for a leaner carcass than previously and you should avoid over-finishing the lambs.

How to market
The lambs are marketed either through a live auction market or direct to meat wholesalers on a dead weight basis. Both methods have their advantages, but over a long period of time there is little to choose between the prices obtained. The only exception to this is with new or unusual breeds and crosses which are not readily recognized by the buyers, who are consequently not prepared to pay a full price. In these cases the lambs should be sold on a dead weight basis so that the price obtained is directly related to the quality of the product.

Milk

Although the production of milk is no longer the primary reason for which sheep are kept in Britain, North America and Australasia, in many parts of the world milch sheep play an important part in the rural economy. Specialist breeds have been developed in the Mediterranean countries in particular. In France the manufacture of Rocquefort cheese has been documented since the eleventh century, and 80 per cent of the total production (more than 16,000 tons/tonnes) is derived from the milk of Lacaune sheep. The Lacaune is milked twice daily during a six month lactation, but its lactation yield of 30–35 gal (140–160 l) is considerably less than the yields of the Fries Melkschaap. The Fries Melkschaap has been exported to Great Britain, where a demand for sheep milk products is beginning to develop, particularly for cream cheese and yoghurt. The latter is very rich and creamy, and currently the demand for milksheep ewes far exceeds the supply. In Australia a Silesian order of monks near Melbourne has bred a strain of Merino-Border Leicester crosses and used the milk to make a local version of Peccarino cheese.

Breeding

The first choice to be made is between pure-breeding and cross-breeding. With pure-breeding programmes the improvement of the genetic potential of the flock lies largely within the owner's control. With cross-breeding systems it is necessary to rely on other breeders for the supply of flock replacements and thus the owner can exert relatively little influence to ensure that succeeding generations are better than their predecessors.

The attractions of pedigree stock and its attendant kudos entice many flock owners into the realms of pure-breeding for the wrong reasons. Very few breeders of pedigree stock achieve the fame and fortune that is sought by many, and the aspiring studmaster should remember that the objective of pure-breeding is to enhance the desirable qualities of the breed by means of a constructive and carefully planned breeding programme. Any glory that results is a bonus earned by improving the genetic merit of the flock.

The basic ingredients of a successful pure-breeding programme are relatively simple:

1. A recording system which identifies the characteristics and qualities of each animal.
2. A selection procedure which distinguishes the elite animals from the remainder.
3. A mating programme which ensures that the superior qualities are concentrated in succeeding generations.

Identification methods

A detailed recording system is only necessary if the sale of pedigree breeding stock is one of your major concerns. On the other hand it is difficult to breed better animals unless the qualities of each sheep are recorded.

The identification of individual animals is clearly essential for a recording system, and there are several different methods. The most permanent and tamperproof one is tattooing, and this should be used wherever possible. It works best for sheep with large white ears, as the dark tattoo is sharply defined against the non-pigmented ear.

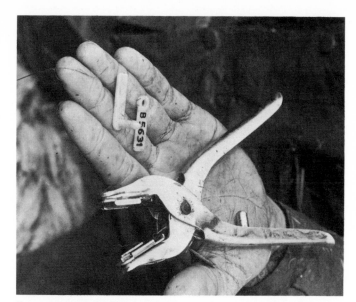

Ear tagging: 1. *Pincers and a two-piece ear tag.*

An alternative method: 1. *One-piece ear tag in its applicator.*

2. *Clipping it into the ear.*

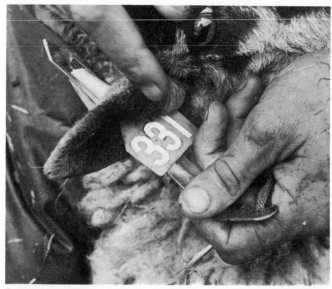

2. *Inserting it into the ear.*

3. *The tag in position.*

3. *The tag in position.*

It is less successful, and in extreme cases completely useless, for sheep with dark ears – especially small, black hairy ears. Thus breeds such as the Dorset Horn and British Milksheep can be tattooed very successfully, while breeds such as Herdwick or Clun Forest present some problems.

For these and similar breeds you will have to resort to other less satisfactory methods of identification. The most usual is the ear tag and there are many varieties available. Some are metal and some plastic. Some are sufficiently large to enable the number to be read at a distance, while others are small and unobtrusive. The loss of the tags from the ear is a continuing problem and there is no type that is entirely reliable.

Notching is an alternative method, but is not widely used. For horned breeds an identification number burned into the horn is as permanent as a tattoo. In most cases the method of identification will be determined by the relevant breed society.

Keeping records

When a sheep can be identified it is possible to make out a record card for each animal on which all relevant information can be entered. Records should not be kept for their own sake. They should be as simple and straightforward as possible, providing only the information you need in order to make decisions on mating patterns and the selection of breeding stock. The records will vary according to the purpose of your breed but in most cases the following items should be included on each record card:

Identification number
Pedigree (at least two generations)
Date of birth
Sex
Lambing record and performance of lambs

A specimen ewe record card is illustrated below.

Adjustments

The main function of records is to enable the breeder to identify the best animals in the flock. However in order to make just comparisons between individuals you will need to adjust and correct the records for certain factors. This applies particularly to the growth rate of lambs. Until it is fifty days old a lamb's growth depends mainly on the milk yield of the ewe, but thereafter its growth rate is increasingly affected by the potential it has inherited from both the sire and dam, and this effect is assessed by its weight at a hundred days old. Usually batches of lambs are all weighed on one day so that the weight of each lamb must be adjusted according to when it was born, as follows:

$$\left(\begin{array}{c}\text{actual} \\ \text{weight}\end{array} - \begin{array}{c}\text{birth} \\ \text{weight}\end{array}\right) \times \dfrac{100}{\begin{array}{c}\text{age at} \\ \text{weighing} \\ \text{(in days)}\end{array}} + \begin{array}{c}\text{birth} \\ \text{weight}\end{array} = \begin{array}{c}\text{corrected} \\ \text{100-day} \\ \text{weight}\end{array}$$

The growth rate of each lamb is affected by the age of its dam, its birth type, method of rearing and sex. Corrections for these factors are as follows (for ewes of average size):

Dam:	ewe hogg (yearling which has not yet been shorn)	Add 15 lb (7 kg)
	shearling (yearling after it has been shorn)	Add 5 lb (2.3 kg)
Lamb:	single female	Add 7 lb (3.2 kg)
	twin male	Add 14 lb (6.4 kg)
	twin female	Add 20 lb (9 kg)

When comparing two animals all factors must be taken into account. For example with a breed which you keep mainly for its maternal characteristics it is necessary to consider its prolificacy, meat yield, ease of lambing and longevity. These factors are combined into an index which you can directly compare with that of any other ewe.

Name _____	Date of Birth _____	Reg. No. _____
Sire _____	G.S. _____	G.D. _____
Dam _____	G.S. _____	G.D. _____

Lambing Record					Weights		Notes	
Date	Lamb No.	Sex	Birth Wt.	Sire	50 day d/wg	100 day	Lamb	Ewe

Typical headings for a ewe record card.

Selecting breeding stock

The selection of flock replacements and the culling of breeding stock should be carried out according to four basic considerations.

1. Pedigree: superior parents can be expected on average to produce superior progeny.
2. Appearance: animals with a visible fault, such as wool blindness in Rambouillets or overshot jaw in Border Leicesters or entropion in Dorset Horns, should not be selected.
3. Performance: the individual record cards are the basis of recording performance data.
4. Performance of progeny: the progeny test is the ultimate measure of the value of a breeding animal. It is applied usually in the evaluation of rams as very few ewes produce sufficient lambs which are then retained as breeding animals, to give a valued progeny test.

Within a selection programme priority should be given to factors which are, first, of economic importance and, second, of high heritability. It is also important not to attempt to select for too many factors, as this slows down the progress made on any particular one. This is one of the drawbacks of general purpose breeds, which pay equal attention to thriftiness, prolificacy, milk yield, growth rate and carcass quality, while specialist breeds are able to concentrate on a much narrower range of factors.

Mating programme

Once the relative merit of each animal has been defined within the selection procedure, a mating programme can be worked out. Various alternatives are open to breeders with pure-bred flocks. The simplest system is to mate the best rams (as evaluated by progeny test) to the best ewes (as identified by the selection index) to produce the next generation of flock replacements. Where an outstanding animal is identified the mating of best to best may be intensified by line-breeding, where the influence of this animal is concentrated. Line-breeding, which is a form of in-breeding, may have detrimental effects, especially if applied too severely.

A good deal of mystery has always surrounded the intricacies of pedigree breeding, and it still remains as much an art as a science. Certainly it is difficult to explain scientifically some aspects of animal breeding. Sometimes the mating of two animals consistently produces better progeny than might be expected – this is known as nicking.

Compensatory mating

This is the most usual system of mating. The ram to be used is selected specifically for those characteristics which are most likely to correct the deficiencies and defects of his mates. If wool blindness is a problem in a ewe flock, the ram used should be open-faced. Likewise shelly, fine-boned ewes should be mated to strong rams of substance, while rams from high-performance families should be used on ewes of poor productivity.

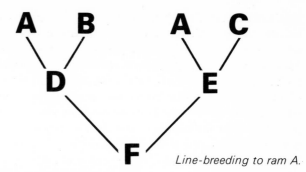

Line-breeding to ram A.

Cross-breeding

Cross-breeding is an extension of compensatory mating. Animals of different breeds are mated so that the combination of their characteristics will produce the desired end product. In the British sheep industry draft hill ewes are mated to rams of crossing breeds. Thus the Milksheep hybrid ewes combine the hardiness and thriftiness of their hill dams with the prolificacy and milk yield inherited through their British Milksheep sires. The high-performance Milksheep hybrid ewes are in turn mated to Down rams, which provide the final ingredient of carcass quality, to ensure the production of large crops of quick-growing, prime lambs.

Preparation for breeding (flushing)

Condition of the ewes

The sheep year, which terminates with the sale of lambs during the summer and autumn, begins with the preparation of the ewes and rams for mating. The condition of the ewes at mating has an important influence on the number of eggs shed from the ovaries and consequently the number of lambs born. It can be assessed by handling the animals over their loins. If there is no flesh on the bones the animals are too thin. If the outline of the bones is lost in excess flesh the animals are too fat. The ideal condition occurs when there is a medium layer of firm flesh covering but not hiding the bones.

Nutrition and health

The condition of the ewes is related mainly to their level of nutrition, and sheep in poor condition will benefit from additional concentrates and minerals. The health of the ewes will also affect their condition. They should have been drenched against parasites before mating and it is especially important to check that they are free from foot rot. All the ewes should have their feet trimmed and treated in good time before tupping commences – excess growth of the hoof should be cut back to the level of the sole. The ewes should also have had all the vaccinations necessary in your area as this will not affect their fertility but will mean that they pass the immunities on to their lambs.

Hormones

Although the nutrition and health of the ewes are essential to enable tham to breed satisfactorily, their actual breeding activity is controlled by hormones and may be enhanced by hormone therapy. The timing of oestrus in a ewe can be determined by inserting a hormone impregnated sponge into her vagina, but I would recommend more natural management techniques. In particular, a vasectomized ram run with the ewes for five weeks before mating will stimulate them to come into season and this will tend to concentrate the mating of the whole flock into a shorter period.

The rams

Rams are often neglected, but without fully fit and functional rams both mating and lambing are likely to be times of great disappointment. Rams should be included in the same programme of vaccination and parasite control as the ewes. Their reproductive organs should be examined for any abnormalities, and their feet should receive special attention as a ram with painful feet will be reluctant to serve his ewes. Any ram which has recently had a fever may be temporarily infertile and a semen examination can be worthwhile.

Mating in the first year

Many breeders are unsure whether they should mate ewe lambs. In general sheep benefit from being bred in their first year, provided that they are well grown at mating and are given special treatment thereafter. If they are too small at mating or if they are not fed adequately during pregnancy for both their own growth and that of their lambs, they are likely to be stunted for the rest of their lives. As a rough guide ewe lambs may be mated if they have achieved 60–65 per cent of their expected mature weight.

Mating and early pregnancy

The best time for mating

The date on which the rams are loosed to the ewes will be determined by the date on which you want to commence lambing. For most commercial flocks, when the end product is the sale of lambs for slaughter or as stores, you should aim to make lambing time coincide with the growth of grass in the spring. An easy guide to remember in Britain is that ewes mated on Guy Fawkes Day (5 November) will lamb approximately on April Fool's Day (1 April). The gestation period of sheep is about 146 days, although this does vary slightly from breed to breed, and in individual animals may vary from 140 to 152 days.

With pedigree flocks lambing may take place earlier in the year, in order that the ram lambs may be sufficiently well grown to be used in their first year. Similarly the lambs from early lambing, being larger, create a favourable impression when standing alongside lambs from later lambing flocks in the show ring, and thereby frequently obtain higher prices in the sale ring. In some cases the time of mating may be limited by the breed. Hill breeds, especially in their natural environment, may not show oestrus until the autumn, whereas ewes of breeds such as the Dorset Horn, Merino, and many European breeds may accept the ram at most times of the year.

The rams should be turned out with the ewes early in the day. This allows the initial excitement to cool down before evening. The danger of a cold night to animals overheated by the stress and exertions of the rams' early enthusiasm when they join the flock cannot be emphasized too much. Rams can be run singly or in groups. Where two or more rams share the same group of ewes, there may be a little fighting initially, but it is unusual for this to be serious, and there is the advantage that the effects of a sub-fertile ram would not be significant as most ewes would be covered by more than one ram. As an added insurance rams should be culled when they show signs of old age or start to lose condition.

Ratio of ewes to rams

If a ram is given too large a group of ewes this may affect the number of lambs born. The optimum number will vary from breed to breed. Solid, heavily fleshed and relatively inactive meat sires, such as Dorset Down or Southdown, should not be given more than forty ewes. Crossing rams from breeds such as Bluefaced Leicester or British Milksheep can cope with sixty ewes, while rams from some hill breeds, such as the Swaledale, frequently cover eighty ewes, and on occasion may exceed a hundred. In general ram lambs should not be given more than twenty to twenty-five ewes and should be given extra feed to ensure that they continue to grow while working.

Raddling the rams

During tupping time the rams should be fitted with a harness which holds a coloured crayon, known as a raddle. When the ram serves each ewe the crayon leaves a mark on the rump of the ewe to show that she has been mated. The crayon should be changed for one with a different colour at regular intervals of not more than a fortnight. This will indicate ewes which have failed to conceive at the first service if they are marked with a second colour. Some flockmasters prefer not to use a harness and instead raddle the rams between their front legs with a coloured paste. If the pre-mating management of the ewes and the use of a vasectomized ram have been applied effectively most of the flock can be expected to be mated in a period of not more than three weeks, in which case the colour of the crayon should be changed weekly.

The fertilized eggs and foetuses are at greatest risk during the first ninety days of pregnancy. During this period stress on the ewe may result in re-absorption of foetuses and a reduced lambing percentage. As prolificacy is perhaps the most important factor affecting profitability, the management of the ewes in early pregnancy is of great importance. Stress may be caused either by adverse climatic conditions or by a poor level of nutrition. The effects of bad weather can be partly offset by avoiding exposed fields and by providing shelter in the form of stone walls, shelter belts or woods. The level of nutrition can be maintained by supplying extra feed as the quality of the grass declines, and the easiest method is to use self-help feed blocks. Provided they are given the benefit of good management before mating and during early pregnancy, many breeds of sheep are capable of averaging two lambs per ewe.

Late pregnancy

From six weeks before lambing onwards the ewes will need a special diet of concentrates and roughage which is described in detail in the *Feeding* section on page 118.

Pregnancy toxaemia

It is important that the condition of the ewes is maintained during the pre-lambing period, as any loss of condition could lead to pregnancy toxaemia, commonly known as twin-lamb disease (see page 139). If the ewes are underfed at this time too much of the energy in the diet will be diverted to the growing lambs leaving insufficient for the ewe which, as a result, is required to call on her own body reserves. This results in the breakdown of fat and the production of toxins. The likelihood of this problem arising is increased if the ewes are permitted to become too fat during mid-pregnancy.

Prolapse

Another problem that occurs in late pregnancy is prolapse. This again occurs usually where the ewe is carrying more than one lamb, but the severity of the problem is increased by bulky feeds, such as roots, and these should be avoided at this time. Slight cases of prolapse, where the protruding part of the uterus is no bigger than a small orange, may be cured by joining the lips of the vulva with a single stitch of cat gut. In more severe cases a special device may be inserted into the vagina after the uterus has been pushed back carefully into place. The device is held in place by tying it to the wool on each side of the hindquarters.

Preparation for lambing

About two weeks before lambing the ewes should be vaccinated against clostridial diseases and Pasteurella. This enables them to pass on their immunity to their lambs. They should be drenched against internal parasites and this acts as a tonic, improving the birth weight of the lambs and increasing the milk yield of the ewes. At the same time the wool should be removed from the area around the tail and hindquarters, otherwise it is likely to become soiled and present a potential source of infection to the lambs at birth.

Lambing time

Prolificacy is crucial to the profitability of most sheep flocks, but it is a term with a variety of definitions. Two measures of prolificacy are the most relevant. The first is the average litter size (ALS). This is the average number of lambs born, alive or dead, for each ewe that gives birth, and is the best indication of the ewe's inherent or genetic ability to produce twins or larger litters of lambs. ALS is the measure generally used to evaluate the breeding worth of a particular ewe. The second factor in assessing prolificacy is lambing percentage. This is calculated according to the number of lambs reared

for every hundred ewes which are loosed to the ram, and it is a measure of the efficiency of your management. It takes into account the number of barren ewes and the mortality of both ewes and lambs, and it provides the most accurate index to your profitability.

The number of lambs reared is determined partly by the genetic background of the ewes, but more by the management of the ewes at two critical stages of the year, namely at mating and at lambing time. The management at mating (already described) determines the number of eggs that are fertilized and implanted successfully in the uterus. The management at lambing time determines the number of lambs that survive. In the national flock in Britain the pre-weaning mortality of lambs is over 20 per cent, which is an unacceptable loss. In well managed flocks, even those comprising several hundred ewes, the mortality of lambs can be limited to 6–7 per cent. This is achieved by round-the-clock observation during lambing time so that no lambs are lost by neglect.

It is important also to ensure that the ewes are quiet and docile. It is not possible to shepherd housed sheep effectively if they are nervous and flighty and crowd into a distant corner when disturbed. Docility varies from breed to breed, but it can be enhanced by the shepherd's quiet and sympathetic treatment which inspires confidence in the sheep.

If the sheep are housed for lambing at least one lambing pen measuring 5 ft × 5 ft (1.5 m × 1.5 m) should be provided for each group of ten ewes. If lambing takes place in the open some shelter should be provided, either straw bale walls or yards, or by wattle hurdle shelters. In either case you should be fully prepared in good time with all the necessary equipment. Here is a suggested list.

Lubricant, soap, towel
Infra-red lamps (or Aga cooker)
Ewe tonic and lamb tonic
Elastrator and rubber rings
Navel dressing
Ear tags
Marker spray
Weigh balance and bucket
Foot shears and spray
Syringe, disposable needles and antibiotic
Feeding bottle and teats
Lambing book
Torch

Abortion

One of the most disheartening problems associated with lambing time is abortion. This may result from stress or mechanical damage, but it is more likely that a disease organism is the cause. Both salmonella and vibrio foetus cause abortion in sheep, but the most common types are toxoplasmosis and enzootic abortion (kebbings). Enzootic abortion is controlled by vaccinating the ewes, but the best way to prevent toxoplasmosis is to build up natural

Lambing: 1. *The early stages of a normal lambing, showing the ewe lying in a characteristic position, straining with her nose pointing upwards.*

2. *The emergence of the 'water bag'.*

immunity by mixing the ewe lambs, which have been selected as flock replacements, with infected ewes. Diseases spread more rapidly among sheep which are housed.

Signs of lambing

A ewe will give warning of her intention to give birth. An early indication is that the ewe will choose a place away from the rest of the flock. She will appear uneasy and pre-occupied. She will repeatedly lie down and eventually start to strain. If the presentation of the lamb is normal a water bag will appear first. This may or may not burst at this stage, but next the front feet and nose of the lamb will emerge. The ewe will continue to strain until the lamb is born. Then the ewe will rise, turn round and lick the lamb to clear the membranes and mucus from its head and to stimulate it. If there is more than one lamb this process will be repeated, and the lambs should find the teat and start to suck within a few minutes. There is a great temptation to assist a ewe which is giving birth, but this temptation must be resisted unless the ewe has ceased to make progress or unless the lamb is incorrectly presented.

Abnormal presentations

Malpresentations in the ways listed below are not uncommon in lambs.

1. Head only – legs back. Push the lamb back into the uterus and carefully bring forward the forelegs protecting the wall of the uterus from the hooves of the lamb. (Make sure also that your fingernails are short.) Then draw out the lamb in the correct presentation.

2. Head and one leg – one leg back. Follow the same procedure as above.

3. Hind legs presented. Draw out the lamb hind legs first, but this must be done quickly otherwise the lamb will swallow fluid and fill its lungs.

4. Breech presentation (i.e. rump first). This is possibly the most difficult presentation. The lamb must be pushed back into the uterus and turned around completely to enable the front feet and nose to emerge first.

5. When a ewe is carrying more than one lamb it is possible that two may become entangled at the time of birth and sometimes the two feet that appear may belong to different lambs. In this case the lambs must be separated. The sorting out of a tangled welter of legs in a difficult task and demands a good deal of experience.

Experience comes with practice and the inexperienced shepherd or flock owner should not be afraid to enlist the help of a more knowledgeable colleague or neighbour, or to call a veterinary surgeon. When a ewe has been assisted she should be given an antibiotic injection as a precaution against infection.

3. *The nose of the lamb just visible.*

4. *The head and forelegs emerging.*

5. *The birth completed, and the afterbirth appearing.*

6. *The ewe licks the new-born lamb.*

A difficult lambing: 1. *Preparing to insert the hand into the uterus to manipulate the lamb. Note the plastic glove and lubricating jelly. Lying the ewe on her back helps the operator to push the lamb back into the uterus, but the ewe may have to be laid on her side if the operator does not have an assistant.*

2. Inserting the hand into the uterus.

After a ewe has lambed it is important to ensure that milk can be drawn freely from both teats.

Rearing

After lambing

When the ewe has lambed she should be moved as soon as possible to an individual lambing pen. If the lambs are carried slowly to the pen in full view of the ewe she will usually follow. The first task should be to check the udder of the ewe to ensure that colostrum can be drawn from both teats and to remove any wool on the udder which might confuse the lambs or which they might swallow. It is essential to make sure that the lambs suck and obtain colostrum as soon as possible and certainly before they are twelve hours old. Colostrum contains the antibodies that the ewe has produced and that give the lambs resistance to disease.

Some ewes do not accept their lambs. This is often because they have no milk – the pressure of milk in the udder is needed to stimulate her maternal instincts. Sometimes the teats of the ewe are too large for the mouths of the lambs and sometimes the lambs do not have the necessary urge to suck. As an insurance you should store a bank of colostrum in the refrigerator.

The first two days
While the sheep are in the individual lambing pens (the time will vary from twenty-four to twenty-eight hours) you should give the lambs the following treatment.

1. Tail with a rubber ring – the tail should just cover the vulva.
2. Castrate, either with a rubber ring or with blunt pincers which crush the cords.
3. Treat the navel with iodine to prevent invasion by organisms that cause joint ill.
4. Weigh.
5. Identify with ear tag (usually only necessary for pedigree animals) and, if you have a large flock, with a field number sprayed on the lamb's side.

It is also a good opportunity to trim and treat the ewes' feet.

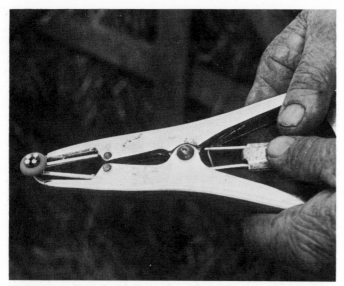

Tailing a lamb: 1. *The elastrator, showing the rubber ring correctly positioned.*

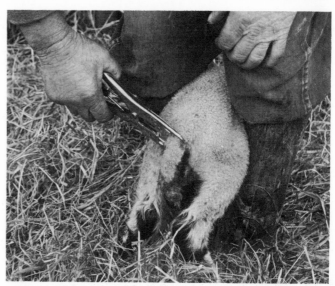

2. *The rubber ring is placed around the tail with the elastrator. The tail should be left long enough to protect the vulva in female sheep.*

When these tasks have been completed and it is certain that the ewe has accepted her lambs, they can be transferred to small group pens.

Problems with new lambs
If the ewe is unable to rear her lambs, either because she has no milk or because she has three or more lambs, they can be fostered by another ewe who has either lost her lambs or has given birth to a single. When a lamb is being fostered its new mother should be restrained in a yoke and not permitted to see or smell her own or her foster lamb. If the ewe's own lamb has died the skin of the dead lamb can be tied to the one to be fostered so that the ewe will recognizing the smell. In most cases the lambs will be accepted within two or three days. While the ewes are still housed and the lambs are young there is a danger of diseases such as coccidiosis and scouring spreading rapidly. If this occurs remedial treatment must be applied as advised by a local veterinary surgeon.

Turning out
When the ewes and lambs are turned out to their pastures they move away from the confined area of winter housing and the lambing area where they could be closely observed all the time, but they must still be inspected regularly and thoroughly. Careful and detailed observation is the kay to successful management of any livestock. Even with those breeds which claim to thrive in 'easy care' systems, simpli-

Castrating a lamb: 1. *The rubber ring is placed around the scrotum using the elastrator.*

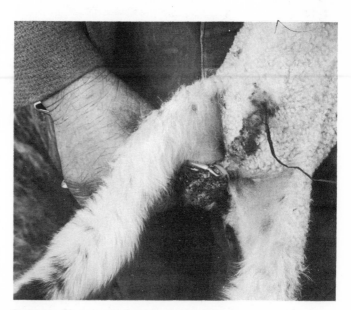

2. *The rubber ring in position around the scrotum.*

A ewe and lamb in a fostering crate. The ewe is restrained by her neck in a yoke so that she can neither see nor smell the lamb, and the creep bars beside her prevent her from lying on it.

city of management must not be confused with neglect!

While the lambs are young predators may be a problem. These may be a neighbour's pet dogs, but are more likely to be foxes, coyotes or dingoes. Ewes of some breeds are more aggressive than others and may have more success in defending their offspring, especially if they have only a single lamb. Electric fencing is a more certain deterrent where possible. Beyond this you can control the problem to some extent by intensive shepherding, but a determined predator will seize any opportunity and is not likely to be denied.

Artificial rearing

If your lambs include some triplets or orphans they will need to be reared artificially, but you must first make sure they receive colostrum, either from their dams or from surplus supplies you have stored earlier. In the absence of colostrum you can give them a substitute – either a specially manufactured one or one which you have made up yourself. (The standard ingredients for each lamb would be: $\frac{1}{4}$ pt (125 ml) milk, some raw egg, and $\frac{1}{2}$ tsp (2.5 ml) each of glucose and cod liver oil.) However these should only be used as a last resort as no substitute will provide the protection against disease given by natural colostrum.

Bottle-feeding a lamb.

Hand-rearing

If you are prepared to devote a great deal of time and hard work to the task, and if you have only one or two lambs, you may want to try rearing them by hand. This will involve feeding the lambs with special milk powder substitute, fed warm from a large sterilized bottle with a rubber teat over its end. Never feed cow's milk to lambs as it is unsuitable for them. For the first four or five days you will need to feed the lambs every four hours, and thereafter every six to eight hours until weaning. The amount you give them will increase gradually as they get older, but they should start by getting about 1 pt (0.5 l) a day, and may be drinking as much as 3 pt (1.5 l) a day by the time you wean them. After a few days you can give them a little crushed cereal with each feed, and when you come to wean them (at about six weeks old) they can be offered as much concentrate feed as they want during the days when you are gradually diminishing their supply of milk. Throughout the period when you are rearing them the lambs should be kept housed and warm and should have clean, fresh water always available (make sure the container is not big enough for them to drown themselves in it).

Other methods

Hand-rearing is, however, messy and time-consuming and you should only undertake it if you are sure you will be able to go through with it. If you have not enough time to spare or too many lambs to make it practicable there are several more labour-saving methods of artificial rearing which have proved successful. The system I describe below is one of these, but others may work equally well, provided that the following essential principles are observed:

1. The lamb must have colostrum before being reared.
2. A high standard of hygiene must be maintained in the rearing pens.
3. The daily routine should be consistent.
4. The lambs should be reared in small groups.
5. They should not be weaned until they are eating solid food as a regular and significant part of their diet.

The problems that occur with artificially reared lambs are often caused by a lack of attention to detail. Variations in the daily routine may cause a lamb to become hungry and gorge itself when food is provided. Variations in the temperature of the milk may cause digestive upsets. (Bloat, for example, is a problem associated with artificial rearing which may be caused by management inefficiencies such as these.)

Careless cleaning routines may give harmful bacteria a chance to multiply – the feeding equipment must be cleaned meticulously each day, while the walls and floor of the rearing pen must be thoroughly cleaned and disinfected after each batch of lambs. It is not uncommon for the elation which the sheep owner experiences after the excellent results of his first batch of lambs to change rapidly to gloom when the succeeding groups of lambs become prone to rampant disease because their pen has not been adequately cleaned.

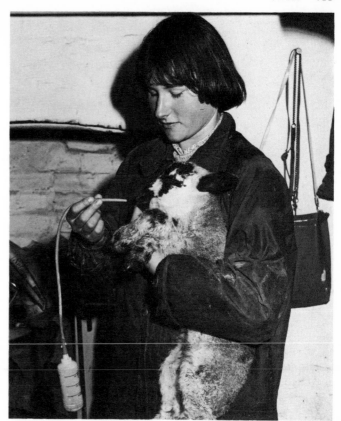

When a weak or uncooperative lamb will not suck, tube feeding may be necessary. **1.** The soft rubber tubing is pushed carefully down the lamb's throat.

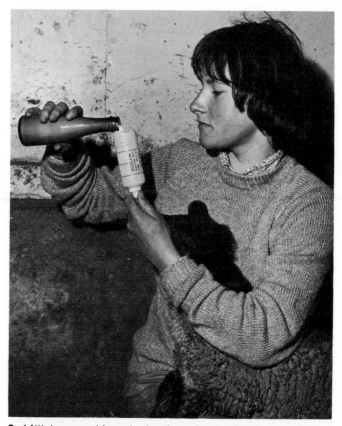

2. Milk is poured into the bottle and runs directly into the stomach. Take care that the milk is at the correct temperature and that hygiene is strict.

Artificially rearing lambs on a bucket. The lambs draw the milk up the tubes by sucking at the teats.

After the lambs have received their colostrum and as soon as they are active they can be moved to a preliminary rearing pen and taught to feed from a special feeder. You can buy a metal container with eight rubber teats which is large enough for up to sixteen lambs to feed from it as and when they wish. A vertical incision in the end of each teat ensures that the lamb can suck out milk but that the teat will not leak at other times. There are also simpler versions of this system available which are fitted onto a bucket of milk and can be used to feed about four lambs at a time (see photograph above).

Initially the lambs are given regular feeds of warm milk every four hours and are trained to suck from the teats. Cow's milk is not satisfactory, as it contains little more than half the proportion of protein and fat found in ewe's milk, so you should use a milk powder specially produced for feeding lambs. As soon as they are able to suck without being assisted, and are active and eager to feed, they can be given access to a limitless supply of cold milk. Care must be taken to make sure they make this transition successfully. They should also have access to solid food (lamb pellets) and clean water.

The lambs should be weaned abruptly when they reach a weight of 30 lb (13.6 kg) – provided that they are eating at least 12 oz (340 g) pellets each day. They can then be kept indoors on a diet of as many pellets as they want, together with clean water and a little hay if they require it, or they can be turned out to grass with continued access to lamb pellets. It is important that at all stages the continuity of the diet is preserved by at least one major constituent so that there is never a drastic change.

Sometimes, when a breed of sheep is kept for milk production the lambs may be taken from the ewe and reared artificially soon after birth as a deliberate policy. In some countries it is more usual to leave the lambs with the ewes for about six weeks. They are then weaned and the ewes are milked for the remainder of the lactation.

Weaning

Lambs which have not been artificially reared are usually weaned when they are between twelve and eighteen weeks old. The actual time will be determined by the level of milk yield of the ewes and the availability of grass. Lambs whose dams can provide large quantities of milk will grow more quickly and can be left with the ewes until a good proportion are fit for slaughter. With ewes of poor milking ability the lambs should be weaned earlier and given fresh grass so they can maintain a good rate of growth.

The ewes and lambs may start to compete with each other for grass as the quality and volume of the supply deteriorates, and this can also precipitate the decision to wean. When the ewe's milk yield starts to fall off the lambs will be unable to compensate for the loss of this part of their diet if there is not enough grass to feed both them and their dams.

At weaning a close watch must be kept on the ewes' udders. If they are still producing an appreciable amount of milk the pressure will build up in the udder and the ewes must be kept on a sparse diet to dry them off. They should be kept without water for twenty-four hours and fed only a minimal amount of roughage feed until milk production has ceased.

Weaning in pedigree flocks

In pedigree flocks the emphasis at weaning will be more on breeding stock than on lambs for slaughter. Flock replacements should be selected and given a permanent identification mark. Any animals with a defect or which are not typical of the breed should not be retained. One or two older animals should be mixed with the flock replacements to provide a stabilizing and quieting influence so that they will be as docile as possible when they join the flock.

At this stage all the ewes in the flock should be examined and ewes with a fault or a deficiency should be culled. Ewes which have had mastitis or prolapse should not be retained and only in special circumstances is it worth keeping ewes with a broken mouth. These are all examples of negative culling. The culling of ewes because of poor performance is a positive process as it is likely to lead to improved production by the whole flock. In hill flocks culling is based mainly on age, and ewes are drafted to farms in the lowlands after producing three or four crops of lambs.

The ewes which are kept in the flock should not be allowed to deteriorate in condition after weaning. (If they have been working hard during pregnancy and lactation they will probably already have lost a good deal of condition.) This is also the time when all sheep on the farm should be dipped against sheep scab (see page 136).

After weaning

Although the period after weaning is not critical in the sheep's calendar, it is a time of recuperation for the ewe when she is steadily building up her resources before flushing. By keeping the ewes in good hard condition

Making a subcutaneous injection over the rib-cage where an abcess damages the pelt.

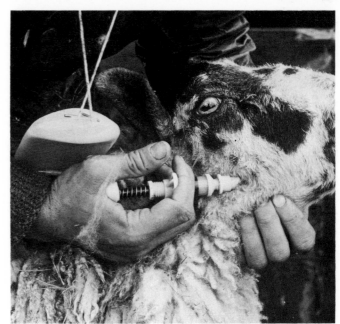

Making a subcutaneous injection on the cheek.

(neither too fat nor too thin) the foundation of another successful year is laid. Although the management of the sheep will vary at different times of the year, and special treatment is required at certain times, it is all part of a single integrated system. Mismanagement at any time of the year will be reflected in problems at a later stage, unless you compensate for the deficiency at the earliest opportunity. The sheep will respond to good management, and its wide range of uses from the production of meat, wool or milk to the maintenance of the fertility of arable land make it a most efficient and satisfactory animal to keep.

Health and disease

Signs of ill health

The flock owner who observes his sheep with sympathy and a deep concern for their welfare will learn to recognize instinctively their normal patterns of behaviour, their appearance, and their reactions to him and to each other. He will know immediately when there are any changes, or signs that something might be wrong. A sheep standing with its back humped, its head down and its ears drooping is an obvious indication that all is not well and the animal should be immediately examined. It may be that the ewe has developed mastitis because she was not yielding sufficient milk and the hungry lambs have chewed her teats, leaving sores and cracks through which the disease organisms could gain entry. A lamb, on the other hand, might be suffering from exposure, especially as a result of hunger or after prolonged and heavy rain.

Other signs are less obvious. If the lambs are reluctant to suck their dam it may simply be because of a dirty udder.

If a ewe moves off when lambs attempt to suck, it may indicate that she has mastitis or that she has rejected her lambs. Lambs may be attempting to suck their dam with unusual frequency which is often a sign that they are hungry – either because the ewe is short of milk or because she is being 'poached' by other lambs.

Vaccination

Ewes being introduced to the flock for the first time should be given an initial vaccination against certain diseases. These may vary from area to area, and local veterinary opinion should be sought regarding details of the vaccination programme, but in most areas ewes should be vaccinated against clostridial diseases, pasteurella and enzootic abortion. The best site for subcutaneous injections is under the skin of the cheek, and for intramuscular injections into the muscle of the neck. Damage resulting from the injection at either site will not reduce the value of the carcass. This vaccination programme will not affect the fertility of the ewes, but it will ensure that immunity from the diseases is passed on to the lambs.

Protection against parasites

Internal parasites, both nematodes and flat worms such as liver fluke and tapeworms, are a problem to which all sheep are liable at some time. A nematode parasite problem is indicated by unthriftiness and scouring, while white segments in the faeces of the sheep indicate the presence of tapeworms. If either of these problems should arise they can be controlled either by drenching with anthelmintics or by one of the other preventive measures which are described below.

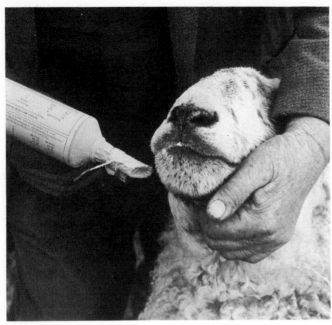

Administering anthelmintics as a paste. The barrel of the special 'gun' is inserted into the side of the ewe's mouth.

Anthelmintics

These are usually given in a liquid form as a drench, although they are also available as a paste. All ewes should be drenched in the spring, as this not only keeps them in good condition during mating but also ensures that a large population of parasites is not carried over the winter to infect the pastures in the following year. Both ewes and lambs should be drenched when the lambs are weaned. Beyond this the lambs should be drenched as necessary. In severe cases this may mean repeating the process regularly every three weeks throughout the season.

Grazing system

The lambs can be protected to some extent by 'forward creep' grazing whereby the lambs, but not the ewes, can go into a forward paddock and so obtain clean grazing. This system is not widely practised. A better method is to mix different species of livestock on the same grazing. Different animals are attacked by different parasites which means that grazing mixed species, such as cattle and sheep, will dilute the effect of the parasites on both of them.

Stocking density

Parasite problems increase when livestock are densely stocked, but the profit-conscious farmer naturally wants to keep the greatest number of lambs he can on his area of land. These two factors can be reconciled by keeping only a moderate stock of ewes but achieving a high output from each one. Thus if your required target is ten lambs from 1 acre (0.5 ha), it is preferable to stock five ewes and reckon on two lambs from each than to keep seven ewes with an average of 1.43 lambs per ewe, and so on.

Dipping

If you have a handling pen it is a good idea to include a dip bath in it (see page 117). Sheep should be dipped about three weeks after they have been shorn to control ectoparasites such as lice, ticks, keds, headfly and blowfly maggots. The dip bath is filled with the dip solution and each sheep made to run through or stand in it long enough for the liquid to penetrate to the skin. An important point to remember is that the dip solution is a pollutant and should never be drained out directly into a water course.

Sheep scab

All sheep should also be dipped in the autumn against sheep scab. In Britain this is a legal requirement and sheep scab is a notifiable disease. When selling sheep it is necessary to produce a dipping certificate proving that they have been dipped in a 'scab approved' dip during the statutory period. A sheep suffering from the disease will lose or rub off much of its wool as the burrowing mites cause intense irritation. The animal's skin will become thickened and rough. If you suspect the disease you should call the veterinary surgeon immediately.

Blowfly treatment

Before the sheep are dipped they may be attacked by blowfly which lay their eggs in damp wool. When the eggs hatch the maggots hidden in the wool eat the skin and flesh of the sheep. The first signs are that the sheep may waggle its tail or attempt to nibble itself. Then a darker damp patch will appear in the wool. Unless the problem is treated at this stage the maggots will cause a severe and unsightly wound.

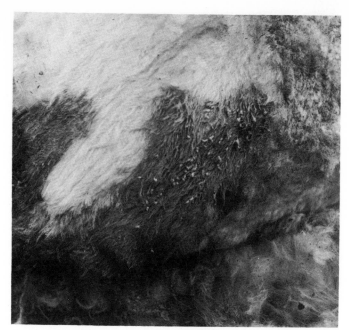

A wet area of wool indicating a blowfly attack. Here the maggots have been disturbed by shearing.

The wool should be cut away from the affected area and the wound treated with dip. The fever caused by the blowfly can affect the fertility of the rams.

Diseases

Clostridial diseases

This group of diseases is caused by a family of anaerobic germs and includes tetanus, lamb dysentery and pulpy kidney.

1. **Tetanus:** This disease, also known as lockjaw, happens when an organism enters the body through a wound. The animal begins to walk stiffly, loses its balance and lies with its legs outstretched, after which it soon dies.
2. **Lamb dysentery:** This is caused by C. welchii type B. It is usually characterized by severe diarrhoea and death in early life, and often affects a high proportion of the lambs to some degree.
3. **Pulpy kidney** in lambs and enterotoxaemia in adult stock are caused by C. welchii type D. Symptoms are rarely seen as the animals are generally found dead. Among lambs it is most commonly the animals which are thriving and healthy that are affected.

Control of all the clostridial diseases depends on preventive measures. Flock replacements are vaccinated initially before they are first mated and the vaccination is repeated every year thereafter about two weeks before lambing. The immunity is transmitted from the ewe to her lambs through colostrum.

Contagious pustular dermatitis (Orf)

Orf is caused by a virus and appears most commonly in lambs either in early life or at weaning. Pustular lesions appear around the lamb's mouth and feet. The lesions blister and form a scab which falls off after a few weeks without treatment when the lesion has healed. The problems are threefold. The lamb will lose some condition, the disease may be transmitted to the udder of the ewe by a sucking lamb, and secondary bacterial infection may occur. This may make the lambs extremely lame or prevent them from eating and it may lead to mastitis in the ewes. Where the disease occurs the ewes should be vaccinated before lambing and the lambs may also be vaccinated early in life. Antibiotics should be used to control any secondary bacterial infection.

Foot and mouth disease

Foot and mouth is a highly infectious disease (notifiable in Britain) which is caused by a virus. Blisters appear on the feet at the coronet and between the claws, and to a lesser degree on the dental pad, lips and gums. Other symptoms are lameness and fever. If the disease is suspected a veterinary surgeon must be called immediately or the police should be notified. The current law in Britain requires compulsory slaughter.

Foot rot

Foot rot is caused by infection of the foot by Fusiformis nodosus which results in suppurative decay of the tissues. As the infection spreads the horn wall of the foot may become detached, and the animal may be so painfully lame that it grazes on its knees. The condition occurs most commonly in areas of high rainfall and where sheep are grazing lush grass or on wet land. However, it will occur under any conditions if steps are not taken to control it. Some breeds, such as Dorset Horn and Finnish Landrace, are more susceptible to it than others.

General control is effected by regularly running the flock through a footbath containing 10 per cent formalin. More severe individual cases should be treated by paring away the outer layers of the hoof to expose the infected tissue, to which 10 per cent formalin solution or antibiotic is applied, and the sheep is separated from the flock until the infection is cured. Where there are no sheep, Fusiformis nodosus cannot survive more than two weeks on the soil. Sheep can be vaccinated against foot rot, but the results are very variable.

Hypomagnesaemia (grass staggers)

Hypomagnesaemia is characterized by sudden death, preceded by symptoms of extreme nervousness and excitability. It is caused by a deficiency of magnesium in the blood, which is often associated with sheep grazing rapidly growing grass. Magnesium is not stored in the body and must be provided at regular intervals during the danger period. It may be provided in a variety of ways, but the most usual method is to feed a magnesium-rich concentrate or to make available a molasses/magnesium liquid feed.

Examining the feet of a ewe for signs of foot rot. Sheep in houses or pens tend to accumulate dirt and straw on their feet, which encourages disease organisms.

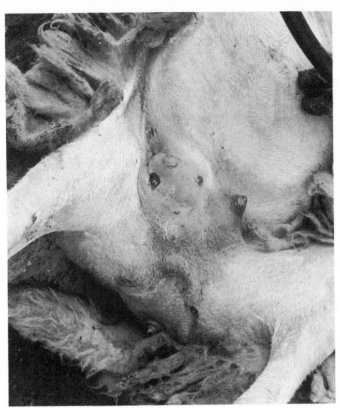

A ewe with mastitis, showing hard tissue and open sore.

Foot rot is found most usually in the cleft between the claws of the foot or underneath the sole.

Joint ill

Joint ill, or inflammation of the joints, is usually seen in lambs. The causal organisms infect the animal through the navel soon after birth or through wounds, including those caused by castration or tailing. The joints become swollen and painful and will remain deformed, while in the worst cases the animal may die. The prevention of joint ill is best effected by strict hygiene, and by treating the navel of the lamb at birth with iodine. Antibiotics are the best form of treatment.

Mastitis

Mastitis occurs when infective agents gain access to the udder either through the teat canal or through an injury. It may be caused by several organisms, of which the most important are *Staphylococcus* or *Coryrebacterium pyogenes*, which cause acute inflammation of the udder. The udder becomes swollen and hot. The ewes appear distressed and will not allow their lambs to suck. In acute cases the animals may die. The affected side of the udder must be milked frequently to empty out the fluid which may be thin and watery, and sometimes bloodstained, but more frequently it is clotted. Treatment with antibiotic drugs may cure the disease if it is diagnosed at an early stage of the infection.

Pine

Pine is caused by a deficiency of cobalt which prevents the manufacture of vitamin B12 in the rumen. This vitamin is essential for all higher animals. Pining animals appear dull and unthrifty. The fleece is dry and lacks lustre, and there may be a discharge from the eyes. The animals lose condition and become weak. The disease is cured by the administration of cobalt or vitamin B12.

Pneumonia

Pneumonia – literally an inflammation of the lungs – may result from various causes, but the most important type is Pasteurella pneumonia. The Pasteurella organism is present in most healthy sheep, but remains inactive until triggered by stress – which may be caused for instance by transport, severe weather conditions or a drastic change in nutrition. Affected sheep may die very quickly, but the usual symptoms vary from a cough and laboured breathing to rapid breathing, high body temperature and a discharge from the eyes and nose. Pasteurella is controlled by vaccination generally carried out at the same time as the clostridial vaccination.

Pregnancy toxaemia (twin-lamb disease)

Pregnancy toxaemia occurs in late pregnancy and is most common in ewes carrying two or more lambs. The immediate cause of the disease is a reduction in the level of nutrition, which is most likely to occur in ewes which have been overfed in early pregnancy, and whose carbohydrate metabolism has been upset. Affected ewes show lack of coordination, often with a nervous tremble. They lose their appetite, then collapse and go into a coma before finally dying. The disease is controlled by ensuring that the ewes are kept in steady condition during early pregnancy, and then given an increasingly high level of nutrition as lambing approaches. Treatment is far from reliable, but injecting the sheep with glucose or drenching them with glycerine may be effective in some cases.

Prolapse

See page 127.

Scrapie

Scrapie is a disease of the nervous system, the cause of which has not yet been identified. In many places it is notifiable (reportable). The disease has a long incubation period and may appear in sheep of any age, but most commonly occurs when sheep are three or four years old. Affected animals are very nervous and excitable, and rub themselves against fences or any other suitable scratching post. They carry their heads high and trot with the action of a Hackney horse. Finally they become uncoordinated and paralyzed and eventually die. There is no known treatment. It seems that some sheep are more susceptible to the disease than others and the best course of action in the light of current knowledge is to cull the close relatives of affected sheep.

Sheep scab

See *Dipping*, page 136.

Swayback

Swayback is caused by the progressive destruction of the nervous system of the lamb before it is born, resulting from a deficiency of copper. Usually swayback is present at birth, but in some cases it develops in the early weeks of life. The main symptom is lack of coordination, especially of the hind limbs. Severe cases never recover and treatment is not possible. The incidence of the disease can be reduced by giving the ewes extra copper during pregnancy. It must however be remembered that sheep are very susceptible to copper toxicity unless the supplement is carefully controlled.

PIGS

Background

The keeping of pigs for food purposes goes back many hundreds of years. There are references to pigs in Neolithic times; certainly Roman writers made many comments about them and of course there are numerous Biblical references to swine. It was not, however, until the eighteenth century that people really became interested in improving pigs. In 1760 in Britain a man called Bakewell began to select Large Whites for breeding purposes and in the latter part of the century the Chinese pig was imported and crossed with native breeds to give earlier maturity and more meat on the carcass. In the nineteenth century pig shows introduced a competitive element between breeders and a National Pig Breeders Association was formed. Thereafter many individual breed societies were established and throughout the twentieth century great work has been done to improve breeding, feeding and marketing.

Pig raising in Australia has grown up largely as a sideline to other kinds of farming, and has often been the province of the part-time or small-scale farmer. For many years the only popular breeds were the Tamworth, Berkshire and Large White, the last of which was particularly well suited to the hot climate. In the 1950s, however, the Landrace took the Australian pig industry by storm. Entrepreneurs at that time made a fortune by importing the new 'wonder breed' which has since become well established in Australia for both crossing and pure-breeding programmes.

It is thought that pigs were introduced into America by Columbus. From this small start they proliferated amazingly quickly and were soon to be seen foraging freely around the countryside in large numbers. In recent years the pig has been intensively developed in America as it has in other countries to become a sophisticated, commercial animal, hardly recognizable as a descendant of its half-wild ancestors.

Breeds

A great change has taken place in breed structure over the last thirty years, following the importation into Britain of the Swedish Landrace in the late 1940s. Today the Landrace and the Large White are by far the most popular in breeding herds and are used extensively in cross-breeding programmes to produce hybrid pigs.

Many of the once popular breeds of pig have now become old fashioned because they do not perform as well as the modern breeds (i.e. they do not gain weight so efficiently) and also because present day commercial breeders prefer to plump for a breed which they know and which has already proved itself. Breeds such as the Tamworth, Gloucester Old Spot and the Berkshire are rarely kept any longer for strictly commercial purposes, although they are still of interest to smallholders because they are hardier than the specialized commercial hybrids and stand up well to tough or outdoor conditions.

Pig breeders generally start with pure-bred stock and breed hybrids from them in the hope of combining all the good characteristics of two or more breeds in the progeny, and of producing a pig which eats less, puts on weight faster and does not lay down a lot of fat. Large companies even make special attempts to develop breeds exactly suited to their own particular requirements. A bacon pig, for example, should have a long back whereas a pig which is going to end up in pork pies or sausages can be more stockily built.

Swedish Landrace

This breed is noted for the quality of the bacon it produces and for its economic performance. In Denmark and Sweden careful breeding and selecting have brought the breed up to a very high standard indeed. The Landrace has also spread through Britain very quickly over the last twenty years.

Large White

This pig has gained a world-wide reputation. The boar in particular is highly thought of and is extensively used for crossing with other breeds for bacon production, and for grading up poor herds.

Other breeds

Of the lesser breeds the Welsh is notably hardy and prolific. The British Saddleback plays an important part as foundation stock in cross-breeding programmes to produce hybrid animals which have excellent prolificacy and milk. The Duroc is a popular breed in North America, and has recently been introduced into Britain.

The special characteristic of the Pietrain, which comes from Belgium, is that it produces a high percentage of lean meat. This virtue is, however, somewhat offset by the fact that the meat is not of a particularly high quality. A testing programme of cross-breeding is under way and the results are being carefully monitored. The Hampshire, which is a North American breed, is likewise being used in testing programmes both as pure stock and for cross-breeding.

Buying pigs

Different types of pig will reach their ideal point for slaughter at different weights so your choice of any particular breed should be determined more by your systems of management and feeding than by differences in maturity.

More important than your choice of breed is to choose

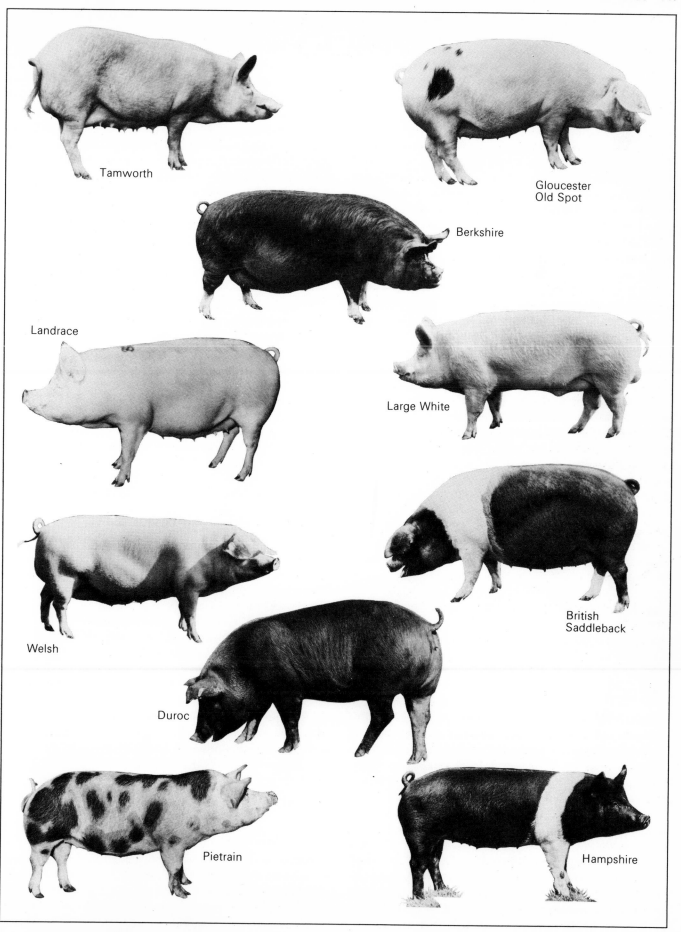

Tamworth

Gloucester
Old Spot

Berkshire

Landrace

Large White

Welsh

British
Saddleback

Duroc

Pietrain

Hampshire

good stock from within that breed. A pedigree animal is generally a better choice than one from unknown ancestors but there is no automatic guarantee that a pedigree animal will be a good one. Pig performance recording is now commonplace and any prospective buyers of new stock, either for fattening or breeding, must first make up their minds on the purpose for which they require the animals. In other words, do you wish to breed or fatten for the pork market, the bacon market, or the pig-meat processing market (i.e. pork pie filling, luncheon meats and sausages)? You can then refer to the records and select pigs that will best suit your intentions.

How to buy

Once you have decided what you want you should seek advice as to the best stock to buy for the project you have in mind. Advisers from the Agricultural Department or local pig breeders should be able to help. In the USA the Agricultural Extension Service is also a good source of information on this.

I would not recommend buying from the local market unless you are highly experienced in choosing stock or can take an expert along with you. Many farmers use the local market to off-load stock which they consider inferior or which they believe will not fatten well. Obviously not all stock at the local market will fall into this category, and this is why you should get specialist help.

I believe that the best source is from the farmer himself and this is where official advisers can help by naming good suppliers and giving a guide to likely prices. The local press will also publish prices after market day and a telephone call to the auctioneers can be a good way to find out whether the general price trends are upwards, downwards or static.

Which to choose

For fattening

If you want to keep pigs to fatten for meat you should aim for animals weighing about 60 lb (27 kg). At this weight the pig has left behind most of the ills that beset the piglet. It will also have been weaned for some weeks, will have been away from its mother for some time and will therefore be quite capable of looking after itself. A reputable pig farmer will only sell you good stock. After all, it is in his own interest – you are not going to go back to him in the future if the pig does not do well, if it refuses to put on weight quickly enough or if it succumbs too easily to illness. However there are certain points which you can look out for yourself.

The pig should look lively and act lively. Its back should have a sheen or gloss. There should be no obvious cysts or lumps on the body, although it is fair to say that if such lumps are present it does not necessarily mean the pig is faulty. I have grown several pigs on to slaughter weight with lumps the size of a tennis ball and had them passed fit to eat by the slaughter house inspector.

The animal should not be too dumpy but should have a fairly long, lean back. The area to the front of the back legs should look full and rounded as this is evidence that the pig is eating well. The rear end of the pig should be clean. If there are signs of dung then it is possible that the animal is scouring (i.e. has diarrhoea). Look out for signs of lameness. A pig which is not four-square on its feet at this weight is going to have problems as it grows and gets heavier.

For breeding

Choosing breeding stock is a near impossibility for the beginner and I personally would never recommend that he should do it himself. It is safer and, in the long run, more economical to go to a reputable breeder and buy through him. He has his reputation to consider and therefore will not be likely to pass on dud stock. A word of caution, though. No breeder can guarantee that a female pig will conceive or that a male pig is fertile. These are points that only time can show.

If you must buy for yourself then here are some points for guidance. The pig must have a straight nose, straight legs, long back and the correct ear position (i.e. for a Landrace, floppy ears; for a Large White, ears that stand up). In addition the female pig must have not less than twelve, and preferably fourteen to sixteen, teats. The boar must have both testicles and very good strong legs. When he is serving the female all his weight is on his back legs.

However I must stress again that it is really quite foolish to attempt to buy breeding stock without advice. To leave it to the expert is bound to be best and cheapest.

Taking the pig home

In many countries there are strict rules concerning the transportation of animals and the penalties for failing to adhere to these rules can be heavy. Your local Agricultural Department will give you details on precisely how these regulations might affect you.

Obviously the type of vehicle you use for transporting your pigs will depend on how many there are and how often you are likely to need to move them. I have regularly used a station wagon to take two 180 lb (82 kg) pigs to slaughter. The same vehicle has accommodated three 60 lb (27 kg) pigs. If you want to transport a larger number of animals your best choices are either to hire a cattle truck of, if you expect that you will have sufficient need of it, to buy a trailer which you can hitch onto the back of a car.

Housing and equipment

Before you buy your pig or pigs it is essential that their housing should already be constructed and ready to receive them. Never buy your pigs and then build their accommodation round them. Do not, incidentally, be influenced by the epithets 'dirty pig' and 'greedy pig' which show the animal in an unduly bad light. In truth the pig, if given the right environment, is by nature a clean and intelligent animal.

Pasture

It is possible to keep pigs on open pasture rather than indoors, but if you choose to do this you should remember the following points:

1. You should allow a maximum of 20–24 pigs per 2½ acres (1 ha).
2. The pigs should be wormed frequently to kill the parasites which they will pick up off the grass.
3. You must ensure that the land is very well drained so that it does not become a quagmire.
4. Strong fencing is essential.
5. It is more difficult to be accurate about feeding with this method and to be sure that each pig has its fair ration than it is with pigs which are housed.
6. If they are left out in cold, wet weather the pigs will need supplementary feed to compensate for the adverse conditions.

Pigs kept on pasture can have a beneficial effect on your land as natural ploughs and the system has some advantages as a low-cost exercise. However, it does generally produce poor performance from the pigs and a lot of extra work for the pig keeper.

Small-scale housing

The small pig keeper will probably prefer to make use of a small shed or building already on his land rather than to build special housing. This should, ideally, have an enclosed yard or small paddock adjacent to it in which the pig can wander. The only two absolute essentials when providing housing for pigs are these:

1. Make sure the pig is warm, dry and free from all draughts.
2. If it is allowed to roam in a yard or field make sure that the enclosure is completely pig-proof. (This is neither as easy nor as cheap as you might think!)

The same rules apply to any type or size of housing. My own pig pens are built out of hollow concrete blocks, with walls about 3 ft (1 m) high. The gates are made of steel bars with a cattle-proof latch. You can make gates out of wood, but the wood must be extremely thick and faced with a metal sheet, otherwise the pig will simply tear it to pieces.

Three different kinds of pig housing.

Larger-scale housing

If you want to keep a larger number of pigs there are basically three alternative methods of housing which are the most commonly used. These are cubicles, kennels or tethering. The principles of each of these systems are described below, but all of them can of course be adapted to suit your own existing buildings and the number of pigs you want to keep.

It is generally a good idea to keep the sows and gilts

(i.e. sows before their first litters) in the same building as the boar, and if possible open-work gates should be used to separate them so that the sows and boars can rub noses. This prior acquaintance helps when it comes to serving (mating). It is best for each pig to have its own individual feeding place into which it can be shut at feeding time. This ensures that each pig gets a fair ration of food and is not bullied out of it by stronger pigs. If individual feeding

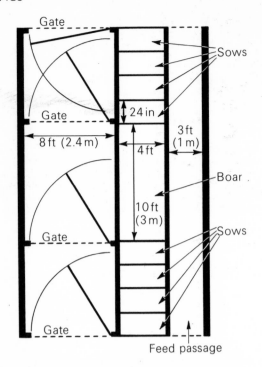

Cubicle housing.

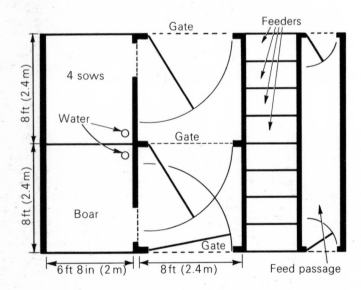

Kennel housing.

places are not possible then the pigs should be fed an extra 15 per cent of food to make sure that the weaker pigs get a fair ration. This, as you will understand if you multiply 15 per cent by the number of days' feed, can work out very costly.

In addition, and regardless of which method of housing you are adopting, you should take special note of the following points:

1. Walls, floors and ceilings should all be insulated. (Floor insulation can be achieved by using fibreglass,

lica, polystyrene or even egg-trays, with a layer of smooth concrete over the top.)
2. It is very important to allow each pig exactly the right amount of space – too little and they will be uncomfortable, will not perform well and may fight; too much and they will muck in the wrong place and create extra work in cleaning out.
3. Adequate warmth, especially for the new-born piglets, is essential.
4. Fire is an ever-present hazard and you should take proper fire precautions, especially in the creeps.
5. Take care to provide adequate feeders and water supply. Pipes or containers for water should be properly lagged if there is a danger of frost.
6. Keep the pigs clean – experience will teach you how often to muck out.
7. All walls and penning must be adequately secured and must be strong enough to take the full weight of the pig to be housed in them.

Cubicles

The accompanying drawing shows how cubicle housing should be planned. The sows or gilts live in their individual cubicles into which they are shut at feeding times. They muck in the muck passage which should, ideally, be cleaned out once a day, but if you have not got time for this you must at least make sure that muck and urine are not allowed to build up. (The diagram shows a straw set-up. The slurry or slatted floor system is discussed later on page 148.)

The area for the sows and the boars should be raised approximately 6 in (150 mm) above the floor of the muck passage. A fall of 1 in (25 mm) from head to tail in the sow lying area ensures that urine will flow into the muck passage. Any dung made in the lying area can be pulled out with a hand-hoe or similar scraper.

Water is provided for the sows by means of an individual nipple drinker. This type of watering system is best constructed out of polythene or steel tubing. The nipple must be placed over the feeding trough so that any spillage will go into the trough rather than onto the floor. It is essential to use steel tubing at any points where the pipe is in reach of the pigs, and it should be fixed very firmly to the cubicle. A pig would take only a few minutes to gnaw through polythene and the resulting mess and damage would be colossal.

The dunging area should be filled with straw – barley straw is best, and you will be able to judge how much to use with experience, depending largely on whether or not you clean out every day. The boar should have a good bed of straw in its sleeping pen and this can be pushed into the dunging passage and clean bedding straw provided as necessary.

The boar's feeding trough should be firmly fixed either to the floor or to the wall to make sure he can't get his snout under it. Pigs love to play with things and given half a chance the boar will have its snout under the feeder, bending it, breaking it or generally causing trouble. Try to

An automatic drinker.

put the trough somewhere where it can be easily filled from outside. Boars grow to a very good weight and can sometimes be extremely disagreeable. You should not go into their pens except when necessary, and then take great care.

Kennels

In this system the sows have a lying area on one side of the muck passage and their feeders are at the other side. Floor space in the kennel should be 12–15 sq ft per sow (1.2 sq m). Larger areas can encourage the pigs to muck in their sleeping quarters. The floor of the muck passage should be 4–6 in (100–150 mm) below the floor level of the kennels and feeders, and the gates should be hung 6 in (150 mm) above the wide kerb. You will see that this is a fairly large and elaborate style of housing and does tend to be expensive unless you can convert an existing building.

Insulation for the roof can be made from straw bales placed on top of wooden or iron battens about 4 ft (1.2 m) above the floor of the kennel so that the pigs cannot either damage or eat it. The floor should be covered with bedding straw and there should also be straw in the dunging area. Group feeding can be arranged but a system of individual feeders and waterers like the one used in the cubicle housing method is better.

The gates between the boar and the sows should be made of vertical bars to allow maximum contact between them. If you are planning to use the boar pen as a service area the pen should be built between the sow pens as this will also allow maximum contact.

Tethering

One other method of securing dry sows is to tether them. Each pig has a very thick leather strap buckled around its middle. Attached to the strap is a short length of stout chain and on the end of this is a strong clip. Embedded in the concrete floor of the house is a staple or eyed-bolt onto which the clip is fixed. The pig is then securely tethered.

The sow can also be tethered by a steel chain partly encased in polythene or rubber which is placed around her neck like a collar. This is then fastened to the floor of the

pen as before. You must be careful to ensure that the chain is not too short and that the pig can stand and move from side to side comfortably.

At the sow's head there is the feed trough and water supply, and at the rear end a muck passage. Straw should be provided for her bedding. This is a very economical way of housing sows but seems to be going out of commercial use. Its greatest disadvantage is that the fastenings can break and the sow will then just wander about the house until she is caught and re-tethered.

Each sow is kept apart from her neighbour by a barred divider, set into the concrete floor. The bars must be close enough together to stop the sow putting her head through, as it is a terrible job to rescue a pig which has got its head stuck. The divider should be about 4 ft 6 in (1.4 m) long and 4 ft (1.2 m) tall, and the bars should be about 9 in (228 mm) apart. The house should, of course, be thoroughly insulated and draught-proof throughout.

Cleaning out with this system is simplicity itself. Any dung in the lying quarters is simply pulled into the muck passage with a hoe and the passage is then cleared out.

Farrowing house

It is a good idea, if you can manage it, to have a special farrowing house where the sow can be moved to have her litter and nourish the piglets during their first three weeks of life. Environmental conditions are particularly crucial at this stage if the young pigs are to survive as their first few days are the most critical of their lives.

The house must have an insulated floor, roof and sides so that heat loss is reduced to a minimum. It must also be draught-free and have good ventilation – if the natural air circulation is not good you should instal a fan. If the air is allowed to become stale or moisture-laden the piglets could develop pneumonia and die. Small pens called farrowing crates are almost always used by commercial pig farmers as they prevent the sow from lying on her piglets and killing them, but some people consider the crates to be inhumane and smallholders, who can attend to their farrowing sows more closely, often prefer to allow the sow to farrow in her pen and use a crush bar to protect the piglets. Details on both these systems are given below on page 146. When the piglets are born you will need to provide them with supplementary heat, and they should be put in an insulated box known as a creep (see next page).

If you do not want to use a crate the sow must be provided with plenty of room to move around and plenty of chopped straw and woodshavings for bedding. You will have to remove each piglet as soon as it is born and keep it warm until the sow has finished farrowing so as to ensure that she will not lie on it.

Temperature

The temperature in the farrowing house should be between 60°F (15°C) and 66°F (18°C) if the floor is slatted. The temperature in the creep should be 70°–80°F (21°–27°C). These temperatures can be maintained by adjusting the

Washing down crates with a power washer.

ventilation system. In more costly pig units the temperature can be controlled by thermostat, but it is quite possible to get excellent results without such assistance.

Ventilation

If you are using fan-assisted ventilation the fan must be capable of 15,000 cu ft (400 cu m) per hour in summer and 1,500 cu ft (40 cu m) per hour in the depths of winter. Take care that the indrawn air is directed away from the stock, and especially from the new-born piglets. Remember that draughts must be avoided at all costs.

If you are using natural ventilation the recommended outlet and inlet sizes for each sow and her litter are as follows:

    Outlet: 40 sq in (0.025 sq m)
    Inlet: 80 sq in (0.05 sq m)

I would advise you also to build a fan shaft in case you found it necessary to instal a fan at a later stage. Baffles should be constructed over all the inlets to protect the pigs from any direct air currents.

Interior

Natural light is not essential in a farrowing house but it must be light enough for you to inspect the stock properly. In the creep area the extra heating must be insulated against the risk of fire. Heating units should have moisture-proof wiring which must be of a high enough gauge to carry the current without getting warm. At least one farrowing unit is lost through fire every year – a very costly business.

The roof, walls and floors should all be insulated. For this you could use a fibreglass blanket – ideally 3 in (75 mm) thick. The trouble with this is that it must be contained in a polythene sheeting bag to prevent any moisture from getting in. Polystyrene sheets are as good a form of insulation as any. These should be at least 1–1½ in (25–35 mm) thick. They are extremely light and therefore easy to handle, which is a particular advantage when you are doing the roof. Local agricultural suppliers and pig farmers should both be

able to give you useful advice on what insulation to use and how to instal it.

Hygiene

In the farrowing house it is absolutely essential to maintain the very highest standards of cleanliness, and the building should be regularly and thoroughly disinfected. If you have more than one farrowing pen in the house it is best to empty the pens at the same time and do a complete clear out. All the floors, ceilings, walls and pens must be thoroughly washed down with a good disinfectant. Any agricultural merchant will be able to help you choose the right one. On large units a power washer is used for the job but this is uneconomic for the small-scale pig keeper. All solid matter such as woodshavings and straw must be removed and the whole place scrubbed with a hot solution of washing soda in a 4 per cent dilution (i.e. 4 parts soda to 96 parts hot water). Then you should douse the house with disinfectant. I have found that the old-fashioned stirrup pump is a good tool for this job. It gives a strong spray and is useful for directing the liquid into tricky places. After it has been left for the prescribed time the disinfectant must be washed away – be particularly thorough over this in any areas where the pigs or piglets could lick it.

Farrowing crates

If you want to use farrowing crates you have a choice between those which are permanently fixed to the floor or an alternative design which has a wooden floor constructed in such a way that the crate cannot move around once the sow is inside it. Designs vary but average dimensions would be:

Length	84–96 in (2.1–2.4 m)
Height	44 in (1.1 m)
Width at top	20 in (0.5 m)
Width at bottom	32 in (0.8 m)

The size of the pen around the crate will depend on how long you intend to keep the sow there and whether you are using side creeps or front creeps (see below). An average size would be about 120 in × 60 in (3 m × 1.5 m).

Crush bars

If you prefer to let your sow farrow in the more spacious surroundings of her pen, a good method of stopping her from lying on her piglets is to build a strong bar about 1 ft (0.3 m) above ground level dividing off one part of the pen. The piglets, but not the mother, can get under the bar and into the separated area where an infra-red lamp will provide them with the extra warmth they need and will attract them back there after they have suckled their mother.

Creeps

An alternative is to use a creep or pig-brooder. This is an insulated box with extra heating into which the little pigs can creep after they have fed from the mother. The extra

A farrowing hut which allows the sow to go outdoors.

Young piglets in a side creep.

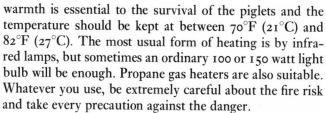

A pop-hole separating the young from their mother.

In this farrowing house, converted from a stable, the creep has been constructed from a hay rack.

warmth is essential to the survival of the piglets and the temperature should be kept at between 70°F (21°C) and 82°F (27°C). The most usual form of heating is by infra-red lamps, but sometimes an ordinary 100 or 150 watt light bulb will be enough. Propane gas heaters are also suitable. Whatever you use, be extremely careful about the fire risk and take every precaution against the danger.

The creep may either be placed at the side of the pen or in the front of it. Many people believe that the front creep is the best as the sow can lie and watch her piglets and is therefore more contented.

The creep area should be well lit so that the piglets will be attracted to the warmth by the light. The front of the creep should be closed off for about two thirds or three quarters of its length so as to leave a pop-hole by which the piglets can go in and out. This helps to keep up the temperature level.

Weaner pool

After weaning the sow goes back to her pen ready to begin the whole breeding process again and at this stage the piglets should be separated from her and kept in what is generally known as a weaner pool. They stay here until they

are approximately 60 lb (27 kg) in weight when they are moved on to be 'finished'. As with all housing for pigs the important things are insulation, good ventilation and freedom from draughts.

You can either have the weaner pool indoors in which case the young pigs are grouped together in pens and a large feeding trough and a water supply are provided for each pen, or, more simply, you can build an outdoor system. This should consist of a kennelled sleeping area with a roof made up of well supported straw bales. The front of the kennel is closed except for a pop-hole door. Typical dimensions for the house would be 8 ft × 6 ft (2.4 m × 1.8 m). The feeders are placed at the rear of the kennel, and you should make a trap door through which to fill and empty them. This system gives very good results. If you have a large herd of pigs you can construct several of these outdoor weaner pools side by side and put gates between them so that a tractor can drive through to clean them out. Try to clean out at least every other day.

Finishing house

When the pigs reach 60 lb (24 kg) in weight they will out-grow the weaner pool and will need a stronger form of

A weaner pool with self-filling drinker.

Weaners on a mesh floor

housing while you are atttening them up to a weight of 220 lb (100 kg). At this stage they should therefore be moved to some form of finishing house, which will need particularly strong pens designed to withstand the occupation of such big and powerful animals.

Slatted floor systems

All the housing systems I have described use straw bedding for the pigs. There is however another method of keeping pigs which does not use bedding and has instead slatted or wire mesh flooring through which the slurry (i.e. the mixture of urine and dung without any other substance such as straw mingled in with it) can fall into a special channel. From there it drains into a watertight pit. The small-scale pig keeper may be able to make use of all the slurry produced by mixing it with straw, allowing it to rot into compost and then using it to fertilize his land. Otherwise the slurry can be pumped into a tanker and taken to a slurry lagoon but this, of course, is expensive.

If you use slatted floors the slats should be not less than $\frac{1}{2}$ in (13 mm) apart, so that the gap is large enough for the dung to pass through but will not trap the piglets' feet. Steel mesh must likewise be of sensible dimensions, and with this you must take care that there are no sharp edges.

There is a special plastic-coated variety on the market which solves this problem, but it is rather expensive.

Equipment

Pig equipment can of course be bought from any agricultural merchant. However sometimes you can also acquire equipment at real bargain prices from farm auctions. I have bought many items this way myself and saved over 85 per cent on their original cost. However, do remember that if you buy equipment secondhand you should disinfect it thoroughly before using it. My own practice is to clean down all metal with a wire brush fixed to an electric drill, disinfect and then paint with a red-oxide paint. Take care however, if you do this, to ensure that the paint you use does not contain lead or other poisons.

A simple piece of equipment that is an invaluable aid for shepherding pigs around (particularly when you are moving the sows to the boar pen to be served) is a length of plywood board measuring 4 ft 6 in × 27 in × $\frac{1}{2}$ in (1.5 m × 0.7 m × 13 mm) with a hole cut into each end so that you can hold onto it firmly and push the pig along.

One easy way to make a feeding trough is to use a short length of glazed draining pipe about 18 in (0.5 m) long, cut in half lengthwise and concreted into a corner of the pen.

Storing straw for bedding

The small pig keeper does best to use straw both for the muck-passages and for the sleeping quarters in his pig houses. Barley straw is ideal as it is softer and more absorbent than other cereal straws. Make sure that you get hold of an adequate supply at harvest time and arrange either to store it yourself or on a neighbouring farm so that you can draw on it as the year goes by. Store it in as dry a place as possible. If you are forced to store it in the open you will have to allow for the weather to destroy the two top bales and all the bales round the outside. As a general rule, a pen of twenty-five pigs from 60 lb (27 kg) up to slaughter weight will require at least four bales a week, although this can only be a rough guide, as so much depends on the housing system you are using and on your own judgements.

Feeding

It may be stating the obvious but it is a simple fact that no animals will perform well unless they are fed well. Until a few years ago most pigs were fed on an ad lib system, i.e. they were fed as much as they could eat. This led to the laying down of a lot of fat on the carcass, which most housewives did not want and objected to paying for. Today almost all farmers have changed to controlled or rationed feeding. This means giving the pig a carefully balanced diet of protein, carbohydrate, vitamins and minerals, rationed out in specific quantities to ensure that the animal will put on lean meat with only the minimum of fat.

What to feed

Tests have shown that wet feeding is preferable to dry. With wet feeding growth rate is improved as is also the efficiency of converting feed to meat, although the carcass quality is not improved. For wet feeding you simply take the dry meal and mix it with water until it is runny. About 3 lb (1.3 kg) of dry meal to about half a bucketful of water is a good mix. Skimmed milk can be used in a ration if it is available. Any feed company will give you a formula for this if you want to make up your own ration.

Feed generally comes in two sorts: meal or cubes. The cube is exactly the same feed as the meal but has been put through a process to force it into cubes. This means that cube feed is slightly more expensive to buy than straight meal. I would imagine that very few people who read this chapter will have their own mill on the farm and they will therefore have to buy in the feed from a feed company. Many if not all of these companies have their own nutritional experts who are more than willing to advise on the correct feed to buy for the pig at each stage of its growth. Rations normally go by protein value and, as a rough guide, this should be about 18 per cent for piglets and about 16 per cent for fattening pigs. Remember too that the pig kept in the best environmental conditions will need less food than one that isn't. A pig kept in a well insulated house will not need to eat to keep warm and so the food taken in will be converted to meat more economically.

Gilts

The gilt will need about $5\frac{1}{2}$ lb (2.5 kg) per day of a dry sow ration. This should be kept up until just before farrowing. After service and during gestation the ration could be slightly reduced to 5 lb (2.4 kg) per day. Do be careful. Remember a maiden gilt is still growing as well as feeding her unborn young, and she needs enough feed to continue to perform these two functions. One school of thought believes that about three weeks before serving, the gilt should have her rations increased to $8\frac{3}{4}$ lb (4 kg) per day. I feel that this is the best policy as it helps to keep the gilt in peak condition. At no time should the animal receive less than 4 lb (1.8 kg) of food per day.

After farrowing

Once she has produced her litter the mother sow has to take in enough food to keep herself healthy and to provide enough good milk for the piglets. The feed should therefore be changed to another type and here again your agricultural merchant will advise you on the best one. If you are going to wean at three weeks then the sow should be fed about 6 lb (3 kg) per day. If you are weaning at five to six weeks then $3\frac{1}{2}$–$4\frac{1}{2}$ lb (1.5–2.0 kg) should suffice. After weaning, the sow must be dried off as soon as possible. There is no need to attempt to do this by withdrawal of food and water as the production of milk in the sow will tail off naturally as the piglets stop taking from her. At this point the sow will be taken off the second or milk-making ration and put back on to the dry-sow ration.

Piglets

When the piglets are just over two weeks old they must be fed on what is termed creep feed. They should be given about one handful every two days in the special creep feeder. As the piglets grow, their appetite will increase and so should the amount of creep feed. The rule here is quite simple: as the piglets empty their feeds, keep increasing the ration. Once the piglets are weaned and put in the weaner pool they should be kept for the first week on the creep feed.

At around ten weeks of age the feed should be changed to a fattening ration, again keeping a high protein content. There are two schools of thought as to how much to feed. Some people go for a restricted ration, others for an ad lib ration. I like to feed ad lib up to about 110–120 lb (45–50 kg) live weight and from there feed a ration of 9 lb (4.1 kg) per day. The ration should be given twice a day, half morning and half evening, and trough space should be such that all the pigs can feed at the same time. This helps to prevent the stronger pigs from pushing the weaker ones out of the way to get more than their fair share of feed. The pigs can be kept on this type of feed until slaughter.

Group feeding from a trough.

How to feed

One method of feeding pigs is simply to put the feed on the floor and allow the pigs to forage for it. Obviously if there is any dung or urine on the floor this system is unhygienic and should not be used, but even if the floor is quite clean there is one other great drawback – waste. Inevitably the pigs will tend to walk in the food and it will stick to their feet. This problem can be reduced if you use cube feed rather than meal, but the cubes do break down when stood on and so you will still get some waste. The system I favour is to put the feed into a properly constructed adjustable feeder which allows you to regulate the speed at which the feed comes out and so prevent all wastage. As the cost of feed accounts for some 80 per cent of the total pig-keeping expenditure, the more you can cut down on waste the better.

A sow being served.

Another method which I have not yet mentioned is swill feeding. Swill is any household waste food such as potato peelings, left-overs from the table etc. which are boiled in water, mixed with meal and then fed to the pigs. There are two main disadvantages to this type of feeding. One is that there is no easy way of calculating the content of waste food in terms of carbohydrate, protein and vitamins – the other is that very stringent regulations apply to the preparation of swill before it can be fed to the pigs. These regulations insist that no waste at all, from any source, which might contain meat can be fed to pigs unless it has first been sent to a properly licensed processing plant. This applies to ordinary household waste as well as to waste from institutions such as schools and hospitals. The reason for the regulation is that some diseases which are shared by humans and pigs can be transmitted through meat. It is therefore essential that you should only feed your pigs on scraps which you are certain contain no meat. In North America the law requires that all garbage or left-overs must be cooked before being fed to pigs. In parts of Australia including New South Wales, swill feeding of any kind is banned.

Breeding and rearing

Signs of heat

The gilt is taken for its first service when it is six to eight months old. Some breeders believe that natural light and a long day length are necessary to promote a strong 'heat', but there is no good evidence to support this. One thing which will certainly bring on strong heat is close proximity to a boar – hence the advice in the *Housing* section to pen your boar where it can have nose to nose contact with the female pigs. Ovulation takes place forty to forty-two hours after the beginning of the heat and the best time for mating is some sixteen hours before ovulation.

When the female is beginning to come into heat her vulva begins to swell and becomes red, and sometimes a white mucus is also apparent. The sow may mount the other sows in the pen. She then makes obvious signs to the boar that she wants his attention and will stand as firm as a rock. This is called 'standing to the boar'.

Serving (mating)

The first service should take place twelve hours after the standing begins and an essential second serving some twelve to eighteen hours later. Sometimes, in larger herds, a different boar is used for the second service if there are any doubts about the fertility of the first one.

Serving should take place in the boar pen or somewhere else familiar to the boar, so that he will get on with his job without being distracted by his surroundings. It is best to supervize the whole serving process so that you can be quite sure that it has occurred successfully. Stockmen on larger farms who are unable to do this have to trust to an inspection of the scratch marks on the female's shoulders as proof that serving has taken place.

Artificial insemination

If you keep only a few sows for breeding purposes, then it will probably not be economic for you to keep a boar and it is here that artificial insemination (AI) comes into its own. The main problem is keeping an eye on the sows so that you will know as soon as they are on heat, and will be able to contact the local AI people straight away. If you fail to spot the sow on heat she will not come on again for another three weeks, and signs of heat will get progressively less pronounced. One of the great advantages of AI of course is that you will be breeding from only the best stock.

Using your own boar

If you decide to buy a boar you are well advised to buy the best one you can afford. Prices can vary by as much as 300 per cent. Go for the boar with the best genetic potential you can get. This should provide good progeny and, if you are going to keep some of the female piglets to use as breeding stock, will help improve your herd.

Boars are generally ready to begin work when they are six months old. At first they should be used only twice a week for two services, and preferably on quiet sows who are well on heat. Over- or under-use of a boar can both produce disappointing results, and once the boar has had some experience four to five regular servings over a weekly period will probably be the ideal.

It is important to ensure that the boar's pen has a smooth floor as a rough one could make him lame and unfit for work. Footbaths of formalin or copper sulphate, and regular check-ups will help prevent lameness. Boars should also be treated with a mange and lice dressing at the same time as the females.

Pregnancy

It is possible to buy pregnancy testers which give a very accurate guide as to whether a sow is in-pig, but they are rather an expensive luxury. Another method of finding out whether she is in-pig is to squeeze her udder slightly and, if she is pregnant, it will secrete milk. This, however, can only be done about one day before the pig is due to farrow so does not help save any of the time which has been lost when she could have been put back to the boar.

Immediately after the service has taken place the sow should be taken back to her pen and left quietly there. She will farrow three months, three weeks and three days (i.e. 115 days) after the serving. Two weeks before farrowing she should be dosed against worms.

Farrowing

Before she is moved to the farrowing house the sow should be scrubbed with a mange and lice wash, but be careful to ensure that you have rinsed off all the residue from the udders and teats. Gilts, who are farrowing for the first time, should receive maximum care and attention.

The floor of the farrowing pen should have a covering of woodshavings or chopped barley straw which will give the piglets added warmth and make the floor a little softer for their knees and feet. Some farmers use shredded newspaper for bedding with very good results, and others have found that a thick rubber matting is effective. Ideal temperatures for the farrowing house are given in the *Housing* section on page 145.

The pig is moved to the farrowing house and (if you are using one) into the crate, about three days before she is due to farrow. This period is essential in order for the pig to become familiar with her surroundings, particularly in the case of a gilt, or of a sow which seems nervous. Once installed she should be left quietly alone, but you should visit her regularly during the day just to ensure that all is well. For twenty-four hours before farrowing and twenty-four hours after, the sow can be given a mash made up of bran and warm water.

On the day that you are expecting the pig to give birth you should keep a close eye on her to make sure that all is going according to plan – how close an eye will depend on you. One of my colleagues makes his last visit at about 11.00pm and then goes to bed. Another stays up all night with the pig until a successful birth has been accomplished and then goes to bed if there is time left, or else simply gets on with the next day's work if there is not. The choice is entirely up to you.

It is often a good idea to remove the pigs as they are born, rub them down and keep them warm until farrowing has finished. Check that the piglets' nasal passages are free of mucus and, once the sow has cleaned everything up and settled down, reintroduce the piglets to her. You can either remove the afterbirth or leave the sow to follow her natural instinct and consume it herself. The piglets will quickly find the mother's teats and begin suckling, and will then be attracted towards the warmth of the creep (see page 147).

Only experience will really tell you whether the birth is normal or not, but if you are in any doubt always call the vet. If you are lucky you may know an experienced pig-man

A sow being suckled by her young.

New-born piglets being kept warm in a box.

who lives just down the road and would come and give you advice, but if it is 2.00am on a cold, wet night he might not appreciate being called out!

Rearing new-born piglets

Suckling

It is essential that new-born piglets should suckle their mother as soon as possible after birth. The milk from the sow over the first few hours is very high in colostrum and contains antibodies which will give the piglets resistance to disease. Their ability to absorb these antibodies is lost after their first few hours of life, and the level of colostrum in the mother's milk will soon begin to decline – all of which underlines the absolute necessity for the piglets to start suckling as early as possible.

One problem to watch out for at this stage is mastitis, as if the sow's teats are infected the piglets may be unable to feed or may pick up disease from them. However if you follow the highest standards of cleanliness and make sure

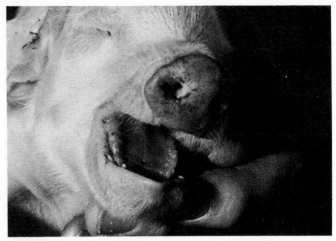

Teeth clipping: 1. *A young piglet, showing teeth in need of clipping.*

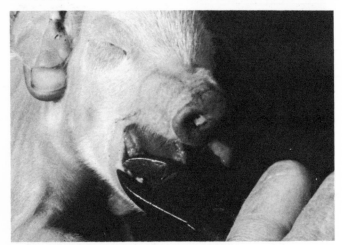

2. *The teeth being clipped.*

the environment is warm you should be able to avoid it.

Fostering

Another problem which sometimes arises is that a sow may give birth to sixteen piglets and not have enough teats for all of them. The easiest way out is to transfer some of the piglets to another sow which has farrowed at much the same time and has had a small litter. The foster sow will accept the new piglets quite happily. If you have not got another sow you can either hand-rear them yourself (see below) or simply allow the litter to take its chance. Obviously if you opt for the latter course of action you are going to have a number of casualties as the piglets will not be able to feed as they need and the weakest will soon die.

Hand-rearing

If you go in for hand-rearing you will need a good deal of time and patience. A friend of mine has taken piglets into her house and fed them from a dropper syringe similar to the type used for eye-drops. The first thing is to ensure that they get colostrum. After that they must be fed every two hours, day and night, with milk taken from the sow or with a sow-milk substitute. They must also be kept warm and free from draughts. My friend puts them into the cool

bottom oven of her stove, but great care must obviously be taken that the temperature is not too hot for the piglets, and of course the door must be left open!

If the piglets survive and grow on they can be put back with the mother after about ten days. This should be done as quietly as possible and when the sow is asleep. All pigs from the same pen have their own particular scent and in this case if the mother smells a stranger she will be disturbed and could quite easily kill the piglet. If she is asleep the introduced piglet has a chance to acquire her scent and thus the trouble can be overcome.

Tending the piglets

Teeth clipping

Once the pig has farrowed and the little piglets have survived, there are still some jobs to be done. Within twenty-four hours of birth the piglets' four canine teeth are generally clipped level with the gum, because they are long and sharp and can cause severe damage to the mother's udder. The job is done with a special clipper which is dipped in antiseptic for each pig. It is quite a painless (though noisy) operation. I would advise the beginner to do it the first time under the supervision of an experienced stockman.

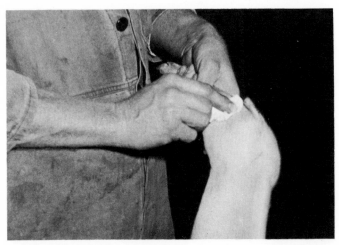

An intra-muscular injection: 1. *Ensuring that the skin area is sterile.*

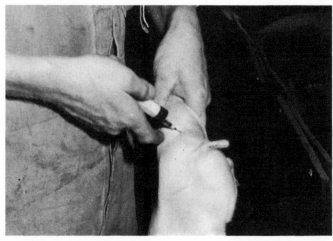

2. *Giving the injection.*

Iron injections

The sow's milk does not contain enough iron to prevent the piglets from getting anaemia, so before they are four days old the piglets should be given extra iron. The most usual method is to give them an intramuscular injection of a proprietary iron medicine, which can be obtained from agricultural suppliers. A word of warning is in order here. There have been numerous reports of salesmen touring farms and selling medicines of all types, some of which are under strength or even quite useless. It may save a little money to buy at the door, but this saving could prove very costly in the long run and my own policy is always to buy from a reputable merchant.

Tail docking

Tail biting is a favourite pastime for pigs and leaves the victim with a painful and bleeding tail. It can also cause fighting among adult pigs, sometimes with fatal results. (For details on tail biting see page 156). One way to prevent the problem is of course to separate the pigs, but if you are not in a position to do this you should dock the pigs' tails within one week after their birth leaving a short stump. This should prevent tail biting when they are older. The two regulations on tail docking are:

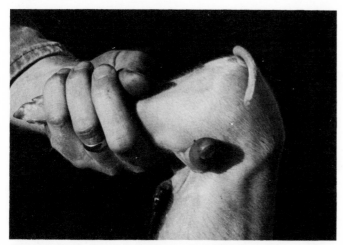

Tail docking 1. *Young piglet before having its tail docked.*

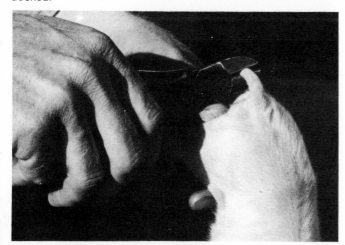

2. *The tail must be clipped neatly and quickly.*

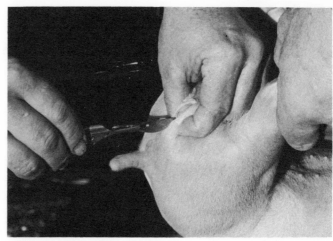

Castrating a piglet: 1. *An incision is made in the scrotum.*

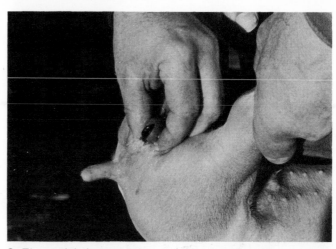

2. *The testicle is exposed.*

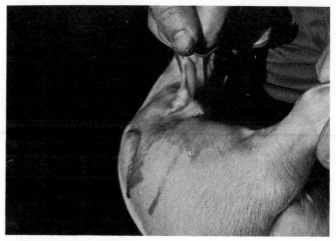

3. *Then it is removed.*

1. The part of the tail to be removed must be severed quickly and completely.
2. Piglets more than seven days old can only have their tails docked by a qualified veterinary surgeon.

Castration

Boar piglets are sometimes castrated, and again this is something which should be done within one week of birth.

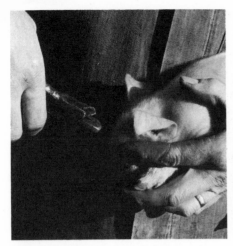

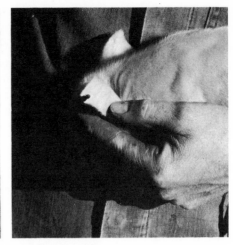

Ear punching: 1. *The piglet must be held firmly.*

2. *The ear being punched.*

3. *After punching.*

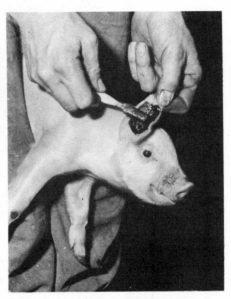

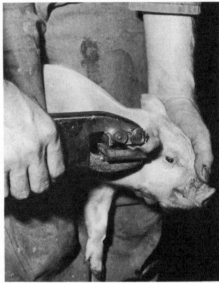

Ear marking: 1. *Ear mark number being selected.*

2. *Ear being painted with dye.*

3. *Number being punched into ear.*

After the age of two months the operation must be done under anaesthetic, and after six months may only be done by a qualified vet. There is a school of thought which holds that the meat of an uncastrated boar has a rather tainted flavour, although as yet there is no scientific evidence to prove this. If you are planning to sell the young boars on to a fattener or to fatten them yourself and then market them, it would be as well to check with your intended buyer to find out if there will be a reduced price offered for un-castrated boars, as some buyers make a fairly big deduction for boar meat.

There is to my mind one very valid argument against castration, and that is that it does set the pigs back, even if the operation is performed when they are very young. A pig which has been castrated soon after birth loses about two weeks' growth by comparison with its uncastrated brothers from the same litter.

This operation is another one which I would advise the beginner to carry out only under the close supervision of an experienced stockman for the first few times. It is of para-mount importance that the boar piglet should be caused the least possible discomfort.

Ear marking

If you are rearing pigs for breeding purposes you may want to identify them by tattooing, tagging, punching or clipping their ears. This too should be carried out as soon as possible and again you should take care to avoid causing any unneces-sary pain or distress. The best age to do it is at about two weeks old, and certainly before weaning.

Weaning

For many hundreds of years pigs have been left with their mothers until they were weaned naturally. However in these cost-conscious days it has become important to get the sow pregnant as soon as possible after she has farrowed, with the aim of producing an average of two and a half litters a year. The following table illustrates the economic reasoning for this change in policy towards curtailing the weaning period:

Days to weaning	18–22	32–38	56
Days in gestation	115	115	115
Days in lactation	20	35	56
Days between weaning and next conception	10	8	7
Days between farrowing	145	158	178
No. of litters per sow per year	2.5	2.3	2.0
No. of pigs reared per litter	9.3	9.5	9.5
Pigs sold per sow per year:	**23.2**	**21.8**	**19.0**

From this it can be seen that weaning later means more piglets per litter survive, but it of course allows fewer litters per year, and the bottom set of figures proves the point that in fact early weaning does produce the highest number of pigs in the end. However, even though these figures prove me wrong, I still hanker after a later weaning date than three weeks and would certainly strongly advise the non-expert to go for a longer period than this until he feels confident enough to start reducing it. These figures are taken from the very best herds under the most expert management, and you must remember that the earlier you wean the more important the stockman's care and expertise become.

The following table gives a guide to the weights you should aim for in your piglets over their first two months:

Birth: 2.9 lb (1.3 kg)
1 week: 6.0 lb (2.7 kg)
2 weeks: 8.8 lb (4.0 kg)
3 weeks: 13.2 lb (6.0 kg)
5 weeks: 24.2 lb (11.0 kg)
8 weeks: 44.0 lb (20.0 kg)

Pigs for meat

After all the work and worry the day eventually comes when your pig is ready for slaughter. Slaughter weights are generally as follows:

Pork   140 lb (63.6 kg)
Cutter 180 lb (81.8 kg)
Bacon   220 lb (100 kg)
Heavy   260 lb (118.1 kg)

I prefer to slaughter at 180 lb (81.8 kg) as at that weight the pig is not carrying too much fat and the killing-out percentage (i.e. the ratio of meat to bone and offal) is good – about 75 per cent meat.

Slaughtering

Before they are slaughtered pigs should not have been given any drugs or medication for a specific period, and you should check up on the regulations on this in your area. Slaughtering should always be done by stunning the pig first and then cutting its throat. (In Britain this is enforced by law.) The best way is to use a humane killer in the form of a special type of gun. The muzzle is placed against the pig's forehead and a steel spike is shot into its brain bringing instantaneous death. Another method is to hold a steel spike against the forehead and strike it one sharp, heavy blow with a large hammer, but I cannot recommend this system as there is grave danger that one blow may not be enough. You could then be faced with having to try a second blow on a pig that is maddened by pain – a very nasty prospect indeed.

Once the pig is dead the jugular vein is cut through, allowing the blood to pour out. If at all possible, a block and tackle should be rigged up in the killing area before the pig is killed so that when it is dead the animal can have its back feet tied to the tackle and then be hauled up until it is clear of the ground. The throat is then cut and the blood can be collected for use later on. The whole operation must be done with the utmost speed.

When the pig has stopped bleeding a very sharp knife is inserted in the throat and a cut is made the whole length of the body. All the innards, including the intestines, heart, liver, lungs and stomach, are then extracted, leaving a clean carcass. With the help of a butcher's cleaver the head is severed from the body and the carcass is split into two halves. Remember that with a pork pig you are dealing with 140 lb (63 kg) and with a cutter pig 180 lb (82 kg) dead weight. These are heavy weights for the average man and it may well be best to seek help at slaughter time.

Using the carcass

It is said of the pig that when it is slaughtered nothing is lost except the squeak, and this is almost true. The blood can be made into black-pudding; the intestines can be cleaned, boiled and used for sausage skins; the head can be cooked, cleaned and made into brawn; the heart, liver and kidneys are used as meat. Only the lungs and stomach are of little value.

Before it is cut into joints the outside of the carcass must be cleaned and scraped using boiling water, to remove any dirt and the pig's bristles. From here on I would recommend that you use a butcher to do the jointing. I myself use a local butcher for slaughtering, preparing the carcass and cutting it into joints. It is not expensive and as he has the staff to do the job properly from slaughter to end-product I feel it is the best policy. It would be advisable to check with your local butchers to see if they provide this service.

From your pig you should get between 70 and 75 per cent of the live weight as meat and sausages. In other words from a 180 lb (82 kg) live pig you should get between 126–135 lb (57–62 kg) of usable meat.

If you want to have your pig used for bacon you will have to find a curer. Some curers will take the pig and do the whole job from slaughter to end-product, others only take the carcass and proceed from there. Curing is a fairly skilled job and I feel it is better not attempted by the amateur. From a 180 lb (82 kg) live pig you can expect about 35 lb (16 kg) of bacon, plus of course two hams, meat joints, sausages and the usual offal.

It is an offence to sell meat for human consumption unless it has been passed by a veterinary inspector. If you are using your pig-meat for your own consumption then you are in the clear, but if you want to dispose of it outside your immediate family you must get the carcass passed by a qualified inspector.

Health and disease

No matter how hard you try to create the best environment for your pigs you are almost sure to come across some illness in the end. My two basic pieces of advice in this situation are:

1. If in doubt call a vet. It may at first seem expensive but if you economize on veterinary advice and end up with a dead pig the expense will be far greater.
2. If you are giving your own medication, be sure to buy it from a reputable source.

You will learn to cope with various diseases by experience, but there is a good deal of truth in the old proverb which says 'experience keeps a hard school, but fools refuse to learn in any other' and I would always advise you to seek expert advice when dealing with any medical problem for the first time, or whenever you are not entirely confident in your own ability to cure it.

General problems

Maintaining the well-being of your pigs may sometimes involve more than simply ensuring that they are free from disease, and the following three problems in particular are ones which the pig owner may have to face and which he should know how to tackle.

Fighting

Sometimes a pen of pigs kept under the best conditions will suddenly turn on one or two members of their group and savage them. This can happen at any time and so far no one knows why. You will soon know if fighting is going on in a pen as the attacked pig or pigs will scream in the most awful way. If you do hear this screaming *go at once* to the pen. To delay for any length of time could prove disastrous as the attacked pig will be killed.

Once you get to the pen you will easily identify the pig under attack, not only by the noise it is making but also by the severe scratch or bite marks made by the other pigs along its shoulders and sides. Once you have identified it,

get it out of the pen at once. You may be on your own or you may feel daunted by the task, but it is essential for you to get the pig out if it is to survive. If you are able to get a neighbour along to help, do so, but while you are awaiting help you must go to the pen and prevent the attackers from getting at their prey. A savaged pig, if found in time and removed from the pen, has a good chance of survival. Do remember though that once a pig has been taken out of its pen it can never be put back again. After it has spent even just a few hours out of the pen the rest will consider it an interloper and will simply attack it.

Tail biting

This can be a major problem. I have seen pigs with the whole of their tails bitten away leaving only a bleeding open wound where the tail joins the body. A pig which is having its tail bitten will lose performance and may be attacked more severely by the others in the pen and even be killed. When you see a pig that is having its tail bitten simply paint a weak tar-oil onto the wound. The other pigs will then be put off by the taste and smell and will leave the wounded pig alone. In many cases it may be necessary to rescue the pig as described in the paragraph on *Fighting* above. Tail biting only develops if the pigs are bored, which usually happens when they have access to plenty of food without enough effort or exercise. If your system leads to tail biting you should try to provide more interest for the pigs, or consider docking their tails (see page 153).

Heads stuck through bars

When a pig gets its head stuck between the iron bars of the pen gate it is obviously essential to free it otherwise it will die. With small pigs it is fairly easy to lift the animal up with one hand, twist its head sideways and pull it out. The pig will squeal like the devil, but once freed will suffer no after effects. With larger pigs the process is rather more difficult and may be a two-man job. The only way is to turn the pig's head on its side and pull. A liberal smearing of oil, grease or washing-up liquid on the head will be a tremendous help as it will act as a lubricant. With very large pigs you may have to distort the bars rather than manoeuvre the animal.

Diseases

Aujeszky's disease

This can become a major problem, or even a disaster. It strikes at the sow and her piglets. The sow will abort and any young pigs which catch the disease will die very quickly. There is a preventive vaccine available in some countries. In Britain however it is at the moment an offence to import it, and British farmers are greatly worried by Aujeszky's disease, which is decimating a number of herds.

It is a notifiable disease and anyone who knows or suspects that a pig in his possession is suffering from it must keep all his pigs on his premises until a veterinary inspector has inspected them, and given advice on control measures and assistance.

The disease can occur in cattle, sheep, dogs and cats as well as in other animals, but such cases are comparatively rare. The first symptoms are signs of general nervousness. Sucking piglets suffer from incoordination and convulsions, and if they are under two weeks old they generally die. Older piglets often recover. In fattening pigs weighing about 60 lb (27 kg) and over the symptoms are different. Only one or two pigs in a pen are generally affected and they appear dull and stupid, stop eating and drinking, are constipated and show signs of nervousness, trembling, incoordination, circling and head pressing. They usually spend most of their time lying down. However they do generally recover and deaths are uncommon in this type of pig.

In the case of sows, nervous signs are generally missing. They go off their food and water and become very dull and constipated. After about four days they appear to recover, but may abort some ten or twelve days later. If the piglets are born at the correct time some may be black and mummified.

The infection is usually transmitted to a herd by pigs which carry the disease. It may also be introduced by people who have been in close contact with, or lorries which have transported, affected pigs. A very high standard of hygiene is of paramount importance in trying to stop the disease entering an unaffected herd.

Cramp

Some pigs may develop cramp, which is similar to human cramp. The muscles of the legs (generally the hind ones) seize up and the pig is unable to walk properly. In fact it may only be able to get around the pen by dragging itself along the floor by its front feet. In my experience this, although incurable, is not a serious ailment and the pig will normally grow quite well. However, there is a danger that if it is in a pen with other pigs they may turn on it and savage it, sometimes fatally. My advice is to take the pig out and put it in a pen on its own.

Erysipelas

This disease is fatal if it is not treated in time. The first symptoms are a form of raised pattern on the back and sides of the pig, which can be felt by sliding the fingers over the pig's body. These raised patterns then become pink and square- or diamond-shaped. Within a few hours the pink colouring turns to purple and within forty-eight hours the pig will be dead.

Most pigs carry this disease but it only seems to break out under certain conditions – particularly in hot, sticky weather. Heavy feeding, such as that often given to the pig when it is a few days from slaughter, can also bring out the disease.

Treatment is quite simple and a shot of penicillin will effect a cure. Better than this, however, is to eradicate the disease from the pigs by a vaccination programme. Young piglets should be vaccinated when they are three months old. If they are being fattened for meat this should last them until they are slaughtered. A gilt which you are keeping for breeding purposes should be vaccinated when she is three to four months old and then given a booster once a year thereafter. The easiest policy is to give the vaccination regularly about two weeks after she has farrowed.

Foot and mouth disease

This disease, which attacks cloven-hoofed animals, is highly infectious. Infected animals become feverish and have ulcers in the mouth which make them salivate very freely, and ulcers on the feet which cause lameness. It is not generally fatal, but the animal loses production rapidly and the disease can spread over large areas in a very short period of time.

Foot and mouth is a notifiable disease and the authorities must be told as soon as any case is suspected. Control is by a rigid policy of slaughter and burial or burning, and a quarantine standstill order is placed on the premises. The standstill order means that nothing may move off the infected premises for a period of several weeks. There are vaccines available in some countries (but not in Britain) which help fight the disease.

Lice

Biting lice live on scurf and hair, and sucking lice live on blood. Lice cause intense irritation and may give rise to large areas of bare skin on the pig. Treatment is by a lice wash which you can buy from your vet or agricultural suppliers. While lice are not dangerous, the pig will lose condition and not perform well if left untreated.

Mastitis

This is an inflammation of the udder, usually caused by bacteria which get into the udder tissues through bites from the young pigs' teeth or through the milk channels in the teats. The chief sign of the disease is a painful swelling of one or more of the milk glands or of the whole udder which, when the sow is producing milk for up to a dozen young pigs, is a very large mass extending over the whole area of her belly. The sow is likely to go off her food, run a high temperature and neglect the piglets. Treatment is by antibiotic and anti-inflammatory injections. The occurrence of the disease can be reduced if the sow's sty and bedding are kept very clean and if the little pigs' canine teeth are cut off as soon as possible after birth (see page 152).

Meningitis

A pig suffering from this will be very obviously ill. It will not eat and will simply lie on the floor. If it does get up it will move around in small circles with its head on one side. An injection of antibiotic will produce a rapid cure. It is given intramuscularly, generally in the shoulder.

Pneumonia

This disease is the same in pigs as in humans. The pig goes off its food, has a high temperature and breathes very quickly and shallowly with a kind of pumping action which can be noticed when the pig is lying down. An antibiotic injection will help and will probably save the pig if it is given early enough. It is best to remove the pig from its pen

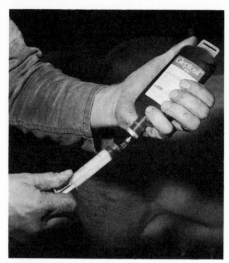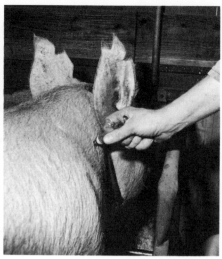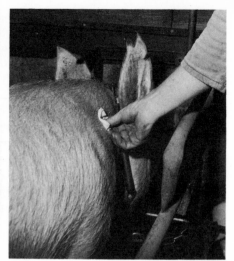

A sow being given an intra-muscular injection.

to a clean, warm, draught-free place, but remember that once it has been taken out it can never go back to the pen.

Scour

Scour, or diarrhoea, mainly occurs in young pigs up to about 100 lb (45.5 kg), although it can occur in a pig of any age or weight. The sign of scour is liquid dung, and you will easily notice this, either in the dunging area or on the hindquarters of the pig. You may also of course actually see the pig passing liquid dung. If you see signs of scour in the dung passage check all the pigs in the pen by looking at their hindquarters for signs of staining, or look for any pigs with thin (rather than fully rounded) flanks. In young piglets the disease can be fatal and will certainly set them back if it is not diagnosed and cured. There are many causes including bad management, poor hygiene, sudden changes in feeding or infectious diseases such as swine fever, gastro-enteritis, bowel oedema, paratyphoid and worms. Reduced feeding and a course of antibiotics will cure some cases, but if you are in any doubt consult your vet.

Swine fever

This is an infectious disease which only occurs in pigs. It is notifiable and if it breaks out the pigs must be slaughtered. Symptoms are a high fever, nervousness, lack of appetite and a catarrhal discharge from the eyes. In its acute form a grey, evil-smelling diarrhoea then develops and the animals become terribly thirsty. Some parts of the body may show a reddening of the skin. A pig with swine fever rapidly loses condition and dies within a few days. The infection can be transmitted from pig to pig, or by the stockman, who can carry the disease from one pig to another.

Swine vesicular disease

This is a highly infectious disease and can decimate a herd quite quickly. It is nearly always caused by feeding the pigs with swill which has not been properly treated (see page 150). It is a notifiable disease and immediate slaughter followed by burning or burial of the carcass is the only answer.

Transferable gastro-enteritis (TGE)

This is a disease which generally attacks small pigs, and it it highly contagious. The piglet scours, soon goes off its food and gets extremely thin. If not treated it will be dead within hours. Treatment is by intra-muscular injection. Medication can be obtained from a vet or agricultural supplier. If your pigs do develop TGE you must take the following precautions:

1. All footwear and car wheels must be thoroughly disinfected. This applies to everyone who goes anywhere near your pigs and to any vehicle coming onto your farm. Most pig farms put a concrete trough full of disinfectant in the drive-way so that all vehicles entering or leaving the premises must drive through it.

2. Do not allow anyone (not even the family) to go to the pig buildings unless absolutely necessary. Any people who do have to visit them must wash their footwear thoroughly in disinfectant before entering your premises, and again on leaving.

Worms

These parasites take their nourishment from the pig, which means that you will be paying for food which the pig then passes on to the parasite. The more serious result is that the carcass of a pig which has suffered from worms will be condemned as unfit, so it is very much in your own interests to keep your pigs free from the condition.

There are several wormers on the market, each proclaiming its own advantages. A reputable merchant or vet will prescribe the one you should use. I prefer the kind which is simply mixed with the pig food so that the pig takes it in quite easily as it feeds. The sows and gilts should be wormed about one week before they are due to farrow, and the young piglets wormed at around 50 lb (23 kg). This should suffice, unless they are put out to graze in which case it may be advisable to worm again about two weeks before slaughter.

CATTLE

Background

Not so many years ago man raised thousands of cattle on the pampas of Argentina and the coastal plains of the Gulf of Mexico and the seaboard of California solely for their horns and hides. They were too wild to milk. There was no way of shipping the meat. It is sad to reflect that in some areas we have progressed so little that even today some wild species are endangered because their horns or tusks or antlers find a ready market. The rest of the animal is left to rot.

Those of us who keep cattle today do so because they provide meat, or meat and milk. The hides and hooves are still used, as are the horns of the minority of cattle who have them. Most of us have become far less wasteful of our heritage and animal resources. Some of us who keep cattle in suburban areas even collect the manure and sell it at the roadside to enthusiastic gardening neighbours who don't have the space or the inclination to keep cattle.

In developed countries only a very small minority of people keep cattle as beasts of burden although in other places thousands of cattle and their close relatives are kept as working animals. Possibly we are as bound by progress as people in those countries are bound by tradition.

There is yet another reason for keeping cattle. We like them. And they like us. Many make admirable, responsive and intelligent pets.

So why don't we all have a cow? First, they are relatively expensive. The novice can easily embark on a small poultry venture without a visit to the bank manager, but deciding to keep cattle involves a major financial outlay. Many of us don't have the pasture or the sheds – in many places people reckon you should have an acre (half a hectare) for each cow. In some parts of North and South America and Australia the figure is closer to 40 acres (16 hectares) a cow. Then too, cattle are large creatures. Some are just plain awkward. Some can be downright dangerous. If you define a dangerous animal in terms of your chances of surviving an encounter with it, then the Jersey bull must rank as one of the most dangerous of all domesticated creatures. Even a small one can weigh a ton (almost 1000 kg)! Much of that weight is concentrated in its lethal head and shoulders. One idle swing can render the strongest man unconscious; a couple of purposeful butts, a bit of a stomp and the gentleman has become a statistic. But, of course, few of us these days keep a bull. More of that later in the *Breeding* section.

The improvement of cattle

The history of some of the common British breeds is both interesting and instructive, and is a good guide to the ever-changing complexion of the world's cattle population.

In the eighteenth century the only 'breed' commonly recognized throughout Britain was the Longhorn. It was as common as the Friesian is today. It was essentially a triple-purpose animal used for milk, meat and draught. The first farmer who tried to improve the type by scientific breeding was Robert Bakewell who was born in 1726. The British have always been carnivorous. To cater for this demand he selected the best beef types of Longhorns and by breeding father to daughter and brother to sister and so on, fixed the type. In the process much vigour and vitality as well as some of the ability to produce milk was lost, but the resulting animals were undoubtedly better for beef. His methods were quickly taken up by other farmers. They tended to concentrate on the Shorthorn and were so successful that for almost 150 years it dominated the British cattle industry. Two definite types were evolved, the beef and the dairy. The best beef types were not really good milkers but even the best dairy types produced acceptable beef – although of course neither as quickly nor to the same high quality as the beef types.

Meanwhile both the Hereford and the Aberdeen Angus were being improved in the same ways by other enthusiasts. They were, as they are today, strictly beef animals. That is not to say that one cannot milk a Hereford or an Angus – lots of people have done and will continue to do so. But if you want lots of milk then they are not the best choice.

It is astonishing how much influence those three breeds have had on cattle breeding throughout the world. Herefords were first exported to North America in 1817, to Australia in 1839 and the Argentine in 1858. Shorthorns made their mark about that time and the Angus a bit later in all those countries. They were so successful in these countries that beef was exported back to the insatiable British, who in their turn began to take a renewed interest in milk production. The Channel Island breeds, the Ayrshire, the dual-purpose Shorthorn and the Red Poll were not able to fill the gap and the influence of the Holstein-Friesian began. Although the breed society was only formed in 1909 and had few members, today it has almost 14,000 members in Britain alone. The Dutch breed dominates Britain and the dairy industry of much of the world.

During the last twenty years there has been a remarkable development in the beef industry. Continental breeds which had formerly been excluded from Britain and North America and Australia because of the fear of foot and mouth disease and possibly the lack of any compelling demand, are now assuming a more important role everywhere. A few years back one could drive through the countryside and say with confidence, 'There's a herd of Herefords (or Angus, or Friesians)'. Today even an experienced farmer will say, 'What kind of bull have they used on that lot of cows? That calf over there looks like a Charolais, yet that one looks like a Limousin.'

Breeds

Aberdeen Angus

This breed is almost a synonym for quality Scotch beef and its breed society is now the biggest in the world. Early maturity and a high proportion of high- to low-priced cuts are its great virtue, and the marbling of its fat is an aspect of the breed keenly appreciated by butchers the world over. It is also hardy, adaptable and generally long-lived. Towards the end of the last century Hugh Watson (who is considered to be one of the founders of the modern breed) had a cow called Old Granny, who produced twenty-nine calves before her life was ended at the age of thirty-five by a stroke of lightning.

The main reasons why they are not popular on small units are that they produce insignificant quantities of milk (just enough to rear their own calves) and, to put it mildly, docility is not a characteristic of the breed. A calf barely able to stand is quite likely to try to move a couple of big men forcibly out of its area. The remarkable thing is that it can sometimes succeed. However it must be said that much, of course, depends on how the cattle are handled and in the right hands I daresay the Aberdeen Angus may be as docile as any other breed.

Semen from Angus bulls is widely used on dairy cows on smallholdings because:

1. The calves are small at birth and seldom cause calving problems even in the smallest heifers.
2. The polled characteristic of the breed is always transmitted. It is a distinct advantage to have calves that don't need to be dehorned.
3. The colour too is often transmitted (either all black or, in some strains, all red). This easily recognizable characteristic usually means a premium in the auction ring.
4. The beef production potential of the Angus will often be transmitted even through a scrawny creature like a Jersey.
5. The hardiness and vigour of the Angus always seems to come through.

Angus cows may not be the small farmer's best bet, but semen from Angus bulls is odds-on favourite and they are very useful in a cross-breeding programme.

Ayrshire

This is the only native British dairy breed. Although many are white and red or brownish-red like mahogany, some are white with a few red spots over the neck and head and a bit of body speckling. The practical Scots laid the foundations for this breed in about 1750 and, like good horsemen, declared that 'no good cow is a bad colour'. They wanted to achieve good milk production despite inferior food and inclement conditions. For some reason they promulgated the unique horns which sweep sideways, upwards and backwards. Describe them as you will they are awkward and dangerous encumbrances and even enthusiastic breed supporters practise debudding as early as possible.

This breed has many advantages for the small farmer. It is incredibly tough. It has proved itself in conditions varying from the sub-arctic to the tropical. It produces lots of milk of extremely high quality, and it does it on varying diets. Although its calves will never take any beef championships they are superior to Jerseys and Guernseys in that they fatten earlier and don't have the yellow fat. But the real reason you should consider the Ayrshire is its longevity. Most cows are productive for eight or ten or twelve years. Many Ayrshires are merely stretching into healthy middle age at twelve and go on producing good calves and lots of rich milk into their late teens. People who consider their cow as a member of the extended family might consider this a characteristic that cannot be measured in money.

Beefalo

There have been several sorts of buffalo crosses. Most have turned out to be sterile or not worthy of promulgation. Ideally, they are $\frac{3}{8}$ American Bison, $\frac{3}{8}$ Charolais and $\frac{1}{4}$ Hereford. They have small calves weighing under 65 lb (30 kg). They are vigorous, hardy and efficient beef producers and do well on low-quality roughage. Canada has done a considerable amount of experimental research on this breed but this has now been discontinued.

Blonde d'Aquitaine

Quercy, Garonne and Pyrenean strains as well as Shorthorns from Britain have been used to produce this breed whose progeny still varies from the ideal. Mature cows weigh about 1600 lb or 750 kg. Average calves at birth weigh 104 lb or 47 kg. In colour they vary from near white to wheat-coloured to light brown. Outside their native France they have not made a great impression except with optimistic importers. No one claims they can produce appreciable quantities of milk, and most beef producers think that other breeds can do the job better.

Brangus

This is another new American breed – developed by a judicious cross-breeding of Aberdeen Angus and Brahman. They have the heat- and insect-resistance of the Brahman and the superior marbled beef of the Angus – and of course they are polled. Like pure Angus they are easy calvers. The bulls can safely be used on females of any breed including maiden heifers. The only thing to watch with the Brangus, as with any other Brahman cross cattle, is their temperament which can vary from the most docile to the wildest – depending of course in part on how they are handled.

Brown Swiss

The Brown Swiss is another dual-purpose breed which has found ready acceptance in areas as diverse as western

Aberdeen
Angus

Ayrshire

Blond
d'Aquitaine
bull

Charolais

Brown Swiss

Canada and tropical central America. Its distinctive, uniform greyish-brown colour and squarish body dominate much of eastern Switzerland in the summer time. They are usually housed during the winter except for brief exercise periods at midday. Their average milk production is about 8800 lb (4000 kg), and the average weight gain of steers about 2 lb (1 kg) a day. They are no longer widely used as draught animals. Archaeologists claim that its predecessors lived in the area as early as 1800 BC. A beef type of Brown Swiss is now being developed in western Canada and will be used in cross-breeding programmes.

Charolais

Until comparatively recent times this breed was hardly known outside the fertile provinces of central France,

where it is considered the prime beef breed. To this day some French farmers fatten Charolais cows indoors for three years! Their prime virtue as beef producers is that they don't lay on a lot of unsaleable fat just under the skin. The fat tends to marble uniformly throughout the high-priced cuts.

In Australia the Charolais enjoyed a boom from about 1969 to 1974 when high prices were paid for pure-breds imported from New Zealand, but some breeders were put off them because of their tendency to have difficult calvings. Today they are widely used throughout the world to improve the beef quality of other breeds. Their medium-length, soft hair – either white or cream – and their short, thick heads and dark eyes are, like the distinctive head of the Hereford, considered a sign of quality and command a premium everywhere where people demand good beef.

Chianina

This is the largest breed of cattle anywhere and is about twice the size of a Jersey. The males average 5 ft 6 in (1.7 m) at the withers and can easily look over the head of a standing man. The cows are only slightly shorter. The males weigh almost 1½ tons (1400 kg) and the cows about 1800 lb (800 kg). Fortunately they are gentle giants. The famous bull Dunetto who in 1955 was acclaimed the largest bull ever, weighing 3834 lb (1740 kg) and measuring 5 ft 11 in (1.8 m) at the *withers*, was reputed to be as gentle as a lamb, although some of the Chianinas bred in Canada have proved less docile.

The breed was largely developed in Tuscany, an area of highly intelligent and, wherever possible, highly intensive agriculture. To this day they can still be seen pulling carts and heavy farm machinery but enthusiasts have started now to concentrate more on their beef potential. They are good animals for the ever-expanding hamburger trade, although they do not fatten easily. Because of their heat tolerance (attributed to their dark skin and smooth white coat) they have found a ready market in hotter climes abroad. In Australia they are crossed with Brahmans to produce a leggy, efficient and extremely hardy breed of cattle. Small units in many different areas have found them ideal beef animals in that they are docile and don't put on a lot of unwanted fat even when they reach their enormous maturity. This one is certainly a breed to impress the neighbours.

Devon bull

Friesian

Devon

This is a hardy beef breed, docile and an easy calver. Devon cows have distinctive cherry-red coats – hence their nickname of Red Rubies – and have been developed for centuries for production of high quality beef. The Pilgrim Fathers took these red cattle with them when they sailed to America. A century later many were exported first to Tasmania and then to the rest of Australia. In the early days of settlement on Australia the Devon was widely used as a draught animal. Its docility and stamina made it valuable as a member of a team of twenty or more, hitched in tandem and two abreast, hauling waggonloads of wheat, wool and timber.

Their prime qualities are early maturity and the ability to fatten on grass. The polled variety now has its own register. It has only recently been stabilized following the use of, among other breeds, red Galloways, red Aberdeen Angus and North American Poll Devons. Both sorts of Devon cattle can easily fit into a small beef herd in many climates.

Dexter

It is thought that this diminutive breed (mature cows of 500 lb (225 kg) or less are not uncommon) was developed by crossing red cattle from north Devon with the native Kerry of Ireland. They do produce reasonable amounts of milk

and acceptable beef. However, it has become quite a rare breed because some individuals are of uncertain temperament and, more significantly, an undue proportion of calves seem to suffer congenital abnormalities like 'bull-dog' heads. The Dexter is either all black or all red. Ideally it is a compact, short-legged, dual-purpose animal. One would have thought an enthusiast with a bit of money, a bit of land and twenty years of scientific breeding could revive the breed.

Droughtmaster

Zebu cattle have been used in Australia since the 1830s but it took almost a century before they were tested on a scientific basis. The modern breed is considered to be 50 per cent Indian and 50 per cent Shorthorn. They are red in colour and carry a small hump. Docility, resistance to insects (due to their loose hides), allied with a high fertility rate and the ability to withstand drought yet respond to high levels of feed, are the characteristics of this aptly named breed.

Friesian

(Holstein-Friesian; Black and White; Dutch Friesian)
If the world was run by computers this would be the only dairy cow ever kept. Certainly this enormous breed – bulls

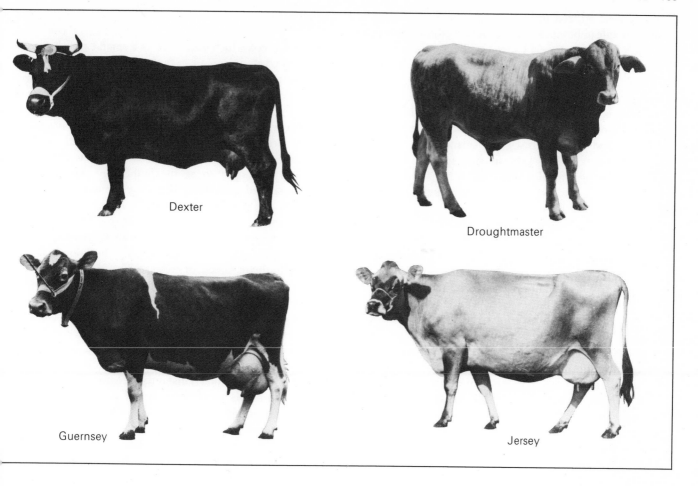

Dexter

Droughtmaster

Guernsey

Jersey

average over a ton (1000 kg), cows about 1500 lb (700 kg) – dominates the dairy scene in its native Holland and in most of the English-speaking world. In Britain, for example, 90 per cent of the total milk supply is provided by Friesians. And to prove that it is an effective dual-purpose animal, figures show that 60 per cent of British beef production comes from dairy herds which are 80 per cent Friesian.

Even among professional dairy farmers there is some slight confusion about classifying the breed. Large black and white cattle have been bred in Holland for a couple of thousand years. As early as the seventeenth century there were exports to North America and to Britain. The first pure-bred Holstein-Friesian herd was established in Massachusetts in 1852. At about the same time in Britain black and white herd books were kept and selective breeding started in Canada.

The three types (or four if you agree that Canadian Holstein-Friesians have made their mark) vary but slightly. The Canadian and American breeds tend, like the people in those countries, to be larger and possibly more rugged than their European cousins. All four types contain some cows that can produce 80–100 lb (40–50 kg) of milk a day, although the average recorded 306–day lactation in Britain is 10,000 lb (4,550 kg). Butterfat averages about 3.5 to 4 per cent compared with Jersey's 5 or 6 per cent.

All in all this is really the commercial farmer's cow. People who keep only one or two cows may find them too big,

too awkward, too demanding and, anyway, who needs all that milk when the kids are screaming for ice cream?

Galloway

The Galloway is a polled black but occasionally dun-coloured breed not unlike its distant cousin, the Aberdeen Angus. It is claimed that the Galloway has never been crossed with outside blood and hence is the purest and oldest beef breed in Britain. This may or may not be true. It is true that they are extremely hardy animals with thick coats and tough constitutions. Although in good conditions they don't finish as early or as well as the Angus, breed devotees declare they do better in bad conditions even on the coarsest grasses.

Guernsey and Jersey

These two distinct and separate breeds originate from two tiny adjoining islands in the English Channel. It is thought that the original stock came from France many centuries ago but was modified by island inbreeding and possibly British modes of husbandry, allied to the inhabitants' need to utilize every bit of land. For the last couple of centuries outside breeds have not been allowed to enter the islands. Hence they may be the purest pure-bred cattle in the world.

They are remarkable in many respects. Both breeds are

universally recognized and respected. The females must be the most docile of bovines. What other breeds can be invited into one's home to share a banana? The aggression seems to reside in the male – the novice cattleman is well advised not to try to keep a Channel Island bull.

Both breeds reflect the influence of the climate of their native islands. This can be tropical at the height of summer but most inhospitable in the winter, which means that both the Jersey and the Guernsey transplant easily to other places. However, when kept in intemperate areas like Saskatchewan, they are almost invariably housed during the dreary months when the outside water troughs are frozen right down to the bottom.

They also reflect the islanders' meticulous utilization of land. It is not unusual to see a Guernsey cow in milk tethered in a fallowing garden patch between a tomato-laden greenhouse and a row of potatoes. A few hundred steps away there may be a small herd of three barren cows or in-calf heifers browsing on the sparse seaside herbage.

Neither breed produces a lot of milk if one thinks in Holstein-Friesian terms. But the milk they produce is very rich indeed. Devotees declare that a good cow should produce ten times her own weight in milk each year and that the milk must contain at least 5 per cent butterfat.

The average Jersey weighs about 900 lb (400 kg). Guernsey cows are bigger, about 1100 lb (500 kg). Apart from size, the other main difference is the fact that the muzzle of the Guernsey is cream-coloured whereas that of the Jersey is dark.

Guernsey people often enthuse about the golden colour of the milk, cream, butter and ice cream. They also boast about the Golden Guernsey donkey which apparently no longer exists and the Golden Guernsey goat which does but is hard to find. The outside observer might find it difficult to tell the difference between Jersey and Guernsey milk.

All butchers and many knowledgeable members of the public are distinctly unenthusiastic about the golden or yellow coloured fat in the beef. Indeed it is usually considered a sign of inferior meat, and that is the major drawback of the breeds – they produce indifferent beef very inefficiently.

If you have only one or two cows you may consider asking the artificial insemination man to use Aberdeen Angus semen for the first calf and Charolais for subsequent calves. The cross-Angus calf will probably be smaller and hence cause less trouble at first calving. Both sorts of crossings will produce acceptable beef animals. Many smallholders however want to build up their small herds and use Channel Island semen and gamble that they will get a heifer. The bulls are not worth much. You must debud the calves when they are a few days old unless you use an Angus or other bull which transmits hornlessness.

This is possibly the ideal milch cow for the small farmer – docile if not downright friendly, hardy, able to cope with wide variations in climate and food, widely known and widely available. It produces high-quality milk but inferior meat.

Hereford

This beef breed which originated in a small English county has stamped its image all over the world. The distinctive white, woolly face is passed on from generation to generation and in beef markets everywhere is considered a guarantee of quality. It is vigorous yet docile, early maturing but productive for many years. Herefords were originally bred as draught animals and only slaughtered for beef when their working life was finished. One Benjamin Tomkins is thought to be the founder of the breed. In 1742 he started selectively breeding for beef allied with hardiness. He was little concerned with draught or milk. His son and others followed his policies. The breed has made its greatest mark in the American West where it has largely supplanted its predecessor, the buffalo. It is also the predominant breed in western Canada with its long, cold winter – no other breed thrives as well on grass alone in sub-zero temperatures. Where better conditions are provided either by nature or by man other breeds compete successfully with the Hereford. In most places it is not the ideal choice for the small farmer unless he is running a calf-rearing operation and selling the weanlings through an auction. Hereford blood is instantly recognizable and almost always commands a premium. But if you are fattening or finishing you might find other breeds suit you better.

There is a polled variety which in all respects is equal to its cousins born with horns. The Australian Poll Hereford is today one of the most numerous of the pure breeds in Australia and it is interesting to note that in October 1979 the council of the Royal Agricultural Society of Victoria decreed that, from a date to be selected in the future, no horned Hereford cattle would be shown at the Melbourne Royal Show.

Highland or West Highland

In parts of western Scotland and the neighbouring isles some people call them Kyloes. Whatever you call them they must be the most instantly recognizable breed of domestic cattle anywhere. Their shaggy coats and enormous horns are unmistakable. Their virtues (among others) are vitality, thriftiness and the ability to rear good calves in climates where other breeds would perish. They will impress your neighbours but your bank manager on the other hand might like to see the figures. In most places other breeds make more economic sense.

Jamaica Hope

It is easy to denigrate the whole British colonial system but, in its favour, let it be said that this breed was developed over a period of half a century from 1910 by dedicated civil servants. Local cattle, European pure-breds and Zebu or Brahman (*Bos indicus*) were used. Today it is a rather rough-looking Jersey type which has the same temperament as that marvellous breed and much of its potential, together with the resistance to heat and insects carried by its 15 per

Hereford bull

Jamaica Hope

Kerry

Limousin bull

cent of Indian blood. If the only thing preventing you from retiring to an island in the Caribbean is the lack of real milk, Jamaica Hope may be the answer.

Kerry

The Kerry has been known in Ireland for about 3000 years as a distinct breed. This figure is widely quoted because it is a nice round figure and easily remembered but scientific evidence is not so readily available. It is a small, black, dairy breed (cows average about 800 lb or 365 kg) which is noted for its ability to produce rich milk and adequate beef on relatively poor hill pastures. Until the 1800s it was the only widely known breed in Ireland and until quite recently was known as the poor man's cow. It is relatively rare today having been supplanted by more modern, larger breeds. However, as it has many obvious advantages for the smaller unit it would not be too surprising if it enjoyed a revival in Ireland and adoption elsewhere.

Limousin

It is claimed that this French beef breed produces more meat and less bone than any other. Its province of origin has a harsh climate and poor soil. Limousins transplant easily and possibly their greatest virtue is that they can be sold as veal at three months, as finished calves at less than a

year of age or can be slowly fattened over a period of three years. These last, known as *Chatrons*, are bulls castrated at about ten months of age, kept on grass for two summers and fattened on grain. Some gourmets suggest that the result is a product that makes other breeds taste like soya.

Like that of its distant cousin, the Charolais, the Limousin's dark yellow-brown coat is now recognized as a mark of quality in many places. However, it is never claimed of this breed as it is of the Charolais that the English Shorthorn was used in its development.

Lincoln Red

These are thought to be of Scandinavian descent, crossed with improved British breeds. In fact it was only in 1960 that 'Shorthorn' was dropped from their name. Today the Lincoln Red is considered to be primarily a beef breed that can be used for early fattening or very large carcasses depending on demand. It is a hardy breed, able to adapt to many different climates from severe Canadian winters to tropical heat. They are probably the only British cattle in recent times which, by selective breeding, have had the horns bred out of them, and today the majority are polled. Although it is not a good dual-purpose breed it is still in demand in some areas where superior beef cattle are required, and is gaining a measure of favour in the more temperate districts of Australia.

Meuse-Rhine-Issel

Luing bull

Dairy Short-horn

Simmental

South Devon

Luing

The Luing is the newest British breed, created by the Cadzow family on an island off the west coast of Scotland within the last thirty years. They used West Highland and beef Shorthorns to produce a hardy, long-lived cow with innate mothering ability that could produce calves who are demonstrably efficient converters of roughage to muscle, and later of grain to marbled beef. The Cadzow name is well known and respected wherever top beef commands a premium. The fact that the breed comes in all colours from red, white and yellow to brindle and roan does not deter enthusiasts in places as diverse as New Zealand and Canada. The only thing wrong with the breed as a beef animal is, to my mind, its name.

Maine Anjou

This one is another French breed which has benefited from the use of English Shorthorns over a century ago. Although it is considered to be a dual-purpose breed its milk production isn't too impressive. Recorded averages are about 6600 lb (3000 kg) of 3.8 per cent butterfat. It does however produce prime hindquarters and for this reason

has been extensively used in breeding programmes in North America and Britain. The type carries a relatively long red and white coat although breeders discourage too much white. Maine Anjou cows fatten well indoors or on grass. Breeders are also aware that they have more than their fair share of calving problems and are trying to select bulls who produce smaller calves.

Marchigiana

The climate of much of southern Italy leaves a great deal to be desired. Certainly many of the parts of Italy where this breed originated are very hot, dry and dusty in the summer and cold, harsh and wet in the winter. It finds most other places salubrious by comparison and thrives accordingly. It was and is used as a draught animal but during the last fifty years the emphasis has been on its unique beef characteristics. These include the fact that the male can weigh 350 lb (225 kg) at five months of age – and that high quality beef. Bulls can reach 1200–1300 lb (550 kg) at fourteen months. Mature bulls go to 2500 lb (1100 kg), mature cows 1400 lb (650 kg). The latter reproduce well up to ten years of age. The breed is docile and finds a ready niche in many intensive farms.

Meuse-Rhine-Issel

The fact that this breed is second in numbers only to the Friesian in its native hard-headed Netherlands is sufficient testimony to its worth. Shorter in the leg and more heavily muscled than the Friesian, it is considered to be more of a dual-purpose breed. Breeders aim for 400 lb (180 kg) calves at four months of age. Current practice is to finish bulls in groups at about eighteen months of age. Strangely enough most farmers in Holland don't dehorn them. Markings are similar to those of the black and white Dutch Friesian but the colours of the Meuse-Rhine-Issel are red and white.

Murray Grey

There are arguments as to the origin of this breed, which sprang early this century from rugged country on the head-waters of the River Murray in Australia. Its recognized founder, Mr Keith Sutherland, says the breed originated from a 'sport' bull calf of Angus parentage which stamped its silver-grey colouring on all its progeny. The other theory, not widely accepted in Australia, is that it developed from a Shorthorn-Angus cross.

Whatever its origins the breed, with its characteristic silver-grey to dark grey coat has proved itself all over the world for its ability to produce prime beef at an early age. Even over-finished steers show a high proportion of lean to fat in the high-priced cuts. This not only appeals to the Australian market but has convinced cattlemen in South Africa, Britain and North America to give them a try. The smallholder must consider himself fortunate if he has a source in the neighbourhood.

Santa Gertrudis

In 1940 this became the first new recognized breed in a century. It was developed by the King Ranch in Texas through the judicious crossing of Brahman and Shorthorn. It is a large, hardy, top-class beef animal which has found acceptance in countries including Australia where, however, it is not generally thought to be as good at coping with droughts as either the Brahman or the Droughtmaster. It is about $\frac{5}{8}$ Shorthorn and $\frac{3}{8}$ Brahman, is generally a rich red colour and carries the early maturing and fattening qualities of its British ancestors, as well as the resistance to heat and insects of its Indian progenitors.

Shorthorn

It is a bit trite to state that no cow can be better than the man who looks after it but it is completely true that in the right hands no cow can better a Shorthorn. There are beef Shorthorns and dairy Shorthorns and both are exceedingly good.

If you want a small herd you must consider this breed. They are hardy, viable, easy calvers, long-lived and can convert grass or grain into top quality beef. They rear their own calves easily and possibly also an adopted orphan and, when fed properly, produce a surplus for home use as well. At one time in Britain this was the only breed that could command a premium both for quality beef and quality milk. In Australia the breed is renowned for its hardiness and there the Poll Shorthorn has had markedly greater success than did its horned predecessor. Although it has not maintained its world-wide popularity it is a breed that will always find a place on the small farm.

Simmental

Medieval records extol the virtues of this breed, which was admired for its size, its working abilities and the abundance of its milk. People were not so concerned then about the quality of the beef – if a beast produced plenty of it, that was good enough.

The Swiss Red and White Spotted Simmental Cattle Association was formed in 1890. With characteristic Swiss thoroughness, meticulous records were kept, which, together with the absence of invasions, has helped the Swiss develop the Simmental to its present premier position not only in its native cantons but in many other continental countries from parts of Hungary and Russia right through to France and Germany. There are reckoned to be more than 35 million Simmental in Europe of which almost a million are recorded pure-breds in Switzerland alone. This single breed comprises 50 per cent of the Swiss cattle population.

Some strains of Simmental are prone to difficult calving which is not a problem under the intensive conditions practised in Switzerland, but which is a real drawback in a country like Australia, where an intensive progeny testing project is being carried out on the breed to try to eliminate sires which cause problems.

Although they are not widely used as draught animals today their strength and hardiness still stand them in good stead in the mountain pastures. Their coat colours vary from shades of yellow to red with white markings. The head is white. Bulls average slightly over 2200 lb (1000 kg) cows about 1650 lb (750 kg). Progeny testing of males and recording of cow's milk production is widely practised. This is a high quality dual-purpose animal which has quickly found wide acceptance outside Europe. It is a good animal to consider in a cross-breeding programme and works particularly well with Herefords.

South Devon

The South Devon is certainly the largest British breed and possibly one of the most underrated except in the corner of England where it originated. The cows can weigh 1500 lb (700 kg) and many bulls are twice that. At one time it was considered the prime dual-purpose animal in that it qualified for premiums both for the quality of its beef and for its abundant production of butterfat rich milk. It is thought to be related to the German Yellow. Both do well on small units in kind climates. Because of their massive size they may not have the refined features that appeal to many

Swedish Red
and White

Welsh Black

people looking for a family cow. But you may well be advised to look for the solid gold under the crags. If there is a breeder in your neighbourhood buy him a drink and ask him if he has got any for sale.

Sussex and Polled Sussex

As recently as 1940 a team of draught Sussex were to be seen working in that county but the fact that nowadays you can spend a lot of time in Sussex without seeing any of these red cattle indicates that most farmers don't find that they possess any outstanding virtue. It is rather sad in a way because it is claimed that they were well known in Sussex and Kent in 1066. Even their most enthusiastic devotees don't claim that they can produce more milk than it takes to raise a calf. They do however state that they produce marvellous beef. Some also claim a relationship with Zebu cattle and on that account they have exported many breeding animals to tropical and sub-tropical areas. One would have thought that it is the sort of breed the small farmer should only buy if there is a commercial breeder in the neighbourhood who is willing to share his enthusiasm.

Swedish Red and White

This is the most popular breed in its native land but is not often seen elsewhere. It owes a great deal to an admixture of Ayrshire and Shorthorn blood and is today considered a dual-purpose breed. Milk yields average 9500 lb (4300 kg) of 4.1 per cent butterfat. They grow rapidly and fatten well. Mature bulls weigh a ton (950 kg), cows about 1200 lb (550 kg).

Texas Longhorn

Descended from sixteenth-century imports from Spain, this breed soon replaced the wild bison in much of Mexico (which in those days included Texas); concurrently the pampas of South America were being populated by similar imports from Spain and Portugal. Their only real virtue was hardiness as in all other respects they were inferior to northern European stock. Their main colours were light brown, or dark Jersey with a dark stripe down the back and a rather heavy cross at the shoulders. Later, red colours were to dominate. At all times they carried their characteristic horns which were 4–7 ft (1.2–2 m) from point to point. An idea of their numbers can be gleaned from the fact that as early as 1796 over one million hides were shipped from Buenos Aires and Montevideo to Europe. In Texas the breed was supplanted by crosses with Shorthorns and Herefords – a process which has also taken place in Spanish and Portuguese America but at a slower rate. In North America the breed was almost extinct but a few enthusiasts have revived it. Some are now also being used again in western Canada, especially on first-calf heifers as they are very free of calving problems. They are, however, not really suited to the small farm.

Welsh Black

An ancient Welsh breed with long spreading yellowish-white horns with black tips, this was once considered a dual-purpose breed. Today it is considered to be a beef breed noted for its hardiness and its ability to nurture calves for many months in harsh conditions.

Buying cattle

Which to choose

There are four basic questions to ask in choosing a breed or cross-breed. First, what do you want it for? Second, is it readily available in your neighbourhood? Third, does it suit your particular place and circumstances? And, fourth, do you like it?

Purpose

Having decided to keep cattle you must decide which of the three basic sorts you want – beef, dairy or dual-purpose. Do remember, however, that the demarcations are not absolute. There may be some overlap. There is no doubt that the Ayrshire, the Jersey, the Guernsey and the Friesian are dairy breeds. There is no doubt that Herefords and Aberdeen Angus are beef breeds. Yet Friesian-Hereford and Ayrshire-Shorthorn crosses, among others, are profitably reared for both beef and milk. In Britain and in much of America and Australia the best known pure-bred dual-purpose breed until recently was the Shorthorn. Today the Holstein-Friesian has supplanted it in many places, although in Australia it is being challenged less by the Friesian than by the Red Poll and the Swiss strain of the Simmental.

Many newcomers cannot tell the difference between beef and dairy types. To them a cow is a cow. They would say that the subtle shifts in body shape cannot really be that significant. You may rest assured, however, that many people spend their working lives assessing those subtle differences. To understand the distinction one should spend an hour or two behind the counter of a butcher's shop. The expensive cuts like sirloin, porterhouse and fillet come from the hindquarters; brisket, stewing meat and most of the mince from the fore. The expensive cuts can cost over three times as much as the forequarter parts – as far as the butcher is concerned the main difference between a dairy animal and a beef animal is that the latter has more of its weight concentrated in the expensive hindquarter cuts. Dairy beef also has a tendency to be tougher and to have more yellow fat than the meat from beef breeds.

If you want to raise cattle for meat you have plenty of pure beef breeds to choose from. The ideal beef animal is one which matures early and rapidly converts food into muscle. Over the decades well established breeds have changed. A generation ago butchers would regularly buy three year old steers weighing well over half a ton (500 kg). Today they are bidding for exactly the same breeds at eighteen months or two years of age weighing just under the same weight. The main difference is that the public now don't eat the fat, nor do they want huge joints. If the butchers can't sell it they won't pay for it so farmers have responded by selecting for and producing leaner, more compact animals.

If you are buying and rearing a calf which is going to end up on the butcher's counter or in somebody's oven or deep-freeze it is a matter of common sense to have some idea of what it is going to look like when it is fully grown, for a pure beef breed is, of course, an investment which will take one or two years to mature.

The ideal dairy cow is essentially a large stomach (actually they have four) linked to a large udder. It produces more milk than any calf needs – the beef cow also produces milk, but only enough for its own calf, and it has a tiny, tidy udder. There is no point in buying an Aberdeen Angus calf and expecting it to grow into a cow that will provide your family with milk, cream and butter. It will provide some, but it will be very hard work. Similarly there is no point in buying a Jersey bull calf and expecting it to fill your deep freeze with choice cuts.

If you intend to milk the cow because you have lots of uses for whole or skim milk (for the family, dogs, cats, pigs, calves or lambs) you might choose a breed like the Friesian as it is noted for high production of milk which is relatively low in butterfat. If, on the other hand, you have little use for whole milk but want lots of cream for table use or making butter and ice cream, then you might consider the Channel Island breeds or an Ayrshire. A South Devon or a Dairy Shorthorn might be your best choice if you want some milk for the table while allowing the cow to suckle a calf.

Local availability

Farmers are pragmatists. They must be – unless they are supported by an institution, everything they grow or rear must prove its worth. If a breed is not kept by your neighbouring farmers and hence is not readily available, there may be a very good reason. It may be that over the years, by trial and error, it has been found that the area is too hot, too cold, too arid or too damp for a particular breed. For example, if your small stretch of land borders the Florida swamps, it is unlikely that any of your neighbouring commercial farmers keep Highland cattle.

There is yet another reason for the popularity of a breed. If a dominant farmer or rancher in the neighbourhood is a Hereford or Angus enthusiast, it is likely that many of the smaller units will have followed suit. They may have found it convenient to use his bull or buy some of his surplus female stock. The same advantages apply to the smallholder or part-time farmer. Thus, although you may have decided that you really want an Ayrshire cow but your area is stocked with Guernseys it would be silly not to consider the Channel Island breed.

In this context you must also bear in mind that all farmers are subject to the seasons. Thus if it is a particularly good year for grass and grain, your neighbours will be less inclined to part with livestock. However, if you have a piece of land which you can easily irrigate from the river that borders your property you may well be able to buy quality stock at bottom prices during a dry summer.

Suitability to your environment

There is hardly any point in embarking on a beef cow project on half an acre (under a quarter of a hectare). You might, however, consider rearing a few calves. Similarly, if you have 100 acres (40 ha) and a cabin that you live in during the summer and visit on occasional weekends during the winter, dairy cows are not for you. If your large garden (or small farm) is in a densely populated, urban area you had better avoid breeds like Aberdeen Angus which treat fences and other human constraints as a challenge. Otherwise, at the very least, you will become an unpopular neighbour. You might even wake up one morning without your small herd and little hope of ever seeing it again. In those circumstances you are better off considering a Guernsey (for milk) or even a massive Chianina (for beef) or any other breed which for generations has been reared in the fallow strips between greenhouses or vineyards.

Personal preference

There are two hundred breeds of dogs among which even hard-hearted income tax inspectors have their personal preferences. There are only about one hundred breeds of cattle and many of them are kept by practical farmers because they like their looks or their temperament. They may be marginally less economical to raise than similar breeds but the farmers reckon they would sooner spend their time with cattle they enjoy than with those they merely tolerate. Many people, for example, are perfectly aware that a Friesian can produce more milk than a Jersey, but they know that seven days a week they are going to have to milk the cow twice a day. Quite rightly, they opt for the cow that appeals to them. As George Bernard Shaw said, 'Do what you like before you like what you do.'

Where to buy

Neighbours

There is no question that the best place to buy cattle is from a neighbour. Provided you are on good terms and he is not an out-and-out crook the advantages are obvious. First, the creatures are habituated. They have been born in the area and are probably suited to the climate and may even have resistance to the local bugs and diseases. Second, transport is easier. Third, if you run into problems in the first few days (which is not uncommon) the neighbour can easily drop by and see where you are going wrong. Fourth, if you are in an area well served by agriculture advisors, vets and inspectors, they will know exactly what to expect from a familiar breed or strain. Finally, as anyone who has dealt with livestock will know, you cannot leave your cattle unattended and should personal or business contingencies drag you away it is much easier to ask a neighbour to look after them in your absence if he is familiar with your sort of stock.

Auctions

You may buy cattle at auction. There are several sorts of livestock auctions. One sort deals with the vast majority of livestock. These are held in stockyards or market places everywhere in the world. Some, like the stockyards in Calgary, Omaha, Winnipeg and Kansas City, deal with hundreds in a morning. Their British equivalents in Cheltenham, Aberdeen, or Swindon deal only in dozens.

The poker faces of the professional bidders are identical all over the world. That sort of auction is really no place for the novice. You would do better to find a professional buyer and offer him a box of cigars to bid on your behalf. Yes, there are such things as professional cattle buyers. One of the authors had a dear uncle called Sam who for many years had a contract to buy 29,000 cattle a year for a leading Canadian company. In his ripe eighties, he died with an unfinished Havana in his mouth. I think his commission was a dollar a head and the cut-off was $29,000 for tax reasons. Anyway, find yourself an Uncle Sam.

Another sort of auction is the less professional or specialized kind which deals with all sorts of neighbourhood produce. These auctions traditionally coincide with market day which is most often on a Saturday, or they may be held in conjunction with a local fair or carnival. In the spring you can be offered lots varying from chicks to piglets to bedding plants. In the autumn you may be offered calves fresh off the vanishing grass.

Later in this section we look in detail at signs of health. At this sort of auction you may have time to deal with little more than the obvious. You look for a glossy coat, a clear eye and no signs of discharge either from the mouth, nose, eyes or anus of the animal. Ideally, a smallholder needs to find a confident, alert animal that is neither cringing nor overly aggressive. Both are signs of too little human contact. On the other hand, if you have a large bit of scrubland in which you intend to set the calves loose for the winter those characteristics might be exactly what you are looking for. If at all possible you should try to attend that sort of auction well before time and examine at leisure the animals that interest you. Then, as at all auctions, mark your catalogue with your maximum price. Otherwise, in the heat of the moment, you will bid more than you intended.

Dispersal sales

The best sort of auction for the beginner is the dispersal sale when the whole place is being sold lock, stock and barrel. It usually means that there has been a death, an unfortunate marriage, a decision to move or a tax problem and the lawyers, judges, accountants and relatives decide that the only way to clean up the mess is to auction the whole lot. Never mind that Grandpa broke his back tilling the soil and Pa bust his arm building a prize herd of Jerseys. The whole lot goes. Usually in these sales there is plenty of advance notice. Almost always there is a detailed description of every breeding cow. There is no question that in general the best way to embark on keeping cattle is to buy a pregnant cow. You may pay up to 20 per cent more than the prevailing price for a similar animal in a commercial ring, but the chances are that you will be getting good value.

Age to buy

Newly born calves

Markets that sell newly born calves are usually held once a week. The farmer who sells his unwanted calves through that sort of auction will fill his truck with every creature he wants to sell, even if it was born while he was loading. A mixed bunch of calves variously described as 'day olds' or 'nurslings' or 'newly born' may include some that are still not dry behind their ears. They have neither been licked by their dams, nor had the time to suckle. There may be others who are near enough a week old. That seven day gap may not seem significant in terms of a cow's lifespan but in fact it may make all the difference. Its chances of survival, its growth rate and your chances of success with the calf may be determined by those first few days which are the period when the calf receives colostrum from its mother's milk. It is possible to rear calves (or other mammals) that have not had colostrum – but it is much more difficult. The success rate with calves who have not had the benefits of colostrum is bound to be lower.

In a 'day old' auction, one must try to avoid those that look exactly that. Bid for the calves that stand without wobbling and bellow with confidence, and whose eyes and coat glisten with good health. The chances are that they have had colostrum. Another thing to check is the condition of their navels. Also of course, avoid calves which show evidence of scours or cough.

Quite obviously, if you buy young calves from a friendly neighbour, he will be able to tell you their exact age and history.

Weaned calves

Weaned calves and those up to about twelve months of age may be variously described as 'weanlings', 'stock calves', 'stores', 'feeders' or 'yearlings' depending on your locale and local custom. In many places it is traditional for people with abundant pasture but little grain to sell their surplus stock as winter sets in. If you live in such an area and are only interested in half a dozen or fewer, again you will be better off dealing directly with a neighbour. Should that prove difficult then you may have to go to a large market. In most cases you will find that the services of a buying agent (who will charge about 5 per cent of the sales price) are worthwhile. His expertise will more than pay his commission. The signs of health which we describe later in the book are something he will recognize at a glance. He will also be able to tell by looking at the cattle as they move through the ring whether they have been tagged, branded, castrated and dehorned. One must note here that some buyers deliberately buy weanling calves which are quite obviously worm-infested, or scabby, or suffering with scours, or coughing and discharging badly. They buy them because they are cheap and they have isolation or hospitalization facilities and good veterinary advice. Many a smallholder doubles his investment by buying such calves and nursing them through the winter. It is essentially a labour-intensive husbandry operation but it is obviously not the best idea for a novice who is starting out with his first cattle.

Mature cattle

A cow's productive life may be as long as ten or twelve years, or even more, and the younger she is when you buy

Cows housed in indoor stalls.

her the more productive years she has ahead of her. However, the beginner must keep in mind that a young cow (or heifer) is more likely to have problems giving birth the first time than she is in subsequent calvings. You could therefore be well advised to buy a cow that has proved she has little trouble calving or rearing a calf but you must be prepared to pay a premium price for a proven cow in her prime. The smallholder may well consider the advantages of buying an older cow who is past her prime. She may be up for sale at slaughter value simply because she is too old to remain in a large commercial beef herd where they don't have the time to look after the foibles of the elderly. You, on the other hand, might well be able to look after such a cow while she produces a calf or two for your embryonic herd.

In this context one may also be able to pick up very good cows at bargain prices because they have individual psychiatric problems which make them unsuited to 'life in the herd'. A particularly dominant or submissive cow for example may suit a small farmer although it is a problem in a commercial operation. Similarly a cow who has lost a quarter or two or her udder through injury or disease may be culled from a commercial dairy operation at beef prices. Many a very good adult milking cow has become unsuitable for continued production in a large herd because, through illness or injury, she has lost one or two of her teats. Large units simply do not allow a farmer the time to hand milk such a cow, but she can fit quite happily into the one- or two-cow unit and make a valuable contribution to the family economy.

Older cattle

Finally, some small farmers situated among very large units can often fill a real gap by buying aged cows and bulls who are well past their reproductive efficiency. In range areas particularly these animals may be scrawny beasts more akin to scarecrows than to the ideal types portrayed in the breed society literature. This sort of business operation can easily be undertaken by one person on even the smallest place.

Essentially it is a 90 or 120 day fattening programme. One needs very good fencing or really strong corrals and a source of cheap grain. Ideally the animals should be separated but in adjoining pens. It is also important that the operator should own or have access to a mobile loading platform and a well-constructed truck in which to move the animals to market. A fat bull weighing almost a ton (1000 kg) cannot be held by a man or even by a man and horse – it must be directed between wooden slats and steel hurdles. But provided you are physically geared to it and have suitable premises, there is no reason why you should not consider keeping cattle of this age group.

How to sell

Auctions

Unhappily it would appear that there is little real justice for the small farmer. The same rules that mitigate against his buying wisely at auction also operate when he tries to sell. Experienced sellers and their agents will get their stock into the ring just after the first couple of lots, while the buyers are still stirring their coffee – well before the stage when people have bought all they require and the auction starts to tail off. The seller must appreciate that although he may have laboured all year his cattle are going to be valued and sold in a matter of seconds.

Should you have no recourse but to put your stock through a large auction, you must again consider the advantages of utilizing a commission agent. He may well earn more than his 5 per cent simply because he knows the buyers' requirements. Even in a large ring he will know many of the buyers who represent the packers, chainstores or restaurant groups. By a nod he can indicate that your lot is suitable for them.

Cattle at auction are sold by the head, by the lot or by weight. Generally calves, stockers and feeders – anything that is not going for slaughter – are sold by the head or by the lot. The health and 'bloom' of the animals is all-important. Although it is illegal in most places, many vendors will give their cattle a massive injection of antibiotic the evening before the auction to mask any illness.

Cattle destined for immediate slaughter are usually sold by weight. Again, although it is distinctly unethical if not illegal, many vendors fill their cattle full of water by the following simple expedient. In the evening prior to the auction they throw a couple of handfuls of salt into each animal's throat, then a few hours later they allow them access to unlimited water.

Selling troughloads of water at beef prices doesn't endanger the health of the public, but a more recent practice is indeed dangerous. The cattle are injected with anabolic steroids and/or hormones, which are rumoured to be commonly used by Soviet weightlifters and act as artificial aids to the deposition of fat and the build-up of muscle. There is no evidence that the meat produced is in itself harmful but there is a great deal of evidence that residual quantities of the drug in the animal's body can affect humans – particularly growing children. For that reason most places stipulate that the drugs must be given at least ninety days before slaughter. Unscrupulous operators however will inject them a mere two or three weeks before slaughter. It is to be hoped that in the not too distant future tests to detect this dangerous practice will be more widely used and the penalties decreed by law be more severe.

Selling privately

Many people with only one or two cows and maybe a few calves find that their best course is to sell them privately in their own neighbourhood. Again, as in the case of buying locally, the advantages are easy transport, minimal disturbance to the animals, no agents' fees, no handling charges and, above all, the fact that both buyer and seller know exactly what is involved. There are many outlets for small lots of even one cow or one calf. These vary according to your neighbourhood. If you have got a couple of Jersey cows, for example, and one produces a heifer calf another small operator may well be interested. In fact, it is not unusual for a large-scale beef rancher to buy a nurse cow to nurture a calf through to championship status. There may be a specialized market locally created by ethnic groups. Italians, for example, are partial to veal. People of Polish descent may prefer old bulls to make traditional sausage. Jews may actually prefer an animal with little meat on the hindquarters – although the surrounding populace may value sirloin and fillet, they place more importance on brisket and rib.

Marketing groups

Finally, every small operator should consider the advantages of joining with others to form a marketing group. Whether you call it a union or a co-operative doesn't matter. Provided it doesn't become too top heavy in administration, the advantages are obvious.

Handling

How do you go about moving a calf that you have just purchased? In our increasingly urban society it is not really surprising that many people have never actually touched a cow or a calf and certainly never handled one.

Halters

Simple halter

Illustrated here is the simplest sort of halter. It can be made in a minute from any bit of rope – and unhappily often is. You simply put a loop over the calf's head, put the end of the rope through the loop to form another loop which is then tightened around the muzzle (see photographs). The disadvantage is that it tightens around the calf's neck and muzzle and can be painful, and that one cannot or should not use it to tie a calf to a post for more than a minute or two. The advantage is that when one has nothing better to hand it is an effective means of temporary restraint. One often sees it used in markets to help move a calf from the sale pen to the buyer's vehicle.

Haltering a young calf: 1. *The halter rope has a small loop knotted at one end. Thread the other end through and put the big loop over the calf's head.*

2. Reduce the big loop so that the knot is under the calf's chin with the small loop to the right of its jaw, and take the loose end of rope over the calf's nose.

3. Put the long loose end back under the rope on the left of the calf's jaw.

When tying a calf up to a fixed object a half bow is a good knot to use as it can be quickly released by pulling on the free end.

The calf is firmly held by the halter and quickly gets accustomed to it.

Rope halter

If you are handy with ropes and knots it is quite easy to make a rope halter. It works on the same principle as the one described above but is far more satisfactory in that it causes no pain, cannot choke the calf and can be used to tie the calf safely for longer periods. Standard sizes for animals from the smallest to the largest are on sale at any place which caters to the farmer's needs.

Although synthetics are cheaper than rope and last longer, most people prefer natural materials because they don't 'burn' the animal or the handler as badly or as readily as man-made fibres.

Slip knot

If necessary, the animal can be tied to a post with a slip knot. The prospective cattle owner should take ten minutes to master this knot (see photograph), the great virtue of which is that, if the animal pulls back and puts all its weight on the knot, you can still loosen it with a simple tug. If you use an ordinary knot and you don't have a knife handy to cut the rope, the animal will certainly get frightened and may injure itself or even choke to death. The principle is applicable to all farm animals, particularly horses.

How to move a calf

Once you have placed some sort of halter on the calf, how do you actually move it? Some calves have been so gently reared that all you need to do is place a couple of fingers in their mouths. Actually one finger would do the job, but it is surprising how much strength even a newly born calf has when suckling. Most people find it more pleasant or possibly

Carrying a young calf. Put one hand over the flank, the other under the neck. Pressure on the neck must be gentle to avoid squeezing the windpipe.

less painful to allow two fingers to share that strength. Once the calf has accepted that teat substitute it can usually be led quite easily for a short distance.

Some calves however have had so little contact with people that they have to be forcibly moved. Experienced cattlemen will, of course, scoff at this restatement of the obvious. Beginners, however, might find these points helpful, both in easing the initial journey and in obviating the chance of injury or pain to either calf or operator.

If you are strong and the calf is still quite small, it will be possible to carry the calf by yourself (see photograph). In the case of a larger calf the easiest thing is to have two people, who should clasp hands under the neck and the belly of the calf. The stronger of the two holds the lead of the halter in case the calf escapes. It is a sort of fireman's cradle.

If your partner simply cannot carry half a struggling calf, he or she may be persuaded to shove from behind while you pull from the front. This is at best an undignified procedure and even if it doesn't move the calf too far can provide endless amusement to bystanders and cannot cause the calf any pain.

Truly recalcitrant range calves may have to have their legs tied and then be moved on something like a wheelbarrow or carried by the local champion weightlifter. A very serious word of warning: many a calf has broken a leg because one hind leg has been left hanging loose. The safest approach for the beginner is to tie the hind legs together with a series of clove hitches ending in a slip knot, before following suit with the forelegs. Use a third set of ropes to tie the two sets together. Use clove hitches for all the knots and finish with a slip knot. Never attempt to carry the calf upside down – it might break its neck if it is restrained while its head hangs loose.

All that sounds barbaric and indeed it is. In this advanced mechanical age it is almost always easier to move a small trailer to the calf than vice versa.

Castration

If you castrate a male calf it becomes a steer, and is generally more placid than a bull. It is thought that steers put on weight faster than bulls and although the meat may not be as distinctly flavoured as bull meat, it is often considered to be more tender. Steers are easier to handle than bulls and one can keep large numbers of steers in a group whereas with most breeds it is difficult, if not impossible, to keep large numbers of bulls together without problems.

For all these reasons bull calves are generally neutered at an early age. The earlier it is done the easier the procedure but the calf is not usually traumatized until it is at least a few days old. The following are the three most common ways of doing the operation:

1. The Birdizzo is a clamp named after its Italian veterinary inventor. It cuts the cords that contain the blood vessels leading to the testicles.
2. The rubber ring method operates on the same principle. It gradually cuts off the circulation to the testicles.
3. The knife quite obviously requires more skill and carries greater risk of infection but is preferred by most commercial farmers. This method is illustrated on the facing page.

Castration of adult cattle is a far more serious operation and in most places is carried out only by veterinarians.

Horns and dehorning

Most cattle have horns. Breeds like the Aberdeen Angus which do not are called polled cattle. Other breeds such as Herefords naturally have horns but a sub-breed has been developed which is polled. In all developed countries it is accepted that horns are a hazard both to the people who are handling the animals and to the rest of the herd. They are not only a means of defence but can also be a very dangerous weapon of attack. Man in his collective wisdom has decided that cattle are best kept without horns and in many countries there is a money inducement to encourage people to dehorn their cattle or to keep polled breeds.

Dehorning is a relatively painless and hazard-free operation when performed on very young calves. Obviously it is best done by a veterinarian but in most places it is done by an experienced stockman. Ideally a local anaesthetic is injected around the horn bud although many people reckon that, like some dental procedures, the operation is less painful than the needle. The horn bud is cauterized and destroyed so that it can't grow. It may be done by applying a caustic chemical or burning the area with an iron heated by fire, gas or electricity.

The calf's head must be held perfectly still during the procedure, otherwise damage may be inflicted elsewhere.

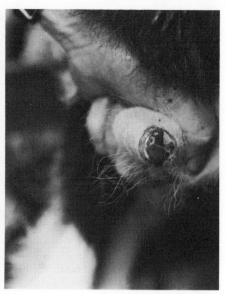
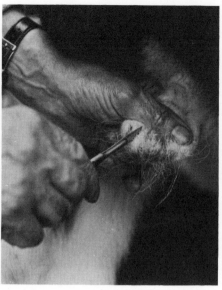
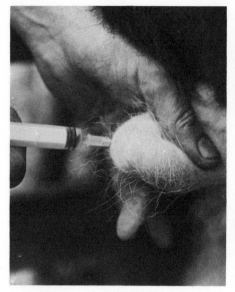

Castrating a bullock: 1. *The scrotum is swabbed with antiseptic, and then local anaesthetic is injected into each testis.*

2. *After ten minutes a testis is gripped so that the skin is tense.*

3. *A cut is made through the skin and the lower part of the testis.*

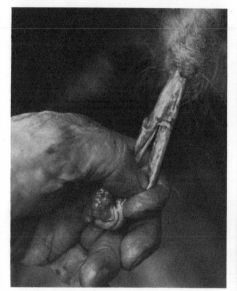

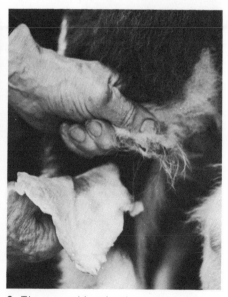

4. *The testis is drawn out of the scrotum.*

5. *The vessels and membranes are cut through under considerable tension.*

6. *The wound is wiped over with an antiseptic swab.*

If one uses chemicals it is best to protect the area surrounding the horn bud with a bland ointment. Calves who have been so treated should not be allowed with other calves lest they spread the stuff.

Adult cattle can be dehorned, but it is a painful and hazardous affair, and should certainly be left to the professionals. In many countries, quite rightly, only veterinarians are allowed to do it.

Identification

In many countries identification of individual cattle is compulsory because of health regulations which involve testing for such diseases as tuberculosis and brucellosis.

Affected animals are slaughtered and a register is kept of the healthy ones which have passed the tests. The efficiency of such controls would be impossible if each animal could not be identified. Cattle resemble each other so closely that written descriptions cannot differentiate between individuals. Identification is effected by tattooing, branding, ear-tagging or by neck straps or chains carrying distinctive numbers.

Tattooing and ear tags
Tattooing is usually done on the inner surface of the ear which is comparatively free of hair. Metal pins are arranged in blocks making letters and numbers. The required blocks are mounted in the head of a pair of forceps. A black or

De-horning calves: 1. *The hair is clipped to expose the horn ends.*

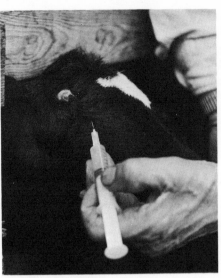

2. *Anaesthetic is injected under the bony ridge that runs from some way below the horn towards the eye.*

3. *After ten minutes the hot de-horning iron is applied.*

4. *It is left to burn around the horn base for half a minute.*

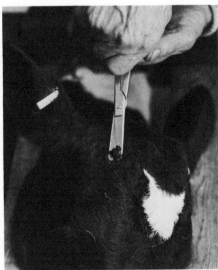

5. *This leaves a projection of burnt horn which can either be left, or snipped off.*

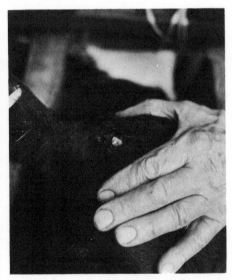

6. *The flat area will soon heal over.*

coloured paste is painted over the inner surface of the ear and the forceps applied so that it punches the pins into the skin. The colouring remains permanently in the skin recording the identifying letters and numbers. The tattooing is difficult or impossible to read in black skinned ears. Metal tags, serially numbered, may be attached by various methods of fastening to the ears but they are liable to come unfastened or to be torn out of the ears and be lost.

Discs

Amenable cattle that are accustomed to being handled frequently may have straps or chains around their necks carrying a disc bearing their identification. This method is unsuitable for rougher conditions under which straps or chains might be torn or broken or, if they are strong enough to resist that damage, could be caught up on fences and lead to strangulation.

Branding

Hot branding with irons to burn identification marks onto the hide is a traditional method in some areas, but it is painful to the cow and the hide suffers some permanent damage. Freeze branding with copper brands cooled in dry ice and alcohol has been used. The apparatus for this method is not as readily available as for the hot irons but the method is painless and the damage to the hide is less serious.

Fly control

Flies are a great nuisance to cattle, and can cause such irritation and restlessness that milk and meat production are significantly lowered and fences are broken. Flies directly spread such diseases as summer mastitis, pink-eye and warbles, and in hot weather positive action against flies becomes absolutely necessary.

Cattle at pasture have little protection from flies. Their reaction is to bunch together and keep the flies at bay by flailing their tails which should not be trimmed from full length. Shelter in the heat of the day is helpful. If woodland is available cattle graze in the open at night and spend the hottest times resting and ruminating in the shade of the trees. They will make the same use of buildings and these should be kept as dark as possible without impairing the ventilation. If cattle must spend their time in the open you can give them some protection from flies by spraying them during their resting periods from a knapsack or a mechanical sprayer which they soon get used to.

In buildings and the farmyard surroundings cleanliness is a helpful fly deterrent. The manure heap should be sited as far away from the cowshed as is reasonable. It should be on a concrete base with a water filled channel surrounding it. Any fly maggots breeding in the heap are likely to move out to pupate and will be caught in the channel. Buildings can be protected with netting over the doors and windows. Cows in the buildings can be sprayed twice a day or more frequently if necessary. Yarded cows brought in for milking can be sprayed in the yard before they enter the milking parlour and a curtain of water from a perforated pipe over the door will markedly reduce the number of flies they bring in with them. The walls, ceilings and floors of buildings used by cattle, and adjacent buildings, should also be sprayed. Fly sprays and other insecticidal preparations should always be used in the strengths advised by the manufacturers.

Milking

When to milk

If you have decided to go in for milk and your cow has had her calf, she must be milked twice a day. For the first three or four days after the birth of the calf the milk is thick and may be slightly pink. This is colostrum, essential to the calf, but not to be used for domestic purposes because it coagulates if it is boiled. A cow of a milking breed will produce far more milk than one calf can use, so from the day the calf is born she will need to be milked by hand or by machine. The calf should be allowed to suckle as it pleases for the first few days, and you should milk out what the calf has left twice a day. After four days the calf can be penned away but brought to its mother three or four times a day for a week. After that you carry on milking twice a day. If the calf is still to be allowed to suckle you should let it have a go for a quarter of an hour before you start milking.

How to milk

The art of hand milking is quite simple. You grasp the teat folding your fingers round it and, drawing gently down, contract one finger after another starting at the top of the teat, so that a stream of milk flows into the bucket grasped between your knees. It helps to have a tutor who will allow you to practise for an increasing number of minutes at each milking. Otherwise it may be some weeks before you can honestly say that you enjoy the exercise.

If you are not enthusiastic about developing an impressive handshake there are machines that will milk the cow by electricity or by engine power. Starting from the cow's udder the machine consists of a cluster of four rubber-lined teat cups connected by piping to an enclosed bucket which is attached by further tubing to a gauge pulsator and vacuum controller mounted with the pump and motor on a wheeled chassis. You can buy an outfit to milk one cow at a time or, at a somewhat higher price, one which will milk two cows at once. Large herds of milking cows are nearly always milked mechanically in carefully designed parlours where the milk is drawn directly from the cluster of teat cups on the udder to a cold storage tank. The small-holder's needs are generally less elaborate.

Feeding

The natural food for cattle is grass and they can maintain themselves, that is they can grow and survive and raise their young, on grass alone. However, on this unsupple-mented fare 'from Christmas to May, cattle fall away' as

One system of feeding cattle is to let them help themselves from a stack of silage. They are prevented from trampling on it by a piece of electric wire.

Teaching a calf to drink: *The calf is always ready to suck a finger.*

It follows the finger down into the bucket . . .

. . . and soon learns to get at the milk without the finger's guidance.

Tusser wrote in the sixteenth century. Stock owners nowadays ensure that their stock do not fall away in the winter by feeding hay and straw and barley. But maintenance is not enough. Good, all the year round productivity is required as well as steady growth from the young stock, milk from the dairy cows and flesh and fat on the beef cattle. This demands generous and careful dieting in calfhood and right through the animals' lives.

The first six weeks

Colostrum

Every calf should be allowed to suckle from its mother four or five times a day for the first four days of its life. The cow's milk for the first few days is known as colostrum, and this valuable substance, as has been stressed before in other chapters, is essential to the new born animal because it contains a concentration of antibodies. Antibodies are minute particles developed by the cow's tissues whenever she is exposed to infectious germs as a protection against the diseases they cause. Through the colostrum the cow can transmit this protection to her calf. A calf that has not received a good supply of antibodies in colostrum is more difficult to raise and is more likely to succumb to disease than one that has had colostrum is good measure.

Milk substitutes

The day the calf is taken away from its mother may involve it in a journey to market, exposure to the shocks of bad weather, crowds, much handling and the close company of other calves with a risk of infection, ending up with another journey to its new home. In such circumstances it can only benefit from the simplest of nourishment and this should be offered in the form of 2 oz (60 g) of glucose powder in a quart (litre) of warm water. This should be repeated in six hours and again first thing the next morning. The calf should then be ready to go on to milk substitute carefully prepared according to the maker's instructions.

The day after the calf's arrival at its new home is the most difficult one because it has been accustomed to suckling from its mother's teats and you are offering it milk (let us forget that it is milk substitute) in a bucket. Each calf has to be taught to drink. The routine is that the calf grabs your finger expecting it to spurt milk. You push your hand down to the bottom of the bucket with the calf still suckling away at one finger. This happily allows the calf to suck up some of the milk. The next move is to persuade the calf to do without your finger. Some learn in a matter of minutes. Some take a week. Two days is the average. It is a tedious business but you really get to know each calf and you end up with a remarkably clean finger.

Mixed diet

For the next eight weeks the calf's diet will be milk three times a day, water, hay and calf pencils. Milk substitute is generally used because real milk is too valuable a commodity on most farms to be fed to calves. Milk substitute, being a dry powder, is very convenient and can quickly be made up into the quantities required. If there is real milk to spare by all means feed that, but for very young calves it is inadvisable to vary the routine.

Calf pencils from a good fodder merchant contain all the necessary foodstuffs not supplied by milk and hay. If this commercial food is not available, the following mixture can be used:

Flaked maize	50 parts (by weight)
Crushed oats	30 parts
Milk powder	10 parts
Soya bean meal	10 parts

changing at eight weeks to:

Flaked maize	25 parts
Crushed oats	35 parts
Crushed barley	25 parts
Soya bean meal	15 parts

The first mixture should be fed in amounts of 4 oz (115 g) daily, gradually increasing to 3 lb (1.3 kg) at eight weeks, when the second mixture can be given at 3 lb (1.3 kg) increasing to 5 lb (2.3 kg) daily at six months.

The new calf will hardly touch the hay or pencils for the first few days but will then become positively interested and will bellow demandingly if his pellet bucket is empty. Be firm about rationing. The stronger calves will eat more than the others and may have their supply increased a little day by day, but they will overfeed themselves if given the opportunity and this may lead to diarrhoea or even to sudden death. Water should always be available.

Weaning from milk

At six to eight weeks of age the calves can do without the milk feed and manage on water, hay and calf nuts. They can be taken out of their pens to run together in groups of a dozen or more. This continues until the spring when they can be allowed out onto fresh pasture for an hour or two a day, gradually increasing the grazing time over a month, at the end of which time the weather may be suitable for them to stay out altogether.

Unless the grazing is very poor they should not need any supplementary feeding. They should keep growing steadily on grass and water. However, it is often found helpful to feed 4 oz (115 g) of pellets a day which keeps the calves interested and gives you an opportunity to feed them any necessary minerals if the local soil is known to be deficient. The concentrate ration can be increased and some hay fed as the colder weather approaches and the grass becomes less nutritious.

Feeding young livestock

Young stock, by which is meant weaned calves and older animals which have not yet been mated or penned for fattening, are fed in much the same pattern. At grass they may need no supplement at all, or just a small feed to keep them interested, and this may include minerals if the pasture

Weaning calves in indoor housing.

is known to be deficient in any particular item. When the grass loses its quality in the autumn or fall, hay may be given in suitable quantities – which can be judged from the amount that the cattle will clear up without wastage.

Young cattle destined for the milking herd or for beef should not be persuaded to overeat. Too much corn or other concentrated foods will make them fat. As long as they keep growing and are reasonably well fleshed their condition is satisfactory. In the winter, the same advice holds good. Whether they are kept out or housed they only need enough concentrates in addition to their grass, hay and clean straw, to keep them in reasonable order.

Heifers for breeding, whether they are for the dairy or the beef herd, will be got in calf more readily and be more productive afterwards if they have not been overfed as youngsters.

Feeding mature cattle
Fattening bullocks
Bullocks lay on condition and fatten very satisfactorily if their feeding is kept to a normal level until they are twelve months old or more, when their rations can be increased steadily up to their killing weight. The resulting carcass will be well fleshed and not over-fat. Housewives, butchers and beef buyers prefer meat that is not over-larded with fat. Farmers who go in for baby beef production have different ideas and feed their young stock generously from an early age. This increases the risk of losses from over-feeding and is expensive but, with selected stock in expert hands, it is usually profitable.

The foodstuffs to be used are, basically, grass and hay, which can always be fed up to the animal's appetite. If too much hay is fed, it will be trampled and wasted, and hay is expensive. If hay is not in generous supply, equal quantities of clean barley straw and hay will do instead. If silage is available that may be substituted for hay and less concentrates will be required.

Concentrates and silage
For beef cattle the simplest way to provide concentrates is to buy them from the millers. They are made up of various grains – wheat, oats, barley and maize with beans, ground-nuts and other imported foods and added minerals and vitamins. They should be bought in reasonable quantities as their feeding value deteriorates with time. Beef cattle may be ready for slaughter at 800 lb (360 kg) in just over a year; at 1000 lb (450 kg) when eighteen months old or at 1200 lb (550 kg) a few months later. Their rations over and above the basic hay or silage can be increased from 3 lb (1.3 kg) of concentrates a day up to 8 or 10 lb (3.5 to 4.5 kg) for the older animals. Silage may be increased from 20 lb (9 kg) a day to 50 lb (22.5 kg) but, with the larger amounts of silage, concentrates may not need to be above 5 lb (2.3 kg) daily.

Dairy cattle
Dairy cattle are fed a basic ration for maintenance of grass or hay and cereals with 3 or 4 lb (1.5 kg) of concentrates for

each gallon (4.5l) of milk they produce. The time of the year affects the food value of grass to such an extent, and the state of each cow as regards milk production and time of calving is so important, that rationing for milking cows must be adjusted to each animal's particular requirements.

Breeding

You can't expect a cow to give milk unless it has had a calf and that of course means that the cow has to be mated. You can postpone the business of breeding for a while if you buy a cow that is already in calf, wait for her to produce her off-spring and then start your milking programme. She will go on producing milk for seven to ten months and then, unless you have had her mated again, the supply will run out. The time to get her in calf again, with a continuing milk supply in mind, is two or three months after the calving.

Nowadays it is not necessary for the cow to be actually mated by a bull. Unless you have a large number of young cattle it is usually more convenient to use the artificial insemination service.

Artificial insemination (AI)

To simplify the process of breeding cattle and to improve their quality by using only very good bulls, government agricultural departments, milk boards and some breed societies have set up artificial insemination centres for both beef and milking types of cattle. A number of bulls of various breeds are kept at these centres and semen is collected regularly from the bulls and divided into single doses. When you have a cow or heifer ready for mating all you need to do is to telephone the cattle breeding centre and one of their inseminators will come along very promptly and inject your cow with a dose of semen from the bull that you have selected or that the centre has recommended as suitable for your type of animal. Your only responsibility is to know enough about your cattle to recognize when one of your cows is bulling, that is, ready for insemination. In areas where distance prevents a prompt service it may be necessary to have these doses in your refrigerator and learn to do it yourself.

Bulling cows

A cow or heifer is said to be bulling, or on heat, when she is showing signs of being ready for mating.

Recognizing bulling

The chief sign of a bulling cow is that she allows other cattle, males or females, to mount her, placing their forelegs on either side of her rump in imitation of mating. In some herds a capsule containing dye is fixed on the cow's rump so that when another animal mounts the cow dye is ejected and stains the area. People familiar with the cow can recognize signs of bulling such as a change in temperament, loss of appetite or a decrease in the quantity of milk being given.

The signs are usually more intense in summer than at other times of the year.

The detection of bulling is important because it only lasts for part of a day, between four and twenty-four hours, and if the cow is not mated by a bull or given an insemination during that time she will not conceive and she will not come on heat again for about three weeks.

Heifers usually come on heat for the first time when they are a year and nine months old, but this can be brought on earlier by intensive feeding and is likely to be later in cases where the animals are in poor condition. Mating at one year and nine months to calve at two and a half years is quite early enough.

The problem of heat detection does not arise if a bull is turned out at grass with heifers of two and a half years or older or with cows. He mates with them as they come bulling. Young bulls under these natural conditions are reasonably manageable. It is the old bulls closely confined in bull pens that may become dangerous and turn on their attendant without warning.

After calving

When a cow has calved and is suckling her calf she is unlikely to come bulling. When her calf is weaned she usually comes on heat within the next ten days.

In dairy herds the newly calved cow is allowed to suckle her calf for four days. It is then raised away from its mother, probably on milk substitute and the cow returns to the routine of the milking parlour. Even though she is producing quantities of milk she is no longer suckling a calf and she comes bulling again, not within ten days, but in a month or six weeks.

The herdsman's objective is to have her mated within three months of calving so that her calves arrive as nearly as possible at intervals of twelve months. Dairy cows that have not come bulling by six weeks after calving should be examined by a vet who will treat any condition that may be upsetting the usual rhythm.

Pregnancy

Cows produce their calves about nine and a half months after a successful mating although the length of pregnancy may vary by a week or two either way. Bull calves are likely to be carried longer than heifer calves.

Recognizing pregnancy

It is not easy to make sure that a cow or heifer is pregnant. A pregnant cow does not, as a rule, come into season but detecting that an animal is in season is difficult anyway. Awkward animals often improve in temperament when pregnant but if you are on good terms with your animals not many of them are likely to be awkward. If it is important in the first few months to know that one of your animals is or is not in calf, your vet will make an examination by feeling through the wall of the rectum. Later on, clear signs of pregnancy develop. Heifers begin to show udder development after five months, and a month or two later

A Friesian heifer about to calve: *The muscles on either side of the tail have lost their usual tension, the vulva is elongated with slightly swollen lips, and the udder is swollen and tense.*

the calf may be seen to 'quicken' or make quite violent movements in the cow's right flank, especially when she has just got up after a spell of lying down. There is also a considerable enlargement of the cow's belly.

Caring for a pregnant cow
The care of pregnant cattle is largely a question of ensuring that they get a good supply of suitable food and clean water and plenty of exercise. All this comes naturally to cattle that are at grass. Animals that have to be housed for any reason should be turned out for several hours a day even if there is no grass for them to eat. They will benefit from walking about to look for it. Remember also, if the animals are housed, that narrow doors and slippery alleyways can cause injury to pregnant cows.

Milking cows need special attention to the quality and quantity of their food which has to cope with three demands – the cow's maintenance of health, milk production and the calf that is moving inside her. Six or seven months after the cow has calved her milk production will go down. This could be overcome to some extent by intensive feeding, but the proper course is to encourage the reduction by giving less food.

Eight or nine weeks before her next calf is due a decision must be taken to dry the cow off. This is done simply by stopping milking her. The udder will swell up for a few days and then subside. It is usual, when drying off a cow, to inject a dose of an antibiotic cream into each of the teats to prevent mastitis. If one or more of the quarters of the udder continues to swell and becomes hot and painful after drying off this is a sign of mastitis and the cow must be treated as described for this disease. If the udder settles down without trouble the cow's rations can be increased again to ensure that she has enough food to nourish the rapidly growing calf she is carrying.

Signs of impending calving (springing)
A month before the calf is due the udder becomes very obvious in heifers and begins to fill out again in cows. In animals that are being overfed milk may run from the udder some days before the calf is born. This is to be avoided as the wasted fluid is colostrum which it is important that the calf should receive. Other signs of springing are that the ligaments on either side of the tail slacken and the vulva enlarges and its lips become full and soft.

Preparations for calving
Cows and heifers may calve out in a field or in a calving box or stall well bedded with straw. Calving out has the advantage of being more natural and healthy but it is less convenient if you want to observe what is going on, and weather conditions may be adverse. One of the risks of calving out is that the mother may choose an awkward place to calve so that the newly born calf falls into a ditch or is born through a fence and can't get back to its mother. In spite of all this the event usually occurs without difficulty.

Animals that calve in a box make extra work and it is difficult to keep a box clean after it has been thoroughly washed and disinfected, if a cow is living in it for several days before producing her calf. There is far more likelihood of the calf picking up the germs of disease indoors than out in the fields, especially if the box is one that is regularly used for calvings. It cannot be over-stressed that the box should be swept out completely after a calving. The floor, walls and fittings must be well scrubbed and then washed over or sprayed with a disinfectant and left for a few days before being bedded down for the next calving. If that is not done disease germs accumulate and calves, apparently healthy at birth, will develop one or more of the calf diseases described. This may mean not only losing the calf but establishing masses of germs around the premises to await the arrival of the next one.

Calving

The actual calving runs to an established pattern. In the preliminary stages the cow appears uncomfortable, perhaps starts to bellow from time to time, and goes off by herself. The next stage is labour pains – muscular contractions of the cow's belly while she holds her breath, that start the calf on its way – and these increase steadily until the calf arrives. A blue-grey bladder appears in the vulva increasing in size until the pressure of the labour pains bursts it and a large quantity of fluid, that has been the calf's water bed and buffer against concussions, pours out onto the ground.

Up to now the cow has usually been standing up but at about this time she lies down. The contractions become stronger and even violent and the cow may moan or bellow in pain. If all is well another bulging membrane appears at the vulva and this is soon seen to contain one or two fore feet and the calf's nose. This membrane also ruptures and a little fluid is discharged. There is usually a pause for a rest before the period of the greatest effort which forces the shoulders of the calf through the cow's pelvic bones and the head and both front legs hang from the vulva. The next stage is easy – the narrower hindquarters are forced out and the calf has arrived.

In a normal calving no human help or interference is needed. The only thing to be sure about is that the calf's nostrils are not covered by membranes that would prevent it from breathing air into its lungs. It is best to leave both the cow and calf resting for a while where they lie. The cow gets up shortly and this breaks the umbilical cord that still joins her to the calf. The cord tears at its natural weak point and may bleed a little but it is best left alone. The cow will then set about cleaning the calf up with her tongue.

The cleansing or afterbirth

The cow still has a bloodstained collection of torn membranes dangling from her vulva. These should not be interfered with, unless they drag on the ground in which case they can be lifted up and tied with string to a higher part of the bundle. They should not be cut or pulled upon. In a few hours, or possibly only after a day or more, they will drop out, pulling with them a whole lot more of the wrappings in which the calf developed. The hanging part should not be cut off because its weight is gradually separating those parts that are temporarily retained in the cow's womb.

Abnormal calvings

Calves are sometimes born backwards, hind feet first. This does not usually cause any trouble though you may be puzzled for a while that there are feet showing but no head.

If the cow is obviously making no progress there is something seriously wrong. The whole process should be steadily moving stage by stage and if, allowing for reasonable rest periods, you think the calf is stuck, veterinary help is necessary. If there are two front legs (with knees, not hocks) and no sign of the head, the head is probably bent back on the neck and no amount of straining by the cow and pulling on the legs by you and your friends will do any good. The vet will push the calf back inside and free the head with special equipment.

The calf may try being born backwards with its hind legs pointing towards its head, really in a sitting down position. You may be able to feel its tail and pulling on that won't do any good at all. That is another problem for the vet.

One of the worst complications arises when the calf is simply too big to be born at all. The vet will give the cow an anaesthetic and bring out the calf through her flank. This can happen as a result either of having used a bull of a big, bony breed on a small breed cow or of breeding from too young a heifer. It is worth taking advice from the cattle breeding centre about the kind of bull whose semen would be suitable for your cows or heifers.

Abortion

Abortion does not occur very often in cattle unless there is a brucellosis infection present. With brucellosis, abortion may occur as a 'storm', with 50 per cent or more of a group of cattle losing their calves. In many countries cattle are now tested for this disease and affected animals are destroyed so the disease is becoming less common and abortions less frequent.

If an abortion occurs it is most important to collect up the foetus and its membranes in a plastic sack. They may be infectious for other cattle which should be kept off the infected piece of ground. Keep the plastic sack for your vet who may want a laboratory test, and isolate the cow until the vet advises that it is safe for her to join the others again.

Barren cows

If you fail to get cows in calf within a reasonable time it becomes uneconomic to keep them and they are sold for beef. There are various reasons why cows do not breed and some of them can be cured by simple veterinary treatments but, unless the cows are particularly valuable, anything more than simple treatment consumes so much time and money that it is not worth it. If a cow has twins, one bull calf and one heifer calf, the heifer does not develop normal sexual organs and cannot be used for breeding. She is known as a freemartin. White heifers of the Shorthorn breed are likely to suffer from a defect of the womb which may make them difficult to get into calf.

Rearing

Details on feeding and weaning young calves are given in the section on *Feeding* pages 177 to 180.

Calf pens

Calves should always be housed in separate pens as they are not old enough for a competitive existence and separation prevents them developing the bad habit of suckling each other's ears. It also ensures that each calf can be given the careful supervision that is so necessary at this age.

The pen wall may be solid or you can use simple open fencing to keep the calves apart. It is convenient if the pens can be dismantled for scrubbing and disinfecting after they are vacated, in preparation for the next batch of babies. The pen floor should be sloped 1:30 towards the front to allow drainage to an open gulley. Some pens have metal or concrete grille floors to allow drainage to a pit below.

Disturbing the bedding creates extra work and unpleasant smells and is not necessary. A little fresh straw is added each day and the whole pad is removed after two months when the floor can be scrubbed clean.

Ventilation

The ideal temperature for calf rearing is 55°F (12.8°C) but ventilation is more important than warmth. A constant supply of fresh air is needed without draughts. This may mean leaving doors, windows and roof ventilators open even in inclement weather, which will lower the temperature considerably, but this is much healthier than what is commonly known as a fug – a warm, damp atmosphere with little air-change. A fug is cosy. Avoid it at all costs. It encourages disease germs and the calves start to cough. Open the doors and windows before they start dying of pneumonia.

An open-fronted shed is not the answer. Even when there is a wind it is surprising how stale the air can become at the back of the building. Through ventilation is one of the most important provisions for maintaining calves in good health.

Construction

Pens of 6ft × 3ft (2m × 1m) are small enough to make handling the calves easy and large enough to allow for adequate movement even after two months' growth. Each pen should be bedded with enough straw to make a comfortable bed, a little clean straw being added each day. The front of the pen, which also is the door, should be slatted vertically, with one cross-bar, so that the calf can put its head out to reach a water bucket hung on one side and a bucket for pencils on the other. Hay can be placed in some soft wire fencing mesh bent over the division between two pens with each end bent up to form a holder for the hay.

Calf pens, for use during the first three months.

Calf vaccination and worming

Young calves may need to be vaccinated against the infections that cause pneumonia, diarrhoea and scours, but if healthy stock is available and good hygiene is maintained, vaccination is not always necessary.

When calves are turned out to grass they cannot avoid being exposed to worm infection – both bowel worms and lung-worms. They can be vaccinated against lung-worms and should be worm dosed in July and in the autumn to deal with bowel worms and fluke.

In many districts vaccination against blackleg is needed and this should be done two weeks before the calves are turned out to grass.

Single sucking

This is a rather pompous title given to the natural raising of calves – when the cow has her calf out at grass and raises it herself. This is the ranch system and apart from general supervision of the herd, growth and progress are left to nature.

Single sucking is also practised in pedigree herds where,

Young stock can be penned between two gates hinged to the same post for purposes such as giving injections.

A calf suckling its mother.

under much closer supervision, each cow is allowed to raise her calf so that it never lacks the benefit of plenty of real milk, soon supplemented with rich pasture and cereal food, and so can reach its full potential of growth. Breeders and feeders of cattle like these say that half the pedigree goes through the mouth – not true! But it is true that when stock are raised in this expensive way they show off the advantages of generations of careful selection and well deserve the high prices they fetch.

Multiple sucking

As an alternative to bucket feeding some calf raisers prefer to let a nurse cow do the work for them. There are amiable cows which will suckle their own calf and two or three others for a month and then, when these are considered well enough started to go on to other rations, accept half a dozen more for a further month and follow those on with yet more batches. One farmer started seventy calves off in a year with the help of such a foster mother. The calves are penned like other calves but are allowed to suckle from the cow two or three times a day, saving the farmer all the bother of mixing feeds and keeping them at the right temperature. It might be worth trying, but bovine paragons are not met with every day.

Calf diseases

Preventive measures
The most important thing about calf infections is that most of them can be avoided altogether if the calves are raised in clean, dry housing with plenty of fresh air and are given clean and fresh milk and water, hay and calf nuts regularly from washed containers. Strong healthy calves are much less likely to fall victims to infectious diseases than undersized and weakly ones.

Diarrhoea or pneumonia or blood poisoning are almost certain to occur if the calves have been born into dirty surroundings or are housed in dark, damp and badly venti-

lated buildings. Mixing with other calves in the market exposes them to infection and the journeys, handling and exposure to cold on their market day undermines any natural resistance they may have. Disease germs get into the calf's body through the mouth when the animals lick each other and from food and drink. They get in through the nose and lungs when the calf breathes in germs with dust or through the navel just after birth. The germs usually come from other calves or older cattle, or from dirty buildings, manure, soil or even from blood and bone meal fertilizers.

If you breed your own calves and allow them to have colostrum from their mothers for a few days, they are off to a good start. If you are buying direct from a farm you should insist that they have the same access to colostrum for several days and be careful to avoid chilling them or mixing them with other calves on the journey home. If you are buying from a market your vet may advise antibiotic injections as a precaution against infection.

Treating sick calves
Calf diseases occur very quickly and unless immediate attention is given to sick calves they are likely to die. Many of the diseases that affect calves during the first few days or few weeks of life are due to infectious germs. In many cases early treatment with antibiotic injections will save their lives. Penicillin was the first antibiotic and is still very useful but there are now many different antibiotics, each suitable for particular conditions. Your vet will know which is most likely to be helpful for the calf problems on your farm. Antibiotics are not the whole answer. If the calf has an infection it is likely to develop a fever and stop eating and drinking. Antibiotics will help but careful nursing is also necessary to keep the calves alive while the antibiotic is taking effect.

Sick calves feel the cold so they must be kept warm without being deprived of plenty of fresh air. They go off their feed so they may have to be bottle fed. Calf diseases spread from calf to calf and if the attendant is looking after healthy calves as well as sick ones he should attend to the healthy ones first and then go on to the ones that are ill. Sick calves should be moved to a building away from the healthy ones if possible. After dealing with the sick calves the attendant should wash well in a disinfectant solution and wash down his overalls and boots before returning to the healthy ones.

For warmth, calves should be fitted with a calf-blanket or a sack cut to fit to conserve body heat. Ventilation should not be reduced but clean dry bedding may be increased and the pen should be protected from draughts.

Feeding sick calves
Feeding sick calves is very important. If they have bowel trouble (diarrhoea or scours) or a fever, they have difficulty in digesting whole milk or milk substitute. Their calf nuts should be taken away for a whole day. They should be given water to drink at blood heat (quite warm to the hand but not hot). They are not likely to take this voluntarily so it should be poured down their throats from a pint bottle a little at a time allowing them to swallow comfortably

without choking. One or two pints (0.5–1.0l), according to the size of the calf should be given four or five times during the day. The next day a warm milk and water mixture can be used, one third milk and two thirds water. On the third day the mixture can be half milk and half water and then full milk (or substitute) can be fed again.

After the first day sick calves should have a small supply of clean water, hay and pellets in front of them to encourage them to start eating and drinking again as soon as possible. Fresh food and drink should be given twice a day. The food they haven't eaten should be carefully disposed of by burning on a bonfire or in an incinerator. It may seem wasteful to destroy good hay and pellets but the quantity is small and there is serious danger of spreading disease if food that has been slobbered over by sick calves is fed to other calves or left around where other cattle may reach it.

Calf diarrhoea
If the calf's droppings are so soft that they run from the anus and dirty the legs the calf is not digesting its food properly. It should be put onto the diet of water and milk by bottle described above and injected with antibiotics. If these cases are not taken in hand immediately they may develop into calf scour.

Calf scour
Calf scour or white scour is an infection that spreads rapidly. The calf's droppings become creamy in colour and consistency and are plastered round the anus and on the wall and floor of the pen. Antibiotics are urgently needed and the water and milk by bottle course of treatment should be applied.

Joint ill
This is usually only seen in calves a few days old though it may appear later. The limb joints swell up and are painful. The calf has a fever and will not feed. The water and milk by bottle diet should be adopted and antibiotics are required. The navel should be washed clean and any scabs bathed off it and the area covered with antiseptic powder, as it is through the navel that this infection usually occurs.

Calf diphtheria
In this disease the calf develops thick grey membranes on the tongue and inside the cheeks. The membranes peel off leaving raw and painful ulcers. If treated early antibiotics will clear up the disease. Food is liable to accumulate in the mouth, bulging the cheeks. This should be carefully pulled out. Mastication is difficult and the calf may be offered sloppy gruel in which a few calf nuts have been mashed up. Dry calf nuts should be taken away, but soft hay may be eaten. The water should be changed frequently to encourage the calf to keep drinking and all normal feeds should be given.

Coughing
Young calves often develop a cough, which is most noticeable when they get up after they have been lying down. This is a sign of lung or throat infection and very often it means that the ventilation is unsatisfactory in the calf house or that they are overcrowded. Attention to these points may be sufficient but a calf that is coughing and off its feed may need antibiotic treatment. Coughing may rapidly develop into pneumonia. Coughing can also be due to husk or other worms and suitable worm treatment should be given if this is suspected.

Pneumonia
A calf with pneumonia breathes very rapidly, usually has a discharge from the nose and coughs frequently. Antibiotics are urgently required as well as attention to warmth, although fresh air is a necessity. Pneumonia is especially likely to occur in calves that are kept in damp and cold conditions. It can appear very suddenly and may be rapidly fatal. The water and milk by bottle feeding treatment may be applied if the calf is not feeding.

Rickets
Calves a few weeks old may show signs of rickets by walking stiffly with slightly swollen but not painful knee and fetlock joints. Sometimes the leg bones appear bent. These calves need fresh air and exercise and a dessertspoonful (10ml) of cod liver oil a day for two or three weeks. Plenty of sunlight will help prevent animals from developing rickets.

Other calf diseases
Calves suffer from a number of the diseases of older cattle described in the next section, such as ringworm, husk and tetanus, and the treatment is the same.

Health and disease

Signs of health
Healthy cattle are alert and active. Even when they are lying down and chewing the cud they take notice of what is going on. Signs of good health, whether the animals are in thin, medium or fat condition, are the same: clear, bright eyes; a moist and clean pink lining to the mouth, lips and nostrils; a moist muzzle; soft, flexible skin that can be picked up in a fold between the finger and thumb, and soft, glossy hair.

Rumination
There is normally a hollow near the top of the left flank just behind the last of the ribs. When the animal is not feeding this hollow can be seen to distend and sink back at intervals of a few minutes. This is caused by the rumen, the largest of the four stomachs, returning a quantity of the partially digested food to the mouth for chewing over again.

Fermentation of food with the production of gas is a natural part of the digestive process in cattle and the gas is discharged by frequent belching – a sign of health in normal cattle. When the animal is suffering from indigestion this activity stops, wind accumulates from the fermenting

1. *Cattle can be cast (made to lie down) by means of a rope tied in a loop around the neck and half-hitched behind the elbows and in the flank.*

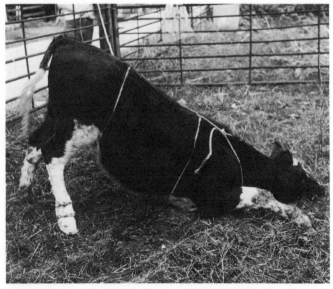

2. *Pulling back on the loose end tightens the half-hitches and the compression causes the animal to lie down.*

Larger cattle can be penned between two gates 2 ft (0.3 m) apart for injections and other treatments.

foodstuff and the flank, instead of being hollow, becomes permanently distended.

Breathing, heart rate and temperature

At rest cattle breathe about twelve to sixteen times a minute. The act of breathing is most easily observed in the movement of the wall of the belly at the animal's flank. The heart beats about fifty times a minute. This is not easy to observe. When the animal has been galloping about or when it is very hot from weather conditions or from fever, the heart may be felt beating a lot faster than fifty times a minute if you press the flat of your hand on the cow's chest just behind the left elbow. The breathing is also increased by exercise or heat, usually keeping to a rate of about one breath to four heart beats.

The temperature of healthy cattle is about 101°F (38.4°C) in the morning, rising to a degree or so higher by evening. The temperature of cattle is taken by sliding an ordinary clinical thermometer, dipped in water, gently into the animal's rectum. Temperatures of over 102°F (39°C) are usually a sign of some disease, although temperatures may be raised temporarily by severe exertion or very hot weather. Temperatures below 100°F (37.8°C) are an indication of impending collapse but, as low temperatures are unusual, this finding should be checked. A low reading may be due to wind in the bowel preventing the thermometer from coming into contact with the wall of the rectum.

Fast breathing, a fast heart rate and a high temperature may be due to some temporary cause such as being chased by a dog or exceptionally hot weather. All these signs settle down to normal in half an hour or so if the animals can be rested or moved to cool conditions. Persisting fast breathing, fast heart beats or high temperatures are signs of disease.

Grazing routine

Cattle are gregarious animals, choosing to move around in a group, filling in the day with spells of grazing interspersed with periods for rumination during which they lie out in the open or, if they need shelter from weather conditions and flies, in buildings or woodland. During rumination the food they have eaten is regurgitated in small quantities for further mastication. They spend about as much time in this process as they do in grazing. When cattle are used as working animals, as they still are in the tropics, eight hours are available for work, the remainder of the time being required for grazing and ruminating. Cattle do best if you arrange for them to work in two widely separated spells of four hours apiece.

You should take note of any animal that separates itself from the herd as it is likely to be sick or injured.

Cattle at rest
Cattle usually lie on their mid-line with the legs on one side close to the body and their heads up and off the ground or turned into the flank. In pleasant summer weather they sometimes lie stretched out on their sides for a short while, but not for long as they cannot bring up wind in this position. Any animal lying flat for any length of time should be checked to make sure it is not ill.

First aid box
The following is a list of the basic equipment and medicaments which you should keep permanently in store for carrying out routine first aid on cattle.

Drenching bottle or doping gun (syringe)
Thermometer
Teat siphon and bougies (for blocked teats)
Trocar and cannula (for bloat)
Antiseptic lotion
Wound powder (in puffer box)
Udder cream
Strong tincture of iodine (for ringworm)
Eye ointment
Eye powder (in puffer box)
Cotton wool
Bandages
Adhesive tape (for poultices)
Worm medicine
Epsom Salts (magnesium sulphate)
Milk fever injections (and flutter-valve)
Grass staggers injections
Medicinal liquid paraffin
Tonic and cudding powders

Two additional items which it is useful to keep, but which you should only use on veterinary advice, are antibiotic cream tubes (for treating mastitis) and general antibiotic injections which will deal with a variety of infections.

Parasites
Ticks
Ticks are short legged spider-like creatures that occur in great variety. They live by burying their heads in the skin of animals and sucking blood. They are harmful in large numbers as they cause the animal to waste from loss of blood but even in small numbers they are a menace as they spread the disease redwater and, in tropical countries, they also transmit heart-water, gall sickness, trypanosomiasis and other serious cattle diseases.

The common castor bean tick is so called because, behind its small head and eight legs, it bulges out to resemble a slightly crinkled shiny grey bean which may reach half an inch (13 mm) in length. This tick is found all over the world in rough pasture.

Ticks attach themselves to animals as they graze and they often congregate around the eyes, ears and anus. In small numbers they may be got rid of by wetting them with turpentine or paraffin oil or by snipping them off the skin with scissors. If they are pulled off they leave an ulcer in the skin. In some climates they are so numerous that regular dipping of cattle or even daily spraying is necessary with tick-killing chemicals but, even then, frequent inspections are necessary for ticks that have survived. One of the most satisfactory ways of reducing tick infestation is to graze cattle on clean pasture instead of on rough scrub.

Lice
Lice are grey or grey-brown wingless insects about a tenth of an inch (3 mm) long. They live by sucking blood and frequently occur on cattle around the head and shoulders, causing enough irritation for the animals to rub the hair off in patches or even to rub the skin raw. Lice mostly affect cattle housed or yarded for the winter and if present in large numbers they can cause anaemia. The lice can be seen moving among the roots of the hair. They are particularly active on warm days. Their eggs (nits) can be seen as isolated objects attached to single hairs.

Small numbers of lice may be dealt with by several applications of derris powder dusted on at intervals of a few days. More extensive infestations can be treated by using a cattle dip mixture either for actual dipping or as a spray.

Warble treatment with an organo-phosphorous insecticide, applied to cattle in the autumn, also kills any lice that are present and this may be sufficient to keep a group of cattle free of lice for the winter.

Warbles
Warbles are lumps up to the size of a hazel nut that appear on the backs of cattle in the spring. They are caused by the grubs of warble flies.

Warble flies resemble bees and buzz loudly as they fly, but do not bite or sting. They are actively on the wing on hot summer days flying around groups of cattle with the intention of laying their eggs on the animals. Their buzzing sets the cattle galloping off with their tails straight up in the air, an activity known as gadding. Gadding cattle are liable to break fences and injure themselves.

The warble flies lay their eggs in hairs on the cattle's legs and bellies. The eggs hatch and the minute grubs bore through the skin and find their way to the region of the

Worming injections.

animal's gullet where they grow and pass the winter. In spring they migrate to the back lying just below the skin where they each produce a lump with a breathing hole at its apex. When the weather warms up they burrow out through the skin, being now half an inch (13 mm) long and as thick as a lead pencil. They take cover in the ground where they pupate to emerge two or three weeks later as warble flies.

The damage associated with warble flies and their grubs is considerable. The gadding causes loss of meat and milk production as well as injuries, while warbles on the animal's back produce spoiled meat known to butchers as 'licked beef', as well as serious blemishes on the hides as leather.

Warbles are most satisfactorily dealt with by applying systemic insecticides in the form of organo-phosphorous preparations. These are supplied as a liquid which is poured onto the backs of cattle at the beginning of autumn. A further treatment may be given in the spring to kill off any grubs not destroyed the first time. (Throughout the winter and early spring the grubs are actively on the move. Killing them at this time may be dangerous for their host.) In countries which have made this treatment compulsory, warbles have been eradicated completely.

Redwater (Haemoglobinuria/Babesiosis/Piroplasinosis)

Redwater is caused by a minute parasite that lives in the red blood corpuscles of cattle, destroying the cells, releasing the haemoglobin they contain and producing anaemia. The parasite is known as Babesia (or Piroplasma) bovis, which produces a confusion of names for this disease. Redwater is spread from one animal to another by ticks which periodic-

ally drop off one animal and attach themselves to another.

The most noticeable sign of redwater disease is the colour of the urine which becomes red or brown due to the release of haemoglobin colouring matter from the blood corpuscles. Affected animals have a high temperature, go off their food, lose condition and become anaemic which is obvious from the pallor of the membranes of the mouth and eyes. Some animals die after only a few days. Others, less seriously affected, may not even show discoloured urine but become anaemic and rapidly fall off in condition. Diagnosis of the disease can be made for certain by laboratory findings of babesia parasites in the blood.

The treatment of redwater is to inject affected cattle with medications that prevent the babesias from multiplying in the blood, but it is just as important to get rid of the ticks by dipping or spraying the cattle so as to prevent re-infection (see *Ticks* page 187).

Careful nursing of redwater cases is necessary. They should be moved quietly from the pasture to a building because of their weakness from anaemia. Feeding should be with nourishing, easily digested food and water should be amply supplied as this disease makes the animals extremely thirsty.

Worms

Worms are constantly present in small numbers and may be causing serious losses of production even though the droppings are normal. In large numbers they cause diarrhoea. Anyone caring for cattle must be aware of worms – roundworms, tapeworms and liver fluke. The roundworms live in the intestines and in the lungs, the tapeworms in the intestines and the flukes in the liver.

Intestinal worms and lung worms produce enormous numbers of eggs which are passed out in the droppings and hatch out on the pasture land, the minute larvae climbing onto grass blades to be eaten and swallowed. Small numbers of worms are fairly harmless. Warm, damp weather conditions and the constant use of the same pasture allow the worms there to reach immense numbers and to cause diarrhoea, wasting, pneumonia and death.

Worm control is a matter of regular dosing and good pasture management. Spring grass triggers off a surge of worm activity and all the cattle should be given worm treatment once spring has turned to summer. This should be repeated in the autumn. If the animals are to be yarded for the winter the autumn dose is best given two weeks after they come in.

Worm medicines can be given mixed in the feed, drenched from a bottle, by a dosing gun, or injected under the skin from a hypodermic syringe. There are medicines which deal with both intestinal and lung worms, and you should ask your veterinary surgeon for advice on the most suitable preparation as he will be familiar with the worms most prevalent in your district.

As regards pasture management, calves should be put out on pasture that has not had cattle over it earlier in the year. Older cattle can follow the calves without much risk of harm. In general cattle should be concentrated on fairly small areas that they can graze right down. The area can then be harrowed, fertilized and rested from cattle completely while the grass re-establishes itself. Distribution of the droppings by harrowing and exposure to sunshine rapidly kills off many of the worm larvae. In winter frosts have the same effect. Using a field for an arable crop or even for hay reduces the worm presence almost to nil.

It is interesting that cattle worms only affect cattle. If the young larvae are eaten by horses, sheep or pigs the worms do not develop. Each of these animals has its own varieties of worms but they do not affect cattle.

Ringworm
Ringworm is caused by a fungus which grows on the skin and hair of cattle, usually when they are calves. Ringworm patches are most commonly seen on the ears, face and neck but they may appear anywhere as grey scaly patches of wrinkled skin. The hair falls off as the fungus spreads outwards forming round bare areas seldom reaching more than 2 in (50 mm) in diameter though if there are many of them they may join up to form large scaly areas. The hair eventually grows again, new growth appearing in the centre of the patches.

To treat ringworm, wash the affected parts with a 10 per cent solution of washing soda in hot water to soften the skin and then scrape off the scabs and dead hair with the blunt edge of a spoon. A gentian-violet solution, strong tincture of iodine or some other ringworm dressing can then be applied. The scabs and hairs should be burnt.

A preparation called Griseofulvin can be fed to the whole group of cattle or calves. This clears up active cases and is also very effective in preventing spread of the disease.

Ringworm is more likely to spread amongst calves that are in poor condition or those that are kept too long in old stabling or in crowded conditions. Improved feeding and management and getting the calves out to grass, even for a few hours a day, is likely to reduce the amount of ringworm occurring on a farm.

Husk
This is the name given to roundworm infection of the lungs because the chief sign of the disease is a persistent husky cough. The treatment of husk is the same as for worms affecting the bowel, but, for husk, there is a worm vaccine, *Dictol*, which is given to calves by mouth to give them a strong resistance to lung worm disease.

Liver fluke
Liver flukes are flat worms that commonly infest the livers of cattle and sheep. The flukes flourish especially where there is damp and marshy land because they need to spend a part of their development in a mud-snail. The signs of liver fluke are not often clearly apparent though massive infestation can occur causing diarrhoea, jaundice and sometimes sudden death. Beef farmers may hear from the slaughterers that the livers have been condemned and it is estimated that even a small number of flukes may lower beef or milk production by as much as 8 per cent.

Applying skin dressings.

Liver fluke disease can be controlled by dosing the cattle in autumn and again during the winter with *Flukanide*, *Zanil* or *Trodax*, and by reducing the number of mud-snails on the farm by spraying damp areas with copper sulphate or other chemicals, but the best control of all is to eliminate wet pastures by land drainage.

Mastitis or inflammation of the udder
Mastitis is an infection inside the udder by bacteria which gain entry through the teat. The large udders of modern dairy cows provide ideal nourishment on which the bacteria can flourish. The bacteria produce poisons which inflame the udder, reduce the quantity of milk produced and change its composition and, in serious cases, lead to illness or even death. Some degree of mastitis is present in any dairy herd. It is easily spread from cow to cow and constant care and attention are needed to keep its effects to a harmless minimum. In parts of the USA it is a reportable disease.

Fore-milk cups
The first sign of mastitis is usually that there are small clots in the milk, especially in the first milk drawn off. These can be detected by using a fore-milk cup before starting milking. The cup has a black plastic plate over which the milk runs, leaving any clots clearly visible on its surface. Other signs may be some difficulty in milking and a slight falling off in quantity. The cow often appears to be in good health otherwise.

Cause and treatment
Before the antibiotic penicillin was produced, the usual cause of mastitis was a streptococcus, *Str. Agalactiae*.

Penicillin, injected in a soft ointment through the teat canal, was so successful that mastitis appeared to be mastered. However, other bacteria, among them *Str. Dysgalactiae*, *Str. Uberis* and *Staphylococci*, have taken over. Treatment is still by antibiotics but these have to be varied from time to time because the infecting bacteria develop a resistance to any medication which is used persistently.

Sub-clinical mastitis and cell counts

One of the problems of mastitis has been that many cows have infected udders but show no clots in the milk and no other sign of the disease, but their milk production is reduced (possibly by as much as 10 per cent) and there is the risk that it will spread to other cows. This situation has now been overcome by routine laboratory examinations for an abnormal increase in the number of cells in the milk. Milk from healthy udders may have up to a quarter of a million cells in each cubic millilitre of milk. Above half a million there is a serious mastitis problem. Purchasers of milk by contract from farmers usually arrange for a monthly cell count on the milk from each herd.

Control of mastitis

The following measures should be taken to keep mastitis within reasonable limits in a milking herd.

1. **Veterinary advice**: Arrange for veterinary supervision and advice on the treatment and control (or prevention) of mastitis. Cell count records are important for this.
2. **Udder cleanliness**: Wear smooth rubber gloves before using the fore-milk cup. Wash the udder with running water and dry it with disposable towels.
3. **Teat dipping**: After milking each cow, as soon as the milking cups are removed, dip each teat in an antiseptic iodophor or hypochlorite preparation.
4. **Milking machines**: Daily maintenance of the machines ensures absolute cleanliness and mechanical efficiency.
5. **Dry cows**: As soon as any cow's lactation is finished a long-lasting antibiotic preparation should be injected into each teat. This destroys any slight infection present and will prevent new infections developing while the cow is dry.
6. **Culling**: Do not retain cows whose records show a persistent liability to mastitis. They are a source of infection to other cows.

Considerable protection against this disease is afforded by injecting long-acting antibiotics into the teats of cows as they dry off and also giving the same protection to pregnant heifers that are at grass during the late summer when flies, which carry the disease from animal to animal, are most active.

Summer mastitis

Summer mastitis, caused by bacteria identified as *Corynebacterium Pyogenes*, produces a sudden swelling of the udder which is extremely painful. The udder is at first red and tender and soon becomes cold and changes in colour to purple or black around one teat or over a larger area. The affected animal is obviously seriously ill and often dies quite suddenly. As this disease may occur in dry cows or in heifers while they are at summer pasture, frequent checks are important to ensure that the animals are behaving normally. The disease is readily detected in milking cows as the milk from the affected quarter soon becomes thick and bloodstained. The udder should be bathed with hot water and frequently milked out. Urgent veterinary attention may save some cases but the affected quarter is unlikely to produce milk again.

Other diseases

Acetonaemia or Ketosis

Acetonaemia is a serious digestive problem that may affect milking cows, especially those housed throughout the winter. Afflicted cows fall off in their milk production, refuse their food, become constipated and lick at the walls or stall fittings. The disease is accompanied by a strong sickly smell of acetone, noticeable in the atmosphere, from the urine and particularly in the cow's breath. Badly affected animals steadily fall away in condition and may die in two or three weeks, unless treated immediately.

The cause of acetonaemia is overfeeding with concentrated food, which means that the cattle do not eat enough suitable roughage, particularly hay. This diet results in a shortage of sugar (glucose) in the bloodstream and the cow's system compensates for this deficiency by increasing the production of ketones as a substitute for glucose. Ketones in excess are poisonous and cause the disease Acetonaemia, which is also known as Ketosis. The ketones are responsible for the characteristic acetone smell.

The affected cows should be given twice daily doses of half a pint (250 ml) of glycerine in an equal quantity of water. A course of dexamethasone or some other glucosteroid injections should be given and gentle exercise should be arranged.

Acetonaemia can be avoided by ensuring that all the cows are actually eating 6 or 7 lb (3 kg) of good hay daily as well as their concentrate ration which will vary from 3 to 12 lb (1.5 to 5.5 kg) according to the amount of milk being produced. The concentrates should be reduced as the quantity of milk falls off and then gradually increased from a month before the next calving date. Molasses or maize meal may be fed as a ready source of glucose. All the cows should be allowed some exercise every day.

Blackleg or Black Quarter

Blackleg results from gas gangrene germs that live in the soil being carried with dirt into a skin wound. These bacteria flourish most actively in deep wounds away from fresh air. They produce a quantity of gas bubbles which cause a swelling on the leg or body that crackles like tissue paper. Treatment of early cases may be successful with blackleg serum and penicillin or other antibiotics, but opportunities for treatment seldom arise as only too often an affected animal is found dead in the field. If the diseased part is cut into the muscles are seen to be spongy and black. The carcass should be removed from the farm or burned to avoid adding infection to the land.

Blackleg is more likely to occur in grazing calves and yearlings than in older animals. Prevention is by vaccination which is usually done before calves are put out for the first time and it may be repeated once or even twice a year in areas known to be badly affected. In some places it is a notifiable disease.

Bowel obstruction and bloat

The most usual form of bowel obstruction is through overloading the rumen with indigestible food such as poor quality hay. Other causes are unsuitable grazing on dried out pastures, a sudden change from succulent to dry food or overeating in fresh green pasture, especially clover.

The signs of bowel obstruction are a dull appearance, a tendency for the animal to go off by itself, a lack of interest in both food and water and the cessation of rumination. The flank becomes distended and firm, the back arched and the muzzle dry.

Treatment is directed at emptying the stomach of its unsuitable load. 1 lb (450 g) of Epsom Salts in 2 or 3 pints (1.5 litres) of water should be given as a drench. If this is administered as soon as a case is detected the animal may recover rapidly. If the distended flank feels full of foodstuffs an oily purgative drench is also needed and a quart (litre) of medicinal liquid paraffin should be given followed by tonic, cudding powders and stimulant powders.

If the flank is distended and drum-like from the accumulation of gases, a condition known as 'bloat', these should be released by using a trocar and cannula. This implement resembles a dagger, $\frac{1}{4}$ in (6 mm) in diameter and 9 in (230 mm) long, in a thin metal sheath. The whole thing, including the sheath, is plunged to the hilt, through the skin and into the distended rumen. The dagger is withdrawn, leaving the sheath (the cannula) in place. Gases escape through the sheath and relieve the animal of the uncomfortable tension, often enabling the rumen to resume its duties, helped along by the oil and stimulants which should have been given by mouth.

The cannula may be left in position to relieve any further gas that may form, by tying a piece of string from the cannula's pierced flanges around the animal's body. As soon as any sign of appetite returns the beast should be offered attractive food such as green grass, thin gruel and milk as well as water. Water may have to be given as a drench to maintain sufficient fluid intake.

Acute and severe cases of bowel obstruction may require a surgical operation to remove the obstructing foodstuff.

Brucellosis (contagious abortion in cattle)

Brucellosis is caused by infectious bacteria that flourish especially in the womb of pregnant cattle, often causing abortion. In some herds in which the disease is active, a large number of cows lose their calves at about the same

The trocar and cannula.

time in an 'abortion storm'. The disease can spread to human beings, particularly to herdsmen and veterinary surgeons who may be handling affected cows or the aborted calf and its membranes. Other people not in contact with cattle may catch the disease by drinking untreated milk from cows with brucellosis. In humans the disease is known as Undulant Fever, giving rise to a variety of aches and pains and sometimes to severe depression, but not to abortion. Horses may catch the disease from associating with cattle or from milk. In horses the disease causes lameness and troublesome abscesses.

Treatment of individual cases in cattle is not effective nor has the vaccination of young stock been successful in controlling brucellosis. There is a very accurate blood test and in the interests of both human and animal health all cattle should be blood tested and affected animals slaughtered. A number of countries have, by following this course, completely eliminated the disease. Blood tests on cattle and regular laboratory tests on milk supplies can ensure continued freedom from this infection.

Cows may lose their calves by abortion from other causes than brucellosis. It is a wise precaution to collect the dead calf and the afterbirth in a plastic bag for a veterinary opinion as to the cause of the mishap, and the cow concerned should be isolated from other cattle in case the cause was something infectious.

Cow pox

This is an infectious disease of cattle in which the animals have a raised temperature for a few days, during which period they lose their appetite and give less milk. The disease is seldom serious but it is a nuisance as it causes a skin eruption on the udder and teats which makes milking painful. The eruption takes the form of nodules that develop into blisters. These burst to leave ulcers that scab over, and eventually the scabs dry and drop off. This blistering may also occur on the vulva, between the hind legs and around the mouth. In bulls the scrotum may be affected. Cases of the disease clear up in three or four weeks but the problem is likely to linger on as it spreads to other cows. It is not now a common disease.

Cow pox can infect people as a mild disease. It is used as the basis for the human vaccination against the much more serious disease of smallpox. People recently vaccinated can cause an outbreak of cow pox if they are dealing with cattle before their vaccination scabs have dried up.

Diarrhoea

Cattle droppings vary in consistency according to the food they are eating. In winter, if they are housed and fed mostly on hay and straw, the droppings are firm. On fresh-growing grass in the summer the droppings are moist and spread out to make a cow-pat on the ground. If the cattle pass motions more watery than this, the faeces run on the tail and hind quarters and down the legs, irritating the skin and causing raw patches with loss of hair. This is a condition of diarrhoea, which may be due to an infection known as Johnë's Disease (see page 193), or to worms or overfeeding.

Eye problems

Cattle whose eyes are injured shed tears down the face and keep the eyelids closed or half-closed. They resent having their eyes examined, especially if they are injured, and they close the lids up tight or turn the eye up into its socket so that only white shows. If you turn the animal's head and grip an ear with one hand and the jaw or the nose with the other, so that the injured eye is facing the ground, the eye almost remains open with the dark cornea visible. In this position it is extremely awkward to observe the eye or to apply a dressing to it, but in some cases it is the only way to get a look at the surface.

A common irritant is an oat-husk, part of the scaly covering of a grain of oats, that flattens itself on the eyeball. You may be able to wipe it off towards the inner angle of the eye using a wedge of cotton wool tipped with an eye ointment. It is useful to have a small tube of eye ointment from your veterinary surgeon or a chemist for use on any animals with eye injuries.

Ulcerated eyes may shed not only tears but thick yellowish discharge as well. Ulcers may be caused by simple injury such as scratching the eyeball on a thorn, or they may be caused by an infection spread from one animal to another by flies. The usually dark cornea in an ulcerated eye becomes clouded with a white haze and there may be a red inflamed area in the centre of the haze which is the ulcer, a shallow wound on the surface of the eye. These are most satisfactorily treated with a dusting powder for eyes, which is supplied in a convenient puffer.

Fog fever

Fog fever is a disease of cattle usually occurring in the autumn in animals grazing on pastures previously cut for hay where some hay has been left and has become mouldy. The signs of fog fever are rapid choking breathing with the tongue protruded, and coughing. The cause is believed to be an allergy, an extreme sensitivity to the spores that occur in mouldy hay, which is also the cause of the human disease known as Farmer's Lung.

Fog fever is usually rapidly fatal. Less serious cases may be helped by frequent fifteen-minute spells of oxygen from a cylinder administered through a tube leading to a funnel held over the muzzle, and by injections of adrenalin. Other cattle in the group should be moved quietly (as exertion is liable to trigger off the disease) to a fresh pasture, not one on which hay has recently been lying.

Foot and mouth disease

Viruses are minute living particles that multiply inside animals' body cells, sometimes with poisonous effects. The foot and mouth disease virus affects cattle more commonly than other animals but the disease can occur in pigs, sheep, deer and hedgehogs and, rarely and mildly, in human beings. The virus spreads from animal to animal in saliva, in the breath or in moist air carried by the wind. It can also survive for long periods in meat and bones and become active again when these come into contact with live animals. For this reason pigswill must be boiled for an hour before it is fed to the pigs.

In cattle the signs of foot and mouth disease are watery blisters that appear on the tongue and around the claws of the feet. In milking cows these also develop on the udder and teats. In a day or two the blisters burst leaving painful ulcers. Saliva dribbles in long strands from the mouth, the blistered feet cause lameness, and the raw teats interfere with suckling and milking.

In some tropical countries foot and mouth disease is not controlled. Outbreaks frequently occur and, while some young cattle die as a result, many of the stock are resistant to the infection and, if they do become ill, will recover in a few weeks with simple nursing. There are vaccines available but their protection is not long-lasting.

In most temperate countries the economic effects of foot and mouth disease on milk and meat production are found to be so serious that, if an outbreak occurs, strict controls are enforced either by immediate vaccination of surrounding farm animals or by a slaughter policy under which all affected cases are slaughtered and the carcasses burned or buried on the spot, while surrounding animals are killed for meat. These measures have proved very effective in maintaining freedom from foot and mouth disease over long periods of time.

Any animals showing signs that might be due to this disease must be reported immediately to a veterinary surgeon or to the police.

Foot rot or foul in the foot

Foot rot, or foul in the foot, is caused by a fungus-like infection from the soil which develops in small wounds and cracks in the soft skin between the claws of the feet. These wounds develop into ulcers and abscesses discharging pus and the part is extremely tender. The chief sign of the disease is lameness with swelling around the foot and up the leg.

Cases of foot rot are most likely to occur in cattle that have to walk through mud, especially in frosty weather when their chapped skin is easily injured by the frozen ground. Animals left untreated may go off their food, become feverish, be unable to rise and die of blood poisoning.

Treatment is by injecting sulphamezathine or an anti-biotic daily for several days. In cases known to be foot rot the infection is usually cleared up remarkably quickly. Lameness with swelling above the foot may be due to other causes as for instance a nail penetrating the hoof or a stone between the claws, and these cannot be cured by injections. Badly injured feet may require to be poulticed.

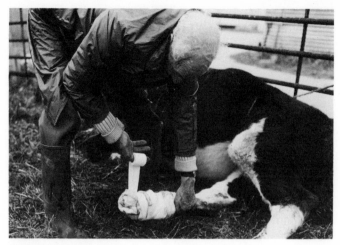

Foot treatment.

Foot poultices
Animals with badly cut or injured feet may require poulticing. Prepared poultices are obtainable from chemists and saddlers but a satisfactory one can be made from cotton wool.

Soak a 12-inch square (300 mm × 300 mm) of cotton wool in a quart (litre) of hot water in which a tablespoonful of flour and a tablespoonful of Epsom Salts (magnesium sulphate) have been well stirred. Wrap the hot poultice round the washed foot, cover with a plastic bag, apply a good layer of dry cotton wool over that and bind the whole lot in place with 3 yards (3 m) of wide adhesive tape. On dry bedding this should stay on for several days. It is inadvisable to leave a poultice on for more than three days before inspecting the injury, cleaning it up and re-poulticing if necessary.

Grass staggers
Grass staggers or grass tetany is likely to affect cattle dramatically a few days after they have been turned out to lush pasture in the spring. Affected animals become excited, fall to the ground in convulsions and die, often in less than an hour after the first signs of illness. There is usually a quantity of froth blown from the mouth and nostrils. Less severe cases show a longer period of nervousness and excitement.

Grass staggers is due to a lack of magnesium salts in the bloodstream. This is brought about by a diet that is too high in protein, a condition found in rapidly growing pasture, especially on ground that has recently been heavily dressed with fertilizer. Excessive feeding of yarded cattle on greenstuffs in the autumn may also produce grass staggers. The disease is most likely to affect mature cows, but bulls, heifers, bullocks and young stock may also be struck down in this way.

Treatment is by injecting 14 oz (400 ml) of a 25 per cent solution of magnesium sulphate under the skin. This can best be administered by using a flutter-valve which slips over the bottle top allowing the solution to flow by a rubber tube to a large hypodermic needle pushed through the skin of the neck or over the ribs. The chances of recovery depend on how soon after the onset of illness treatment can be given. If one case occurs in a group of cattle it is likely that further cases will follow. The herd should be moved to poorer pasture and carefully watched so that any animals showing excitement can be treated immediately.

Grass staggers can be prevented by using calcined magnesite either as a pasture dressing at 10 cwt per acre (one tonne per hectare) or by adding 2 oz (60 g) of it to each cow's daily ration during the flush of spring grass. The risks of this disease occurring can be reduced by avoiding a protein rich diet. This can be achieved by feeding a quantity of hay each day before turning the cattle on to the green pasture or, in the case of autumn yarded cattle, giving them hay before making greenstuff available.

Johnë's Disease
Johnë's Disease is due to an infectious bacillus that multiplies in the bowels of cattle causing a severe and persistent watery diarrhoea. The disease is spread to other cattle grazing over pasture contaminated by droppings from infected animals or through contaminated drinking water. The effects of Johnë's Disease are not seen for a year or more after infection. The droppings become loose and the animal progressively loses condition till it is reduced to a state of skin and bone.

There is no cure for Johnë's Disease. It is very difficult to eliminate from a herd as the calves are likely to be infected soon after birth but do not show signs of the disease until they are fully grown. There is a skin test for the disease, using a diagnostic agent called *Johnin*, but the test is not entirely reliable. Healthy animals may acquire the disease from contaminated land several months after all the diseased cattle have been removed.

Johnë's Disease is less commonly met with when water is supplied in cattle troughs. The practice of allowing cattle to walk into and defecate in pools from which they drink encourages the persistence of the disease more than any other factor.

In most cases the only satisfactory course is to close down the herd and use the land for arable purposes or, since

Johnë's Disease is almost entirely confined to cattle, for other animals. It would be safe to re-stock the land with cattle after two years.

Ketosis
See Acetonaemia, page 190.

Leukosis
Leukosis is a slowly developing disease in which hard swellings occur under the skin and in various parts of the body. The disease is progressive and eventually fatal but this may not be for a number of years. There is no treatment. The cause is a virus infection. Leukosis is becoming more common throughout the world and is a notifiable disease in many countries where animals found by laboratory examination to be suffering from leukosis have to be slaughtered.

Milk fever
Milk fever affects cows a day or two after they have produced a calf. After a short period of staggering about, the cow collapses to the ground and becomes quite comatose, usually in a fairly normal lying position with the head turned round to the flank. Those that lie over on their sides should be propped up with straw bales as the stretched out position encourages gas from fermented food to accumulate in the bowel with possibly fatal results. If they are not treated, animals remain in a state of collapse and die in a few days.

Milk fever is due to a shortage of calcium in the bloodstream and this is associated with the activity of milk production. The disease is almost entirely limited to cows that are heavy milkers.

The treatment of milk fever consists of injecting 14 oz (400 ml) of a 20 per cent solution of calcium borogluconate under the skin. This can best be given by using a flutter-valve which slips over the bottle top allowing the solution to flow by gravity through a rubber tube to a large hypodermic needle pushed through the skin of the neck or over the ribs. In some cases small amounts of magnesium and phosphorus are required in the solution to bring about a cure. Recovery is often as dramatic as the disease, the cow gradually regaining consciousness so that she is able to stand, take food normally and care for her calf within an hour or two of treatment.

The disease is not due to any deficiency of calcium in the diet. There do not seem to be any satisfactory ways of preventing milk fever occurring in heavy milkers after calving, though it is suggested that several injections of vitamin D before calving may be helpful.

Pink-eye (Infectious Keratitis or New Forest eye disease)
This disease causes inflammation of the front of the eye, recognized at first by the fact that the animal keeps the lids of the affected eye closed with tears flowing down the face on that side. Later the eye becomes clouded over with a white film, sometimes surrounding an angry red area. This develops as an ulcer which may penetrate into the front chamber of the eyeball so that the eye collapses and that eye becomes blind. The cause is an infection from an organism known as a Moraxella. This can spread rapidly in a herd of cattle, and is probably transmitted by flies.

Early treatment is very effective in limiting the damage to the eye. A number of antibiotic ointment preparations are available and effective but it is often difficult to apply ointment to the surface of the eye in cattle. A mixture of antibiotic and anaesthetic in powder form is obtainable in a puffer pack and this is much easier to apply.

Snow-blindness
Snow-blindness is not due to the effect of strong white light but to frostbite, as a result of extreme cold. The front of the eye clouds over with a bluish film and there may be a red inflamed area spreading in from the white rim of the eye. Such signs are difficult to see as the eyelids are swollen and are kept tightly closed if any attempt is made to examine them closely. The condition is extremely painful. The remedy is to move the cattle from the exposed conditions into shelter where most cases will recover. Eyes that continue to run with a thick discharge should be dressed with antibiotic ointment or preferably with a dry combined preparation of an antibiotic powder and an anaesthetic from a puffer pack which is obtainable for this purpose.

Sun scald (photo-sensitization)
Cattle which have areas of white hair on unpigmented skin are liable to a condition of sunburn which only affects those areas. In white faced animals, the eyelids are particularly sensitive. The skin becomes inflamed with a moist discharge. In severe cases the skin of the whole white patch may dry up and slough off as a large scab. Affected animals become obviously distressed and may die if they have no shelter from the sun. The condition is related to certain foodstuffs which sensitize the cattle to the effects of sunshine and cause the liver to produce poisonous substances.

Varieties of clover and hypericum (St John's Wort) and buckwheat are known to have been associated with this condition. If cases are detected in time the cattle will recover rapidly if moved from that particular pasture and housed or provided with adequate shade. Laxative medicines such as magnesium sulphate are of help. The offending crop need not be wasted. It may be safely used as hay or silage. If it must be grazed cattle should only be allowed on it for a few hours at a time. It should be borne in mind that if the crop is causing photo-sensitization in white-marked cattle it can also be affecting the livers of dark-coloured cattle and it would help them too to be given alternative grazing for a part of each day.

Teat problems
If there is difficulty in drawing milk from one of a cow's teats the cause may be a narrowing of the teat canal from injury or mastitis. A fibrous thickening can sometimes be felt in the teat. This may be overcome by using a milk-siphon, a metal tube, closed at the tip with an opening at the

side. This is passed gently into the teat canal beyond the obstruction, allowing the milk to flow out through the tube.

Cleanliness is essential in using a siphon as it is liable to carry infection into the udder and cause mastitis. The siphon should be cleaned in disinfectant, boiled for several minutes in clean water and wrapped in a clean piece of gauze each time before it is used, and the teat should be carefully wiped over with an antiseptic solution. An antiseptic bougie (which resembles a small wax taper) can be passed into the teat and left there after you have drawn off the milk with the siphon, to prevent the obstruction closing the teat canal between milkings.

In many cases a better course is to allow a calf to suckle the cow. She may lose the affected quarter but the risk of an active mastitis occurring is considerably reduced.

Sores on a cow's teats are a particular nuisance as they cause trouble at milking time. They may be caused by simple injuries, by the chapping of soft skin in cold wet weather or by cow pox infection or foot and mouth disease. The simple treatment of sore teats is to wipe them over with a mild hypochlorite antiseptic solution and then apply sulphanilamide powder. If the skin is harsh and scabby the powder should be mixed with a little medicinal liquid paraffin and applied to the teats as a soft ointment.

Tetanus or lockjaw

Tetanus is a disease caused by bacteria that are commonly present in farmyard manure. If tetanus germs are carried deep into a wound, away from air, they flourish and multiply producing poisons which affect the nervous system.

The chief sign of the disease is muscle tension. The tetanus poisons cause the muscles over the whole body to contract and remain contracted so that they feel as hard as wood. The animal stands with its head poked forward and its legs propped out. Feeding is difficult because the tension of the jaw muscles makes it hard for the cow to open its mouth. The beast moves with difficulty if it can move at all. In the early stages noise or fright cause further spasms in the muscles but later they become so tense that they can contract no further. If the animal loses balance and falls over it remains on its side with the legs stuck out just as a knocked over wooden toy would lie.

Horses and sheep as well as human beings are very susceptible to tetanus and it is usual for these animals and the people who work with them to be protected against the disease by vaccination. Cases of tetanus are not common in cattle but they do occur and cattle can be well protected by a mixed vaccine prepared against tetanus, blackleg and other infections caused by air-shy bacteria.

Signs of tetanus may appear a few days to a few weeks after dirt has gained access to a wound which may be from accident or from an operation, such as castration. Cases that speedily develop dramatic signs are likely to end fatally. Those in which the disease progresses slowly have more chance to recover with treatment. Affected animals should be kept in a quiet and darkened shed to avoid nerve stimulation. They may be able to take water and nourishment in the form of sloppy mashes from a bucket or else feeding by stomach tube may be necessary. Tetanus serum is given to counter the effects of the poison and sedatives and subcutaneous injections of a 25 per cent solution of magnesium sulphate help to relax the muscle tension.

Tuberculosis

Tuberculosis is caused by the *bacillus tuberculosis* which can grow and poison the tissues in almost any part of the body. This was a widespread disease in cattle in temperate areas of the world but over the last thirty years it has been almost eliminated in many countries by government action.

Tuberculosis may affect cattle for years or even for life without giving any external signs of its presence but advanced and active cases show signs of lung disease and enlarged glands in various parts of the body including the udder. Cows with affected lungs start coughing and thereby spread the disease germs in the sputum. The infection is also spread from diseased udders to the milk and so to calves and other animals as well as to human beings. This last point has made the elimination of the disease an urgent matter in the interests of public health.

There is a very reliable skin test for tuberculosis and by its regular, and in most countries compulsory, use, affected animals are detected and slaughtered before they reach the stage when they might spread the disease. All cattle need to be skin tested at intervals because isolated cases occur from time to time, sometimes traced to the spread of infection from a human being. A few years ago cases of tuberculosis in cattle were traced to a group of infected badgers.

While there are medicines which may cure cases of tuberculosis the consequences of spreading the disease are so serious that treatment of affected cattle is not allowed. Prevention of the disease depends on the regular use of the skin test followed by destruction of cases that show a positive reaction.

Wire in the heart (hardwear disease; traumatic pericarditis)

Cattle frequently swallow pieces of wire, nails and other metal objects. Because of their weight these items are liable to lodge low down in the second stomach which closely adjoins the heart. If the wire is sharp the digestive movements may drive it through the stomach wall and into the membranes surrounding the heart, where it can form an abscess leading to pericarditis. The signs of the condition are, at first, those of indigestion with variable pain and sometimes coughing. The animal may go off its food, run a high temperature and show signs of pneumonia. Sudden death may occur at any time.

If a case is suspected a metal detector may confirm the diagnosis. Placing the animal in its stall on a large door raised 1 ft (0.3 m) higher in front than at the tail end sometimes relieves the symptoms temporarily. Opening the rumen to reach and remove the offending object may sometimes save the animal's life but this is a major veterinary operation that is not always successful. In many cases slaughter is the most economical outcome.

HORSES

Background

There is a lot of hard work to be done on small farms and it is twice as hard if the farm is hilly. Horses or stocky ponies can provide a great deal of energy. A horse with a saddle can save the farmer's legs when he goes around the farm shepherding stock, inspecting fences and attending to his many routine tasks through the worst the weather can produce by way of mud, snow or floods.

A horse with a small cart can take hay and straw and other foodstuffs to stock in yards or fields, transport manure, building materials and fence posts and help in harvesting farm produce. It can also be useful in assisting with local deliveries of eggs, poultry, vegetables, manure, firewood or other saleable products and in collecting incoming supplies that may not always be delivered to your door. A horse and cart may also provide a useful small-load transport service for your neighbours. When conditions are unfit for wheeled vehicles, pannier baskets slung on each side of a horse or pony can carry useful loads.

A horse with the simplest harness, with traces to a cross-bar, can move poultry or pig houses, water troughs, field mangers and tree trunks, draw a sledge or hurdles loaded with heavy objects or be used to pull a plough, cultivator, harrow, rotary digger or cutting machinery.

Buying a horse

In some places, particularly in certain parts of the USA, there are zoning regulations imposed on keeping livestock of any kind. Naturally, before embarking on buying a horse you should first check that it will be legal for you to do so on your own particular piece of land.

Breeds

Horses come in a variety of sizes, from the Shire horse that can weigh a ton (1000 kg) and measures 6 ft (1.8 m), or eighteen hands at the withers (the highest point at the shoulder when the horse has its head down grazing) to the South American Falabella, a toy horse that reaches 28 in (0.7 m), or seven hands at the withers. The smallest working horse is the Shetland pony which is 36–40 in (0.9–1 m) high, or nine to ten hands. For convenience, in this section horses and ponies are usually both referred to as horses to save constant repetition of 'horse or pony'. There is no real difference between them but it has become customary to describe as ponies animals up to 58 in (1.45 m), or fourteen and a half hands at the withers, while those above that height are horses. (Polo ponies are an exception. Polo was originally played on small animals known as polo ponies and although nowadays polo players often ride animals well

above fourteen and a half hands, they still call them polo ponies.) The handiest animals for all the jobs about a farm fall into the 48–60 in (1.2–1.5 m) bracket, or twelve to fifteen hands. The temperamental thoroughbred race horses and their near relatives, the show ponies and the Arabs, are all too lively and delicate for farm work.

There are medium-sized breeds of horses and ponies in most countries in the world suited to providing the horse power needed on small farms. Some of the well-known breeds are listed below.

Europe
Austrian Haflingers.
British Dales, Fells, New Forest, Exmoors, Dartmoors, Welsh Ponies and Cobs, Highlands, Connemaras and Irish Cobs.
Dutch Freisians.
French Comtois, Camarguais, Bretons and Merens.
German Dolmens.
Greek Pintos.
Hungarian Nonius.
Italian Avelignese and Salernos.
Icelandic Iceland ponies.
Norwegian Westlands.
Spanish Balaeric ponies.
Swedish Gotlands.
Yugoslavian Bosnians.

Russia
Terskys, Karabakhs, Akhal-Tekes, Kabardins, Lokais, Kirghiz ponies, small Russian Heavy Draught horses, Viatkas, Kazakhs, Huculs and Koriks.

North America
Sable Island ponies, Mustangs, Appaloosas, Pintos, Palaminos, Morgans, American Welsh ponies, Mexicans, Galicenos and the Pony of the Americas.

South America
Criolos, Mangalargas and Paso Fino ponies.

South Africa
Basuto ponies.

Australia
Walers, Australian ponies, Appaloosas, Quarter Horses and Australian Stock Horses.

Asia
Turkish Native ponies, Iranian Plateau Persians and Darashouris, Burmese Shans, Indian Kathiawaris and Manipuris, and Chinese ponies in great variety.

In Middle Eastern countries, where the hot-blooded Arab and Barb breeds originated, less temperamental horses and ponies share the small farm work with camels, oxen and, above all, with donkeys of many breeds and sizes. In the world's mountainous areas the work is largely put onto mules, of which there are as many local varieties as there are of horses in the flatter lands.

Some of the horse types listed above are pure breeds with recorded pedigrees, but many are not, the name of the country or district being given to the type of horse most commonly used in those parts because it is suited to the prevailing conditions. Unless you intend to make horse breeding a part of your enterprise it is sensible to forget about breed and pedigrees. Acquire a good specimen of the horse that is native to the area where you are farming. There will then be no problem in acclimatizing the animal, and you will have a common interest with your neighbours, who will be familiar with the management and capabilities of that type of horse.

Which to choose

Age
Horses mature at four years of age and can then be worked for twenty years or more. Two-year-olds and three-year-olds need most of their energy for growing but they can be introduced to their duties and do a little towards earning their keep even at these ages.

It is sensible to buy a working horse up to ten years old. After that some depreciation is likely to set in and you don't want to spend your money on a depreciating asset – unless it is a great bargain, and unfortunately great bargains in horse dealing only too often turn out to be useless disappointments.

Appearance
Working horses should look stocky and compact. Long thin legs are right for racehorses and foals but for steady work the front legs from the bottom of the chest to the ground should not be longer than the depth of the chest. The chest should be deep and rounded to give plenty of room for the heart and lungs and the back short so that the hind legs join on strongly and really look as though they belong to the same animal.

Looking at the horse from the front the fore legs should be well separated below the chest and from there they should be straight to the feet without being bandy or knock-kneed. Horses with narrow chests and with front legs which apparently come out of one hole are short of accommodation for their heart and lungs.

Viewed from behind, the hind legs should appear to be set well apart with plenty of muscle between them at the thighs and they too should be straight without deviations to the feet.

Viewed from the side, the back of the front leg should be straight from the elbow to the fetlock without the knee being bent back or bent forward. The slope on the front of the fetlock should continue without deviation down the front of the hoof. If the slope of the hoof is much more upright, the foot is blocky and this may lead to lameness; if the hoof line is more sloping than the fetlock, it may

1. *Draught horse power is applied through collar and chains.*

2. *It only requires a simple harness.*

3. *The chains are kept clear of the horse's legs by a swingle tree or spreader, which may be linked to a harrow or to other machinery.*

4. *A pole is used to spread the pull to the width of the machinery.*

5. *Work can be done with a single horse . . .*

6. *. . . or it can be shared by several horses linked to a single pole.*

be the sign of a weak, shallow foot or it may merely indicate that the shoes have been left on too long: this last condition can soon be corrected by the blacksmith.

The hind leg, from the side, should be straight from the point of the hock to the back of the fetlock. If this line shows an angle 6in (150mm) or so below the point of the hock, this is an indication of weak hocks and general lack of driving power. The slope of the hind feet is likely to be slightly more upright than the fore feet.

Size

A big horse is obviously stronger than a small one but, size apart, a four-square, compact animal is much stronger than a tall, gangling, long-necked, long-backed one whose thin legs are likely to suffer from strains and concussion. Besides this, accumulated experience has shown that big horses are much more susceptible to disease and lameness than smaller ones, and diseases and lameness can lead not only to big veterinary bills, but to the expensive embarrassment of having a horse that is incapable of doing any work. An insurance policy covering vet's fees, permanent loss of use and death of the horse from accident or disease is a good investment.

How to buy

If you mention in public that you are thinking of buying a horse, there will be no dearth of suggestions and advice. Riding school and trekking stable owners are helpful people who may know of suitable animals in your area. When you have discovered a horse that might suit you it will take some time to find out if you and the animal can get along together and, if at all possible, you should establish this by borrowing it for a while for a trial run. If you have no way of getting acquainted with a horse you may have to buy one without an introduction.

Horses may be bought by auction at horse sales or fairs, from professional or amateur dealers, from friends or by answering advertisements. There has always been roguery associated with horse dealing and it still goes on. The amateurs can afford to be less scrupulous than the professionals, and the know-alls will overwhelm you with advice that may well prove expensive.

Horse dealers

The real experts on horses are equine veterinary surgeons and horse dealers. The veterinary surgeon is not usually concerned to find you a horse. When you have selected one he will tell you what is wrong with it for a considerable fee. This may prevent you from buying the wrong horse but it won't find you the right one. Your best course may be to put your requirements to an established horse dealer and ask him to find you a suitable horse for a fortnight's trial. Horse dealers are hard-headed businessmen and you must be equally hard-headed. There are various laws to protect buyers from deception, but going to law over a horse is usually too expensive to be of much help.

If you consider the animal the dealer produces is worth

the money, you should ask him for a written note of the horse's age and a warranty that it is sound (that is, fit to do the work required) and free from vices (vices include kicking, biting, jibbing, bolting and being intractable over grooming, harnessing and shoeing). It must be agreed that he accepts a cheque post-dated to fourteen days after you take delivery. If you allow yourself to be talked out of this arrangement so that you part with your money before the end of a two-week trial, the dealer can lumber you with one of a series of useless animals but you will never get your money back.

To get back to your trial horse. If you are happy with the animal, the dealer cashes your cheque and is happy too. If you find the horse unsuitable for a good reason, tell your bank to stop the cheque. If you are in doubt about your reason, it is worth while asking for a veterinary surgeon's opinion about soundness or vices, or a blacksmith's if it is a shoeing problem. Then let the dealer know you are returning the horse, telling him why. He is entitled to be very awkward if your complaint is frivolous.

If you approach a dealer openly with the proposition suggested above, he is not likely to let you down – but stick to your plan. If he refuses to deal on your terms, you are no worse off and should go elsewhere. If you let him jolly you out of your foolish suspicions and ungenerous approach (which, I lay odds, he will try) your troubles will be of your own making. Holding back your money is the only card you have to play.

Capital outlay and maintenance

Initial costs
Buying a good working horse in the first place will obviously be your greatest single expense, although more modest horse power in the form of a pony, donkey or mule could be a way of economizing on this. Saddlery must be of good strong material and workmanship for safety and will last a lifetime but leather goods are always expensive. As well as a saddle and bridle for riding, you will probably need a breast collar, saddle, breeching and traces for draught work and a small vehicle. All these items, with the help of friends, neighbours and a little shopping around, could be acquired part worn but still serviceable for a good deal less than their normal retail prices.

Maintenance
Maintenance items are stabling, feeding, labour, shoeing costs, saddlery repairs, veterinary items and insurance. There is usually some building on a farm that is suitable for stabling a horse; food is mostly grass, which is a natural product of the farm; and you or your family are likely to be providing the labour, so you must do your own business costing to value those items. In addition, the horse will probably require a ton (1000 kg) of hay and up to half a ton (500 kg) of horse nuts as supplementary feeding and half a ton (500 kg) of straw for bedding.

Shoeing costs vary according to how many sets of shoes the horse wears out. Regular work on roads wears out a set of shoes in a month. Working on grass they may last three months but they should be removed for the feet to be levelled every month – a necessary process, comparable with cutting one's finger nails.

Veterinary expenses and insurance
You should allow for routine veterinary items such as worm treatment, tetanus vaccination and tooth rasping but, in addition to this, veterinary attention over diseases and accidents can be very expensive. This is one item that can be covered by insurance. For about 6 per cent of the value of your horse you can obtain cover for the animal's veterinary fees in connection with illnesses or accident, and full compensation for its death or permanent incapacity or if it is stolen. The theft of saddlery is also covered.

Feeding

Horses at grass

Horses are content at grass. To stay healthy they need food, a water supply, exercise and fresh air. As for food, grass is their mainstay. They are continuous feeders, grazing for long periods during the twenty-four hours with a number of breaks, during which they doze on their feet, slumber on the ground or, in very favourable conditions, sleep flat out on their sides.

As long as their teeth are in order and they are regularly treated for worms they can obtain all they need out in the fields. They vary their grass diet with other herbs that grow in the pasture or on the hedge banks; they browse off the twigs of bushes and the bark of trees for the bitter, spicy taste; they eat a certain amount of earth, and they enjoy eating softwood fences and gates. All these items contribute to a complete diet. Grass supplies nourishment and the bulky roughage which is necessary for proper digestive activity; the herbs add flavour and a helpful variety of vitamins; and the horses get their essential minerals from the herbage and from the soil. Nobody seems to know if the browsing and wood eating are beneficial or just mischievous pastimes. The fact remains that horses do serious damage to trees that are not protected by seasoned hardwood fencing.

Horses are not tidy grazing animals. They are inclined to eat down closely over particular areas, pass their droppings in others and leave some parts untouched so that the field soon looks patchy. They thrive best on short grass, so if there are cattle or sheep on the farm (which graze much more thoroughly) horses may graze with them or, if the pastures are being used in rotation, horses may be turned into a field after the cattle or sheep have eaten it fairly short.

Grazing in groups
Horses like company. They are difficult to keep singly and will break out of any fencing to join other animals. If there are several horses, they will be company for each other

Shoeing: **1.** *Heating a mild steel bar to shape into a shoe.*

2. *Bending the bar to shape.*

3. *Punching the nail holes.*

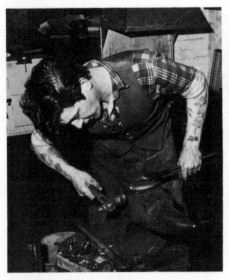

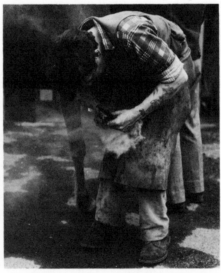

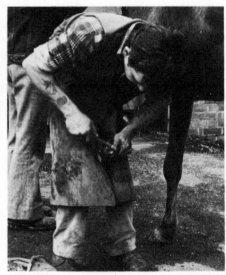

4. *Drawing the toe clip.*

5. *Checking the fit of the hot shoe.*

6. *Nailing on the shoe.*

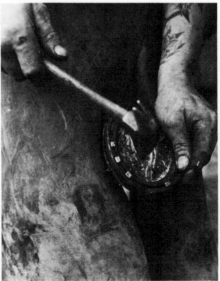

7. *Twisting off the long nail ends.*

8. *Hammering home the nail heads.*

9. *Filing the sharp points off the clenches.*

after they have decided, by dint of a certain amount of quarrelling, on a pecking order (which is their order of rank). This is quite a serious matter, especially when supplementary food is given out in the field. The top-ranking horse gets the most, and the underdog, the last in the line, gets little or none; that is the one most likely to need feeding up. This problem can only be overcome by feeding in the stable or deliberately separating the horses in some other way. As well as having a pecking order, any group of horses pairs off. If there is an odd number of horses, the one at the bottom of the pecking order, having no best friend to protect it from bullying, can really have a bad time.

Condition

The condition of a horse refers to its bodily state (fat or thin) and variations in fitness. Horses with well developed muscles and not too much fat are in fit or hard condition; those with rather less muscle and a little more fat are in show condition; while those that have spent a happy, idle summer on good grazing and look well are likely to be in soft condition, soon tiring if they are put to work. At the other extreme are the overworked and worm-ridden ones, with their bones prominent under the skin, in disgracefully poor condition.

Controlling condition

Horses can vary noticeably in condition in the space of a week or two. This puts a clear responsibility for the animal's condition onto the owner. The variations are controlled by food, work and the state of the horse's health. A horse that loses condition, if it is being reasonably worked and properly fed, must be unhealthy and the cause should be sought. The usual reasons are worms or sharp teeth which should have been dealt with as a matter of routine. If these are not thought to be responsible, a veterinary opinion should be asked for.

A pathetically thin horse, if it is young and healthy, can be improved with good food and careful management to a pleasing and fit condition in a couple of months. It takes rather longer to fine down an overweight, pampered animal that has laid down deposits of fat in its muscles and other body organs and to bring it to a condition in which it is fit for hard work.

Overweight horses

A horse at grass that is not doing much work will probably keep well on grass alone, with the help of some hay in the colder weather; but some horses, and particularly ponies, get too fat on good grassland. This renders them unfit for work, partly because they are carrying an extra weight of fat around, but also because too much fat interferes with the work of the heart and liver and may lead to heart trouble and laminitis.

Horses that get fat at grass may have to be moved to poorer pasture or fenced in on a very small area of land or even confined to the stable except for a few hours' grazing each day or each night (because flies bother horses by day in the summer, and even horses that are living out often prefer to shelter by day and graze at night). Healthy horses losing condition at grass should be moved to better grazing, or supplementary feeding should be considered.

The balance of feeding and work

Horses need exercise. Grazing horses are constantly on the move and this is enough activity to keep all their organs and their circulation active. Work and food need to be balanced. A horse at grass is capable of doing some work without extra feeding. If the work is increased, extra food must be given to keep the horse in good condition. When the work is reduced it is important that the extra food must also be reduced.

Horses' digestions are easily upset, and if a horse is being fed well to meet the needs of hard work on five days a week, its rations should be drastically reduced during the two days' rest. It is best to turn the animal out to grass for the benefit of light feeding and gentle exercise. If a horse is confined to the stable over a weekend on full rations it may suffer from indigestion, which might well develop into colic, laminitis or lymphangitis. If it must be stabled, bran mashes should be substituted for the corn feeds on the days it is kept in.

Grass

The basic food for horses is grass, either as it grows in the field or as hay. It is risky to cut green grass and feed it to horses because it ferments very quickly, especially if it is wet, making it indigestible and likely to cause colic or even death. Lawn mowings are the worst – being compact, they heat up and ferment rapidly. Nevertheless, horses which are constantly stabled benefit greatly from fresh green food such as rye grass and clover or the vetch-like plants, lucerne or alfalfa. These must be fed to the horse very soon after they are cut, and it is an advantage to mix them with the hay ration.

Hay

As a basic ration for stabled horses, hay is usually fed liberally without rationing. According to their size, horses will eat 5–20 lb (2–9 kg) of hay a day. They are happy with it at their normal feeding level on the ground or on the stable floor, but this is extremely wasteful as they trample on it and soil it with their droppings and will then not eat it because it is dirty. Hay fed in the manger is largely pushed on to the floor. It is best fed from hay nets of tarred string which can hold about 10 lb (4.5 kg) of hay. These are tied to a ring in the wall to hang at about the horse's eye level.

Hay is a very varied commodity. Meadow hay is the kind made on farms where animals are regularly grazed and some of the land is left free of stock in the spring so that a hay crop can be taken in the early summer. This is soft hay, usually a tasty mixture of grasses with a variety of herbs. As hays go it is not highly nutritious. By contrast, hard hay or seed hay is grown on freshly ploughed land as a special crop consisting of rye grass and clover or plants of the vetch family. This produces a much less compact hay with long hard stems, making a highly nutritious feed for horses in hard work.

Horses do not like sudden changes of food. If there is to be a change from one lot of hay to another, you should start on the new lot before the old hay is finished so that the horses have a gradual change from one to the other.

Telling good hay from bad

Whatever the composition of the hay may be it should have a pleasant smell at least faintly reminiscent of new-mown hay. Hay may vary in colour from its standard buff to faintly green or to burnt brown. The green hay has been dried and baled quickly without spoiling. The brown has heated after baling but is often very palatable. Hay that throws up clouds of dust when a bale is opened is dangerously dusty. Moulds from this dust can cause severe bronchitis in horses. It has probably lain about too long in damp weather before being baled.

Oats

Oats are for energy. A pony doing light work does not need any oats. If it is doing several hours' regular work, it may need 1–2 lb (450–900g) a day. Horses might need 2 lb (900g) each night and morning, and two or three times that much in the winter, while a heavy Shire horse on haulage work may require 20 lb (9 kg) or more. If more than 2 lb (900g) of oats is being fed daily, the ration should be divided. Horses enjoy oats and will overeat if they get the chance. The penalty for this indulgence may be swollen legs, urticaria, colic, constipation or laminitis. Great care should be taken not to allow horses access to the oats bin.

Oats vary in quality. The simplest checks are that they should have a pleasant smell and they should not be dusty when a handful is picked up and poured back into the sack. Good oats should not be compressible – that is, if you pick up a fistful you should not be able to compress it noticeably by tightening your fist on it. There should be a good fat seed inside.

Poor digestion of oats

Ideally, horses should grind the oats by masticating them with their back teeth so that their intestines can fully digest them. In practice it often happens that there are whole oats in the horses' droppings for the sparrows to pick over. A horse's failure to crush the oats properly may be due to uncomfortably sharp teeth or to rapid feeding because of hunger or jealousy of the other horses.

Obvious remedies are to attend to the teeth and feed the horses separately. Their rate of eating can be slowed by mixing the oats in chopped hay. A lot of people feed oats crushed, first putting them through a crushing machine. It is not usually worth while installing a machine for a few horses but crushed oats can be bought from a corn merchant. Crushed oats should have a pleasant smell. It is not easy to

assess their other qualities. The horses' improvement in condition is your only real guide.

Horse nuts

Horse nuts are often fed instead of oats. They are manufactured from grass, hay, oats and other grains with the addition of some vitamins and minerals. They have the advantage of being made up in a variety of mixtures to suit all kinds of horse, from those that are only requiring a small supplementary feed to those, like steeplechasers, that need a highly concentrated and nutritious ration. Suggested feeding quantities are usually printed on the sacks but, once again, you must make your own assessment of their value from the alteration in your horse's condition.

Laxative feeds

One of the guides to a horse's health is the state of its droppings. These are normally passed twelve to fifteen times a day, consisting of a number of boluses the size of a lemon that break on hitting the ground and are light brown in colour, turning dark quite quickly. On fresh grass the droppings are much looser and are greeny brown. (Watery droppings are a sign of illness.)

Horses that are constantly stabled and are living on dry food are likely to pass droppings of small, hard boluses – an indication that the digestive processes are working too slowly, which is a form of indigestion. This can be countered by laxative feeding. Laxative feeds are fresh grass, any of the cabbage family, apples, carrots and bran mashes. A quantity of the green food or a pound of cut apples or carrots should be mixed with chopped hay to make an attractive feed.

Bran

Bran used to be a regular and valuable addition to horse feeding. The flour content was nutritious and the bran slightly laxative. Nowadays modern methods of milling ensure that all the flour is extracted so there is very little nutrition left. Bran, as a mash, still remains a useful means of warming other foods when horses are chilled or tired and it retains its laxative properties.

Bran mash

To make a bran mash, pour 3 pt (1.5 l) of boiling water on to 3 lb (1.25 kg) of bran in a bucket and stir it thoroughly, adding 1 oz (30 g) of common salt. Cover the mash with a thick folded cloth pressed down well and let it cook itself for half an hour. Now mix in 8 oz (225 g) of crushed oats and feed when cool enough. A bran mash is slightly laxative and with the salt and oats makes an attractive feed for a tired horse on a cold day. Bran mashes must be freshly made as wet bran sours quickly.

Other feeds

Gruel

Warm drinks for exhausted or chilled horses may be made

by mixing 1 lb (0.5 kg) of either barley meal or oat meal in a gallon (4.5 l) of cold water and stirring it until it boils. Allow it to simmer for a quarter of an hour and add cold water to make an appetizing gruel.

Linseed mash

Linseed is a valuable feed for hard-working horses to keep them in condition. It is dangerous to feed the seed uncooked as it is poisonous and extremely indigestible. It must be boiled for three hours to make a nutritious jelly, putting 1 lb (0.5 kg) of linseed to a gallon (4.5 l) of water. Constant stirring is needed to prevent it sticking to the pan and burning it. It is worth while buying a double cooker to avoid all that stirring. The resulting half-gallon (2.25 l) of jelly should be mixed in a bucket with 2 lb (1 kg) of bran and 1 oz (25 g) of salt and allowed to stand until it is cool enough to mix up again and feed.

Sugar beet pulp

Sugar beet pulp is available from some sugar factories to make a cheap and nutritious food for horses. It helps to lay on fat rather than provide energy. The pump takes up water like a sponge, swelling enormously in the process. It is very dangerous to feed it dry because it swells up rapidly when the horses have swallowed it and can kill them by choking them or blocking up their intestines. It must be soaked in plenty of water for eight hours before being fed to horses mixed and chopped with hay.

If you have become accustomed to feed sugar beet pulp to your horses and you are ill or are called away, remember to tell your stand-in feeder about soaking the sugar beet pulp.

Vitamins and minerals

Horses that are at grass throughout the year acquire all the essential vitamins from their varied diet. They only need additional minerals if the land is mineral deficient and this will be well known to the local agricultural advisory organizations who will advise on what is needed as a land dressing or a food supplement.

Vitamins and minerals are essential to life and a deficiency of any of them can be very serious. This makes a convincing advertising point for the people whose job it is to sell vitamin and mineral mixtures, but deficiencies are rare in horses that are living reasonably natural lives and such mixtures are often a quite unnecessary expense.

However, horses that are stabled except when they are at work may well be deprived of some of these substances and the advertised mixtures will certainly do them no harm and are likely to cover any deficiency there may be in the diet.

Storage of feed

Grass should not be stored because it rapidly ferments and becomes dangerous for horses to eat.

Hay

Baled hay keeps in good order for two years or more though its feeding value deteriorates slightly as it becomes older and drier. It is best to use most of it in the winter after it has been cropped. Hay must be kept dry, which means protecting it not only from falling or driving rain but also from moisture seeping up into it from damp ground. If it must be stacked in the open the bottom layer of bales should be raised off the ground on a layer of stones, cinders or timber.

Stacks of bales are difficult to thatch. They are often left flat-topped and roofed with a large plastic sheet lapping a little over the edges and held in place against the wind by an extra layer of bales. This layer may suffer so badly through bad weather as to be useless, although some of it may be fit to feed to animals as the stack is being dismantled early in the winter. Stacks of hay should never be completely enclosed in plastic sheets or in unventilated buildings as this encourages condensation and the development of mouldy hay. Farmers say the hay must breathe.

Straw for bedding

Bales of straw require the same storage arrangements as hay and they keep just as long.

Oats

Oats must be kept dry and inaccessible to rats and mice. Small supplies are usually kept in metal bins with a lid that can be fastened securely. Dampness allows the oats to become mouldy. Rats and mice steal the oats and soil the remainder with their droppings so that horses will not eat their food. The fastening must be horse-proof or the horses will help themselves, which may lead to a severe colic.

Bran

Bran should be kept dry in an airtight bin. Damp bran goes sour and is unpalatable. Bran in sacks or in open bins rapidly becomes infested with maggots. Because of this fresh bran should be obtained at short intervals, weekly if possible, and the bran bin should be emptied, wiped clean and dried before the new supply is put in.

Horse nuts

Horse nuts that are kept dry are fit for horses to eat for a few months. Their nutritive value falls off with storage and it is advisable that they should be used fresh. A month's supply at a time would be reasonable, mixing some of the new delivery with the last bag or two of the previous month's supply so that there is not a sudden change from stale to fresh. There is sometimes an alteration in composition which the horses resent, and they may refuse their food if the change is made suddenly.

Stabling

Farm horses are likely to live out most of the time but you must have a place where you can deal with the horse at close quarters under cover and where the horse can have food and shelter and be available for immediate use. Horses used to be kept in narrow stalls for convenience and economy of space but this restricted their activity and it is now usual to house horses in open boxes 12 ft (3.5 m) square, or more.

Furnishings should be kept to a minimum. A manger for oats and chopped hay and a ring on the wall for a hay net are all that is necessary. Water can be supplied in a heavy rubber bucket. The manger is best in a corner, boarded in from its lip to the ground so that it presents no sharp edges or projections that could damage the horse.

Construction

Floors

The stable floor should not be slippery. Horses use a lot of leverage when getting up and, especially if they are shod, are likely to slip on a smooth floor. Hard-rammed chalk is considered most suitable but it is not widely available. Concrete with a rough finish is the usual flooring and it can be roughed up from time to time by treating it with acids which dissolve some of the cement and expose the rougher grit. Floors should be sloped about 1 in 30 towards the door so that drainage can be to a channel outside the stable to lead water to a trapped gulley, piped from there to a catch pit. Piped drains in the stable or its immediate area should be avoided as they readily become blocked.

Walls

These should be stout enough to withstand both pressure and percussion. Horses leaning or rolling against a wall or kicking it can cause considerable damage. Building blocks with a smooth concrete finish are usually strong enough. Wooden stables must be of very heavy construction if they are to withstand being kicked and chewed at. The main objective in stable construction is airiness which is achieved by floor area, height and ventilation, so the walls should be 7 ft 6 in (2.25 m) with no ceiling. It is also, of course, vital with all accommodation for horses that there should be no dangerous projections – protruding bolts, pieces of wire etc. – which could injure the animal.

Ventilation

The stable door should be made generously wide, 3 ft 6 in (1 m) or more and if possible there should be straight access to the door. Horses turning sharp corners into a narrow stable door can do serious damage to their hip bones. The chief reason for a large door is ventilation. The 7 ft (2.1 m) door should be cut horizontally at 4 ft 6 in (1.4 m) from the bottom with the top section permanently open and fastened back. It is handy to keep the top half of the door in place in case you want to use the stable at times for chickens or dogs that need to be shut in, but as far as the horse is concerned, the top half should stay open.

There should be openings at the top of the walls for ventilation. These may be useful for light as well but they should not be windows that can be shut. Ridge ventilation through the roof is the most satisfactory way of arranging

air extraction and the ridge opening should be generous. If the roof is a single slope to the back of the stable, good ventilation can be arranged between it and the high back wall. If it is the front wall which is high there should be ventilation over its top but the back wall must also supply space for the through passage of air. In countries which have long winters horses are often stabled in barns with a row of boxes on either side facing on to an open central passage over open half doors.

The importance of fresh air

Stale air in dusty stables leads to bronchitis that is chronic, and although at first this interferes only slightly with the horse's ability to work it can extend to a severe condition of broken wind with a deep incurable cough that eventually renders the horse useless. Horses that live out do not get affected in this way, nor do those that are kept in airy stables. Bronchitis is almost inevitable in horses that are kept in unventilated boxes that are only reached by passing through other stabling.

Horses do not mind the cold if they have been allowed to keep their long winter coats and are getting some exercise. Those that have had to be clipped because they are being worked hard and regularly in the winter may be kept warm with rugs and blankets but not by shutting the doors and windows and stuffing the ventilators full of sacks. That way lies coughing and bronchitis. This point is supported by observing that infectious coughing is a constant menace to race horses and show jumpers which spend a great deal of time in the stable and always need to be kept warm so that they can produce their exceptional performance.

Breaking and training

By far the most satisfactory way of breaking a young horse is to accustom it to being led about and handled as a foal, and to continue these frequent or occasional contacts through the yearling and two-year-old stages. A head collar is the really important and only necessary piece of equipment. If a foal can be led behind its mother to and from the pasture and can be tied up for short spells when she is stabled, half the breaking battle is over.

Gaining the horse's trust

Untrained horses are inquisitive but they are naturally nervous of anything unfamiliar. Training a horse is designed to overcome these nervous reactions by establishing mutual confidence. The trainer's first objective should therefore be to avoid appearing as an unfamiliar object. The more often he approaches the horse the more familiar he becomes and if he brings attractive food – carrots or oats are the handiest – his visits are welcome. The young horse should be confined in a stable or a small pen. If it is a foal it will follow its mother into the stable. An older animal can be led in by another horse. With the help of carrots, patience and time the young horse will accept food and become used to having its nose, face and ears touched and rubbed with the hands and eventually with the head collar, which can then be buckled on.

The touching and rubbing process can be extended to the neck, body and legs. In some cases the trainer may find it safer to use a pole to do the early touching and rubbing to avoid being kicked but the pole should be used as cautiously as one's own hands. If the touching and rubbing can be presented as pleasant massage the kick-reaction will not be stimulated and the horse will soon be safe to handle. If a pole is used it may be shortened each day as the young horse becomes less inclined to respond by kicking.

Tying

Tying a horse up may be a problem. If the animal has been tied as a foal it will be accustomed to the discipline. Those that have not been tied by their head harness before being broken will hang back and break the rope or, if it is too strong they may hurt themselves. Tying them up in a pen that is too small to allow them to put any tension on the rope can gradually accustom them to this control. Galvayne's system of using a crupper may be helpful. This is a piece of thin strong rope about 20 ft (6 m) long, which is placed under the tail. The long ends are twisted to make a few turns to the middle of the back and the ends are taken either side of the shoulders and tied in front of the neck. A fairly loose rope girth attached to the side ropes will prevent them slipping down the forelegs. The horse is then tied to a strong fixed point by the head collar rope as well as by the crupper rope extended from the knot in front of the neck. The ropes are tied so that the crupper rope becomes tense first leaving the head collar rope slightly slack. If the horse hangs back or flies back the crupper rope pinches under the tail and he soon gives up. Horses are gregarious animals and the early stages of handling and tying are more readily accepted if there is a companion animal not far away. Individual training which is the next stage is best carried out alone as the horse must learn to be guided by its rider and not be constantly sharing in a group exercise.

Training

Week one

The horse must learn to be led, ridden or driven. Leading is usually simple once the horse has accepted the trainer as a friend and a source of tasty food. Lead the horse with a lungeing rein attached to the head collar and let it graze around you a little. Lead it among strange objects and familiarize it with traffic.

Week two

The next stage is to teach stop and go. With the horse on about 18 ft (5.5 m) of lungeing rein an assistant leads it round. From a halt you say 'walk' and the assistant leads on. The horse may need to be encouraged by the trainer using a touch of the whip. When the trainer says 'halt' the assistant stops and gives the horse a carrot. After a while the

A general purpose pony with all the harness needed for riding.

horse will stop or go without any help from the assistant who should only control the horse if necessary by means of a short rope.

Week three

Once the horse has learned to obey the walk and halt orders without help from the assistant, it can be taught to trot. The assistant goes with the horse, walking on the order 'walk' and encouraging the horse into a trot on the order 'trot'. Again the trainer may need just to show the horse the whip. At the order 'walk' the assistant slows the horse down and when 'halt' is called, he gives the horse a carrot. The horse soon learns to act on its own and should be taught to change from trotting to walking or to an immediate halt from the trot.

During these first three weeks the horse should be familiarized in the stable with the bit and bridle and to having a blanket over its back. After that it can be accustomed to the feel of a saddle and somebody getting on and off, so that these items and activities are already half familiar when the horse is introduced to riding or harness work about a month after serious training began.

Breeding

If you have accustomed yourself to having one or two horses about the place breeding is quite a simple step, and means that, as well as the interest of producing foals, you will have young horses for your own replacements or for sale.

Mares come into season – that is, they are willing to be mated by a stallion – for a few days at a time at intervals of about three weeks during the late spring and through the summer each year. There are suitable horse and pony stallions available to mate with mares in most districts, which are advertised in horse magazines and known to veterinary surgeons, blacksmiths, saddlers and corn dealers.

It is usual for the mare to be taken to the horse when she is ready for mating.

Pregnancy lasts for eleven months. Most mares are mated in late spring or early summer, so that the foals are born at a time of the year when they will have summer weather for the first few months of their lives. Mares usually foal out in a field without assistance, and they supply all that the foal requires in the way of care and attention until weaning at about six months of age.

The time for mating

Signs that a mare is in season
The most important consideration is that the mare should be mated at the right time. The details of her periods 'in season' must be known so that arrangements can be made with the stallion owner or stud farm manager. The mare's season is recognized by her change of temperament and because she stands from time to time with her hind legs extended and her tail raised as if to pass urine with the lips of the vulva opening and closing. She is sensitive to being touched on her hind quarters and she squeals and kicks. This state persists for four to eight days, and she then returns to normal quite quickly.

When to have her mated
The important point is that the foal egg, which is the cause of all these changes, is discharged from the mare's ovary to pass to her womb two days before her season ends, and there must be fresh stallion semen there, ready to fertilize the egg if she is to produce a foal.

As the mare's season is likely to last at least four days the best plan is to have her mated on the third day – that is, two days after she has first shown the signs described. If she is still in season two days later (that is, on the fifth day), she should be mated again, and if still in season on the seventh day, should be mated for the third time. In this way there is almost certain to be fresh semen ready whenever the egg is released from the ovary.

If a mare does conceive, she will not come in season again until after she has produced her foal in the following spring. She will then come into season about a week after foaling. It is not advisable to have her mated at this time as there is an increased risk of the foal dying in the womb. She can be mated when she comes in again after three weeks or on a later occasion.

If the mare does not conceive after being mated, she will come in season at intervals until the autumn, unless she is successfully mated in the meantime.

Mating

Mating behaviour
The mare's behaviour in the presence of the stallion is important. If she is taken to him early in her season (on the first or second day) she will probably kick at him and refuse the mating. A day or so later she will approach him in an entirely friendly way and allow him to mate her. If she is still receptive two days later and again two days after that, mating should be allowed, but most mares are only in season for a few days, after which they revert to being offensive instead of friendly to the horse.

Accommodation
Because mating may be required on several days, mares travelling some distance may have to stay at the stallion owner's premises and mating can then be arranged at suitable times. Valuable mares already in foal from the previous year are sometimes sent very long distances to be mated to a particular stallion and may travel to the stud farm before they foal down and stay there until the stud groom is satisfied that they are pregnant once again.

Pregnancy

Pregnancy is usually quite uneventful. Young mares may not even lose their neat figures. Older mares often bulge considerably but even then there may be some doubt as to whether the bulge is a foal or just good living. Pregnant mares may continue to be worked reasonably until a month or so before foaling.

Foaling

Before the birth
The most certain sign of approaching foaling is that the udder swells and oozes milk, which dries on the teats as a waxy yellow crust. This indicates that the mare will foal in the next few days – usually between midnight and dawn.

A mare likes to foal alone. If she is in a field with other animals she will go off into a secluded corner. If she is stabled she will delay foaling until nobody is about. If you want to be present when the foal arrives, you should make yourself comfortable in an adjoining horse box and peer through an observation slit occasionally. It is a mistake to think that she might appreciate the company of somebody she knows at this difficult time. She may delay foaling so long that the foal is born dead. On the whole it is cleaner and healthier to allow foaling to take place naturally out in a field, regardless of the weather.

Labour
When the foal is on its way the mare becomes very restless with colicky pains, getting up and lying down frequently. She then starts straining and a dark bulging membrane appears at the vulva. This soon ruptures and discharges a large quantity of fluid. A few minutes later another bulging membrane, white in colour, emerges. One or two of the foal's feet can be seen pressing tensely inside. This membrane ruptures next, revealing the fore feet and the foal's nose. So far the mare may have been standing up to do her straining. Now she lies down and with further efforts forces out the head, followed, with the biggest heave of all, by the widest part of the foal, the shoulders and chest. The hind quarters follow much more easily.

Immediately after the birth

It is particularly important that there should be no interference at this stage. The foal has been delivered but is still attached to its mother by the umbilical cord. During the next ten minutes or so the afterbirth, still in the womb, shrinks and delivers nourishing fluids to the foal. Very soon the foal begins to struggle to get on to its feet. The mare gets up and the umbilical cord breaks naturally quite near the foal's navel. The foal has arrived.

The whole birth process takes about an hour, sometimes much less, sometimes more. If a mare is not making much progress after straining for an hour, call the vet.

Suckling foals

For the first six months after the birth the mare and foal can take care of themselves with very little attention. It is a great advantage if the foal can be fitted with a head collar and be caught up from time to time. The head collar should be adjustable so that it can be let out as necessary – a foal's head grows surprisingly fast.

The foal has to be handled for routine worm treatments and to make sure that its feet are growing and wearing normally. The occasional handling accustoms it to people and the first elements of discipline, saving a great deal of time and argument when, at two years old, it begins to be introduced to harness and the idea of work.

Feeding

Foals should grow slowly. Summer grass is nutritious and so mares and foals do not need any supplementary feeding until the early autumn, but condition rather than calendar should be the guide. If the mare begins to lose weight she may need extra food and the foal will demand its share. The mare may be given one or two small feeds a day. These can be put into a high box that the foal cannot reach if it is doing well enough on grass and milk. You must remember that you are raising foals not fat stock. Foals can readily be fattened but they will suffer from it, perhaps for the rest of their lives. Overfed foals develop joint and foot troubles and they grow soft bone so that when they are put to work they become liable to bone diseases that cause lameness and may well shorten their working lives.

Weaning

There is no need to hurry over weaning a foal. Ponies that live on the moors suckle their foals until cold weather and the shortage of food dries up the milk supply. In more comfortable circumstances a mare will remain in milk through the winter. The foal should certainly be weaned a couple of months before the next foal is due, if the mare is in foal again, but this assumes that you have just about been persuaded to become a horse breeder. If you require the mare for work the foal can manage very well without her milk when it is six months old, but there is no reason why she should not do a certain amount of work while she is feeding her foal.

Weaning a foal from its mother at six months old or over is not a dramatic event unless they are the only horses you have. If that is the situation you must put them in adjoining paddocks so that they can be company for each other across the fence but cannot get to each other for suckling. If you have other young horses they will provide the company and the foals will hardly notice being separated from their mothers.

Foals can be sold as soon as they are weaned but there is not much demand for them at that age. Most people, even those who want to break their horses to work themselves, prefer to buy their horses when they are old enough to start their training at two and a half to three years old.

Lameness

Signs of lameness

This is the most frequent problem affecting working horses. It is sometimes quite obvious that a horse is lame when it is walking. If there is any doubt the horse should be trotted in hand without anyone riding it. Lameness can then be seen and heard as a break in the normal trotting rhythm and is confirmed if the horse lowers and raises its head unevenly which it will do when it is lane in a fore leg. The horse raises its head as the foot of the affected fore leg strikes the ground and drops its head or nods as the normal leg comes down. If a hind leg is affected, the point of the hock of the lame leg is seen to go up and down through a shorter distance than the point of the hock of the good leg. In each case the horse is obviously throwing more of its weight on to the normal limb than it does on to the painful one.

Finding the cause

Lameness may be the result of some obvious injury to the limb. If not, the limb should be felt all over for pain or tenderness and then picked up and the joints bent one by one to see if they are sore. Then the foot should be examined. This should be picked up and looked over to see if there are any injuries, wounds around the coronet, broken horn on the walls, stones or other objects stuck under the shoe or discharges from the sole or the frog. If there is nothing unusual to be seen the foot may be tapped on the wall and sole with a small hammer and compressed around the rim with pincers to detect any exceptional tenderness.

If you feel fairly certain that the foot is causing lameness, it is as well to ask the farrier to take the shoe off and make a careful examination. If he detects nothing it is a case for the vet who may find a simple explanation or need to use local anaesthetic injections or X-rays to help him decide what is the trouble.

Treating the feet

The hoof is designed to expand each time the horse puts the foot down and contract when the weight is removed. In spite of this slight elasticity any injury inside the hoof becomes extremely painful because the inflammatory fluids that accompany injury increase the pressure inside the hoof. It is possible to soften the hoof by poulticing or by tubbing. This allows the hoof to expand and gives the horse great relief from pain. The softer horn is also easier for the farrier or the vet to deal with if they have to cut away part of the hoof in treating the cause of the lameness.

Poulticing

Poultices suitable for use on horses are obtainable from many chemist's and saddler's shops. They are dipped in warm (not hot) water, squeezed out and wrapped around the foot, then covered with a plastic bag and held in place by a roll of sticking plaster. A thick pad of cotton wool in the hollow of the heel protects that sensitive part and prevents the plaster being applied too tightly. If poultices are not obtainable, a simple one can be made by taking a piece of thin cotton cloth 12 in × 16 in (300 × 450 mm) and applying a thick layer of flour paste to half of it. Lay a thick layer of cotton wool on the paste and fold over the other half of the cloth. A little mild antiseptic and a sprinkling of magnesium sulphate powder can be added to the paste. When these poultices are dry they will keep until required and only need dipping in warm water and squeezing out before being used.

Tubbing

Tubbing a horse's foot consists of standing it in a gallon (4.5 l) of warm water to which 1 lb (0.5 kg) of magnesium sulphate and the same quantity of washing soda (sodium carbonate) have been added. The foot should be stood in this mixture in a bucket for twenty minutes twice a day. The hoof gets softer and softer with each tubbing to such an extent that the nails will no longer hold the shoe in place. When the tubbing is stopped the hoof returns to normal hardness in a day or two.

Hoof wall troubles

Tread wounds

These are painful wounds at the coronet, the junction of the hair of the leg and the horn of the hoof. Tread wounds are caused by the horse treading on its hoof with the shoe of the opposite foot. They are usually very painful as this is the tender area from which the hoof grows, similar to the quick of a finger nail. The hair should be clipped from the wound and it should be dressed with an antiseptic ointment. These wounds are important because the growth of the hoof wall may be disturbed so that a false-quarter develops.

Sand-crack

A sand-crack is a vertical split from the coronet down the hoof wall. It may be caused by some sudden strain put on the hoof, when, for example, the horse twists or slips on rough ground or makes some violent effort. This is a painful condition and, like a tread wound, it may lead to a false-quarter.

False-quarter

This is the name given to a line of irregular hoof horn growing down from the coronet to the lower edge of the wall. The cause is defective horn formation at the top of the hoof because of a tread wound or a sand-crack. False-quarters can be very troublesome as they are breaks in the continuity of the hoof wall. The gap may allow dirt and infection to reach the tender parts below or these may be painfully pinched between the edges of the gap.

Tread wounds, sand-cracks and false-quarters all need expert attention to ensure that permanent damage is not done to the hoof. It is often necessary to remove a triangular piece of the hoof wall, the apex pointing down, just below the coronet so that the new horn can grow without stresses from below.

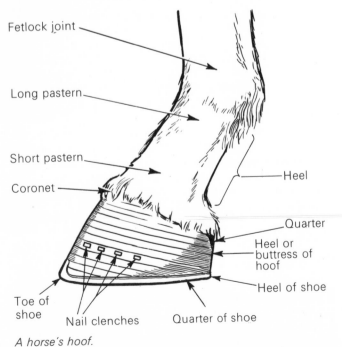

A horse's hoof.

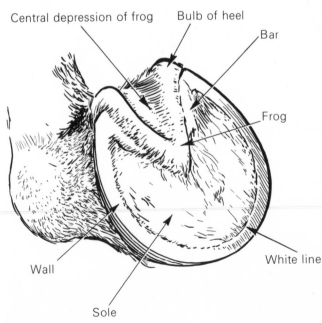

The ground surface of the foot.

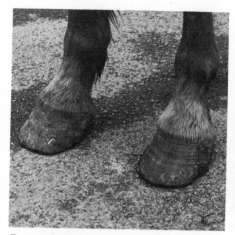

Foot trimming: 1. *Overgrown fore feet to be trimmed back and shod.*

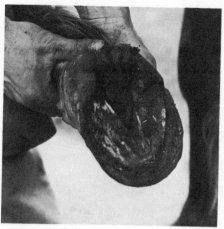

2. *The untrimmed foot, showing the overgrown wall.*

3. *Lowering the excess wall with clippers.*

4. *Further trimming with the foot knife.*

5. *The excess wall horn removed.*

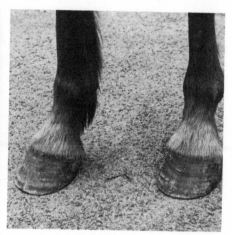

6. *The fore feet trimmed ready for shoeing.*

Broken hoof walls

Like a finger nail the hoof wall grows a little over a third of an inch (10mm) in a month. This growth is worn down by ordinary bare foot usage or is trimmed off by the farrier when he gives his monthly attention to shod feet. Sometimes a piece of the wall breaks off accidentally or is pulled off when a shoe is lost. The accident is only serious if the tender parts beneath are exposed. They should be dressed with an antiseptic ointment and covered with a pad of cotton wool held in place by a strip of plaster bandage. The tender parts soon cover themselves with a thin layer of horn. The wall continues to grow and eventually fills the gap.

White-line abscess or gravel

Under the foot the rim of the sole can be seen to join the inner edge of the hoof wall in a line of white and rather crumbly horn. Grit or small stones sometimes penetrate into the soft horn of this white line and work their way up under the wall giving rise to a painful abscess. The hoof should be softened by poulticing or tubbing and the horn may have to be cut away to open and discharge the abscess. The damaged hoof can then be treated as if it was a foot with a broken wall.

Seedy-toe

In seedy-toe a cavity spreads up from the white line under the hoof wall. Mud and gravel may pack into the space and make it painful but seedy-toe is often quite painless only being discovered when the farrier has removed a shoe. The treatment is to remove the wall from the whole of the face of the cavity, the inner aspect of which will always be found to have covered itself with a thin layer of horn. There remains a hollow on the front of the hoof which disappears as the horn grows down during the next few months.

Hoof sole troubles

Under the hoof is the slightly arched sole at the back of which lies a stout rubbery wedge, the frog. As this is the part of the horse that strikes the ground it is naturally exposed to injury from rough and sharp objects.

Picked-up stones

Small stones may wedge in between the sole or the frog and the shoe causing pain each time the foot is put down. They are usually quite easy to hook out with a hoof pick, but there may be times when the shoe has to be removed to release them. If the horse remains lame after the stone has been extracted it is probably because the sole has been bruised and the foot should be poulticed for a few days.

Bruises and corns
Treading on rough stones or other projecting objects may leave the sole bruised. Rest and poulticing will usually effect a cure in a few days. Corns are bruises in a particular part of the sole. On either side of the frog is a small triangle of sole which is covered by the heel of the shoe when the horse is shod. This area of sole is often bruised by dirt or grit getting under the shoe-heel or by the heel of the shoe itself and this is particularly likely to happen if the shoes are left on for too long. The hoof grows fastest at the toe and this drags the shoe forward so that the heel of the toe is no longer supporting the wall at the heel but digs into the sole. These bruises are known as corns. If neglected they not only cause lameness but may give rise to abscesses in the hoof. Regular and careful shoeing should prevent corns occurring. If they do develop they should be cut out by the farrier and they may need dressing with an antiseptic pad of cotton wool under the shoe heel.

Punctured sole and under-run sole
A wound through the sole from a nail or a flint is likely to develop into an abscess which spreads, under-running the sole. This is very painful and can only be relieved by cutting a fair sized opening in the insensitive horn of the sole to relieve the fluids which are likely to be under considerable pressure. The foot should be poulticed for a few days and then covered with a dry antiseptic pad for a few more days to allow deeper tender surfaces to cover themselves with a protective layer of horn.

Dropped sole
In this condition the normally concave sole bulges downwards. This is a sign of laminitis which is discussed under that disease (page 212).

Thrush
In thrush the frog is found to be wet and pitted, oozing a dark musty-smelling liquid. This occurs most commonly in the hind feet and is an infection associated with standing in dirty stable conditions. It is often accompanied by contracted heels. As much as possible of the diseased frog should be cut away without exposing the sensitive fleshy tissues below. This exposure to air often leads to a cure. An antiseptic powder dressing should be applied. A run at grass barefoot helps to restore a healthy frog.

Contracted heels
The heels of the hoof are kept apart by a large rubbery frog which is stimulated to grow by contact with the ground at each step. Without regular ground pressure the frogs are inclined to shrink and the heels of the hoof contract. It is quite a problem to keep the frogs of shod horses regularly contacting the ground, especially those which are working on hard roads. Allowing the heels to grow long raises the

frogs even further from the ground. Regular shoeing and working on soft ground are both helpful. A run at grass without shoes gives the frogs a chance to fill out again.

Troubles inside the hoof and in the heel
Ringbones, sidebones, navicular disease, laminitis
The membranes covering the bones inside the hoof can become inflamed and the increased pressure causes pain. Ringbones can form inside the foot or bulge in front just above the top of the hoof; sidebones form a swelling above the coronet on either side; navicular disease develops insidiously deep in the centre of the foot; and laminitis causes a very painful condition under the front wall. The earlier these conditions are dealt with by experts the better chance there is of the horse continuing in work.

Mud-fever and cracked heels
The skin in the hollow at the back of the pastern is very thin and flexible and sometimes a combination of wet, cold and mud will chap the skin so that it becomes red and inflamed. This is mud-fever. Some clay soils are especially likely to cause this. The constant bending of this part as the horse moves leads to cracking of the skin and it does not get a chance to heal, so that cracked heels develop.

Treatment of these conditions is to keep the part dressed with an antiseptic powder in vaseline under a bandage which must be changed several times a day. Cracked heels and mud-fever may be avoided by washing off the mud after work in cold weather and bandaging the heels for the night. White heels are much more liable to these complaints than those with dark hair.

Over-reach wounds
These occur when a front foot is not picked up fast enough to avoid being trodden on by the shoe of the hind foot. The result is that the hind shoe cuts off a slice from the heels of the fore foot. If this leaves a flap of skin it is often best to cut off the flap and treat the damage as an open wound. Over-reaches are the result of careless riding or the horse getting off balance. Horses that are inclined to injure themselves in this way should be fitted with protective guards over the heels of the fore feet. These are obtainable in the shape of a rubber funnel that fits around the pastern.

Troubles on the limbs
Strained tendons
Horses concentrate their weight and the thrust of their work on the tendons that run from the back of the knee down behind the fetlock. These tendons and, less often, the similar ones in the hind legs, are sometimes strained and the back of the leg swells and is very tender. Several months' rest is needed for them to recover.

Curbs
A strain that may occur in the hind leg appears as a small bulge on the back line of the hock, 6in (140mm) below the

point of the hock. This is a curb which may be very painful when it happens and some weeks' rest is required. The bulge persists for life but once it has settled it is unlikely to give further trouble.

Splints
On each side of the tendons there is a small bone reaching from below the knee nearly to the fetlock. These are the splint bones. Bony enlargements called splints occur along the lines of these bones. They can be very painful for a while but usually settle down in a few weeks so that lameness disappears. They also persist for life. If they continue to cause lameness over a long period, it may be because one of the splint bones has been fractured and the broken portion will have to be removed.

Spavins
The small bones low down in front of the hock may become inflamed and painful causing a hard swelling, lameness and a dragging toe. This is a spavin. Treatment is aimed at fusing the affected bones together because it is their movement against each other that causes the pain. Fortunately locking these bones together removes the pain and the horse can work normally without further trouble.

Bursitis
The horse's joints and various other prominent parts are lubricated by a substance called synovial fluid whose extent is always limited by a capsule or bursa. On account of bruises and minor strains synovial fluid is sometimes secreted in excess of the normal supply so that the capsules bulge causing unsightly swellings. These cases of bursitis are seen as windgalls which are hazel-nut sized swellings at the sides of the fetlock joints, popped knees on the front of the knees, capped hock on the point of the hock, thoroughpin up to the size of a tennis ball between the point of the hock and the leg, bog spavin about half tennis ball size on the front of the hock, and capped elbow on the elbow. As long as they are only unsightly and not causing lameness, which they seldom do, these swellings are best left alone.

Bursal swellings may also occur on the poll as poll-evil and on the withers as fistulous withers. These two conditions readily turn to abscesses and if they occur it is advisable to have veterinary advice immediately to prevent serious consequences.

Body and back troubles
While lameness is usually traceable to the legs and especially to the feet, it may at times be caused by injuries to the body. Bruises, strains and abscesses may show themselves as swellings or be found as painful when pressed upon. Rest is the best treatment to allow bruised or strained muscles to recover. Abscesses should be encouraged to come to a head and discharged by bathing them twice a day with warm wet cloths for ten or fifteen minutes each time. If the swelling is not an abscess but a bruise or a blood blister the warm applications will help the swelling to disperse.

Spinal trouble due to falls or strains or arthritis causes the horse to move stiffly especially when turning. Veterinary advice may be to prescribe rest for some time and pain killing medicines may allow even arthritic or rheumaticky horses to continue to do reasonable work in spite of their complaints. Persisting spinal troubles may require specialist treatment at a veterinary hospital.

Health and disease

Diseases associated with digestion and management
Any signs of stomach ache or indigestion in your horse are known as colic. A horse suffering from the condition stops eating, looks round at one of its flanks and generally presents the appearance of being uncomfortable. If the pain continues the horse may keep getting up and down and rolling on the ground. The cause is usually unsuitable food or food that has been eaten too fast and not properly masticated – either because the horse is hungry, or because it is bolting its food from jealousy or fear of other horses. The unsuitable food ferments producing wind that distends the bowel and is painful. An early effect is that the bowel movement stops and droppings are not passed. The food in the static bowel continues to ferment and causes more pain.

Treatment is directed to restoring activity to the bowels and easing the pain. Laxative medicine must be given and soft and easily digested food provided. Veterinary attention should be called early in colic cases to prevent them developing into acute colic.

Acute colic
This is diagnosed when the stomach upset has reached such a serious stage that the horse shows extreme distress by rapid noisy breathing, staring eyes and inflamed membranes of the eyes, nose and mouth. The cause may be poisons absorbed from the fermenting food, or the painful bowel may be so distended with gas that it has ruptured or become twisted from its own activities or from the horse rolling in pain. Such cases are often fatal though some may be saved by a surgical operation.

Lymphangitis
In this disease one leg begins to swell at the knee or hock. Within a few hours the one leg has swollen down to the foot and up to the elbow or the stifle and though the swelling may get bigger and bigger, it does not extend any further.

Lymphangitis is a disease of faulty management usually occurring in well fed and hard working horses that are on a proper ration of good food. It occurs when for some reason, such as the weekend, bad weather or an accident, work has to be stopped but (this is the fault) rich feeding has gone on as usual. When a horse's routine of work is reduced, feeding must be reduced immediately and laxative food must be given over the idle period.

Lymphangitis is not very painful considering how

enormously the leg sometimes swells but pressure inside the forearm or under the lower thigh reveals a very tender area. Treatment should consist of a purgative followed by laxative feeding, and exercise should be started as soon as possible in spells of a few minutes several times a day. Affected horses usually have a high temperature and sometimes an infection. Antibiotic injections are often very helpful. It is unlikely that the leg will ever go down completely to normal. It remains slightly thickened and if further attacks occur it is usually the same leg again, so that that limb becomes a hindrance to the horse's full working capacity.

Acute laminitis

Laminitis is also called fever in the foot. In acute laminitis the fore feet suddenly become very hot and painful. The hind feet may also be affected. The horse will not move unless forced to do so. To ease the pain it puts the feet forward so that less weight is on the front part of the foot and more on the heels. The horse's temperature rises from the normal 100°F (37.8°C) to 104° or 105°F (40.2° or 40.8°C).

The pain is due to leakage of fluids into the front of the foot between the hoof wall and the pedal bone. This is a most peculiar result of poisons circulating in the blood. The poisons may arise from some severe infectious disease such as pneumonia or from severe indigestion from unsuitable food such as overfeeding on oats or beans, or eating wheat or uncooked linseed.

The pain must be relieved as quickly as possible. Taking the shoes off and poulticing the feet would be a help but the horse will not allow a foot to be picked up because of the pain. Cold hosing the feet or standing the horse in a pond or stream cools them slightly. Casting the horse with ropes may be possible to get the weight off the feet. Anaesthetic injections to relieve the foot pain are effective for some hours. Water and sloppy mashes should be offered in the hope that the horse will drink. Purgative medicines by tube with plenty of saline should be given as soon as possible. As soon as the horse is willing to move, gentle exercise on soft ground should be given to improve the circulation and so ease the pain.

Chronic laminitis

This, like acute laminitis, causes fluids to leak into the front of the foot between the hoof wall and the pedal bone, but it occurs insidiously over weeks or months and is not acutely painful.

The cause is poisons in the blood from indigestion or overfeeding. It is especially common in native ponies that are accustomed to poor mountain or moorland pasture but are being raised on good grassland. They get fat and their feet become misshapen. The fluids leaking into the hoof displace the pedal bone downwards and the toe of the hoof grows long, the hoof front concave with irregular ridges and the hoof soles become convex and tender so that they are painful to walk on.

Treatment consists of reducing the feeding to a light laxative diet. Grass is such a diet but even it may be too nutritious and the horse or pony may have to be rationed by fencing it onto a very small area of grazing or only allowing it to graze for an hour or two a day. The farrier should be asked to trim back the feet, cut down the heels and fit a seated-out shoe, perhaps with hoof pads. As far as possible exercise should be given, and steadily increased as the animal improves.

To prevent chronic laminitis you should watch the animal's condition and balance its food intake and work. It is often idle animals living on a diet which is too rich (even though they may not be getting any extra feed) that develop chronic laminitis. It doesn't often occur in riding schools or other establishments where the horses are getting plenty of regular exercise.

Azoturia

This is another disease which results from faulty management. It usually occurs in horses that are fit and well and doing fast work at rather irregular intervals, such as occasional trips to the shopping town. If this fit horse is confined in the stable too long (and this may be a day or two in some cases or a week or more in others) without exercise and then really put to work fast it may suddenly stiffen up, like a man with lumbago, come to a halt and be unwilling to go on. The condition is acutely painful due to the sudden demand for energy from muscle cells which are choked with stored foodstuff.

The horse must not be moved. Put it into the nearest field or get a horse box and drive it home. The muscles of the back are hard and painful. The urine is dark coffee coloured and the animal has a high temperature.

Treatment is to keep the horse warm, apply warm wet cloths to the back and hope that veterinary treatment with cortisone injections and sodium salicylate will bring about a cure. Many cases end fatally.

Infections

Tetanus

Tetanus germs live in soil, especially in horse manure, and horses become infected when these germs get into wounds – even quite small cuts or pricks. Horses affected with tetanus move stiffly and, as the disease advances, become unable to move at all. The slightest noise sends their muscles into spasms and they are unable to eat, but can usually suck up water or gruel from a bucket if it is held up for them.

Treatment consists of a special serum and muscle relaxing injections. The death rate is high and horses that go down in the stable lie stiffly like a wooden horse. They are difficult to raise and will probably die. Those that remain standing and show any slight improvement in the first few days are likely to recover slowly but completely. All horses should be vaccinated against tetanus once a year.

Strangles

This is an infectious disease causing a thick discharge from the nostrils and swellings around the throat which burst and discharge a thick creamy pus. Treatment consists of feeding soft food, grass rather than hay, because chewing and swallowing are painful. The throat may be bathed with cloths wrung out in hot water twice a day for ten minutes to help the throat abscesses to ripen and discharge. The disease takes five or six weeks to clear up. It is doubtful if medicines or injections are helpful in treatment. The use of antibiotics may delay the ripening of the abscesses. Recovered horses are immune to the disease. There is no satisfactory method of vaccination.

Stable cough and influenza

Horses are subject to a cough resembling the human common cold. They run at the nose and develop a cough which persists for a few days. Treatment consists of taking the horse off work for ten days or more so that the signs disappear and the inflamed nose and lungs have time to heal up. Unless the weather is extremely bad this convalescence is best spent out in a field. The cough is infectious and is spread when horses are gathered together at horse sales, shows and so on. It also spreads from horse to horse in stables, especially when the ventilation is not efficient. When groups of horses are kept in badly ventilated stables they may suffer from the cough several times a year. Medicinal treatment is not usually helpful.

Influenza shows very similar signs to the cough, but it is usually accompanied by a high temperature and the horse goes off its food. Epidemics of influenza occur at intervals of a few years. Unless the affected horses are rested until the temperature is normal and the appetite is restored, and for a fortnight's convalescence after that, permanent damage to the lungs and heart may be caused so that the horse becomes useless.

Vaccination against influenza is very effective and it is well worth while having this done every year. There is a combined horse vaccine against tetanus and influenza. There is no satisfactory vaccination against the stable cough.

Broken wind

Broken wind causes coughing in horses. The disease is a bronchitis brought on by dusty hay and straw that contains irritating moulds in the dust. Some horses are particularly susceptible to this irritation and you should protect them from stable dust by keeping them at grass all the year round (or in exceptionally well ventilated stables on sawdust bedding instead of straw) and feeding them on horse nuts instead of hay. The early signs of broken wind are a persistent cough and a double lift of the stomach muscles when the horse is exhaling its breath after any exertion. If you continue to keep such horses in a dusty stable they develop a deep churchyard cough and thick discharges are coughed up or run from the nostrils. Such cases are not likely to recover and, though they may live at grass for a number of years, are almost useless for any work.

Lung worms

Lung worms cause coughing which seldom becomes very marked but is a worry because coughing horses are not in a fit state to work. The worms live in the bronchial tubes in the lungs. If the infestation is heavy or the horse is put to hard work severe bronchitis or pneumonia may develop. Treating with thiabendazole worm medicine, using four times the dose recommended for ordinary worming, is the most effective way of clearing up this complaint. Donkeys are frequently infected without showing signs of coughing. Not only does this mean that the donkeys will not thrive but also that they may spread the disease to horses. If you keep donkeys they should be dosed when the horses are given their regular worm treatments.

Sinus infection

Horses sometimes discharge a thick yellow pus from one nostril. This comes from the hollow bones of the head, and may be due to an infection following a cough or influenza, or to a diseased tooth root. Treatment may be by antibiotics but it is usually necessary to drain the sinuses by making openings through the bones of the face or by removing the offending tooth. Both these methods are major operations requiring veterinary skill.

Grass sickness

Grass sickness is a disease which brings the horse's bowel activities to a halt. Food and drink can reach the stomach but can go no further. The disease occurs in spring and summer, mostly on the east side of Britain and also in Scandinavia and Normandy. Affected horses become thin and make a loud snoring noise as they breathe. They swell up around the eyes and mouth, and slime and saliva runs from their nostrils and mouths.

The cause is not known and there is no treatment. All cases die, some a few hours after the first signs appear, some after months of pathetic survival. Once a case is clearly diagnosed as grass sickness, the horse is best put down.

There is a suspicion that grazing close to the ground may oblige horses to consume the cause of the disease and it is advised that if grazing is short the horses should be given a good quantity of hay a day as a measure of protection.

Worms and parasites

Worms

Horses harbour a great variety of worms which are seldom seen. They live in the bowels, feeding either on the partly digested food there or sucking blood from the bowel wall. The female worms lay enormous numbers of eggs which pass onto the ground with the droppings and hatch out in suitable weather so that the young worm larvae may climb on to grass blades and be eaten and swallowed by a horse.

The young worms do not all stay in the bowel. Many varieties go on tour in the horse's body penetrating into the liver and lungs before returning to the bowels to mature and lay eggs in their turn. Some young worms enter blood vessels, damaging the vessel walls and causing blood-clots to form, which interfere with the circulation and may cause colic, paralysis, heart damage or sudden death.

The number of eggs and young worms on a pasture occupied by horses may rapidly build up to a serious infestation, especially in a moist summer, and it is essential that all horses should be wormed at least four times a year, ideally between the spring and autumn. There are a number of useful worm preparations available. One of the bendazole preparations should be used in the spring and late summer, haloxon in mid-summer and dichlorvos in the autumn. It is best to avoid using the same preparation each time because the worms may become resistant to it.

Bots

Bot flies are black and brown, rather larger than a house-fly. They lay their eggs on the hair of horses' legs in the summer and can be seen as a peppering of yellow dots. They are most easily removed with a fine-toothed metal comb. If left they hatch in two or three weeks and the minute grubs bore through the skin and eventually reach the stomach where they live through the winter sucking nourishment from the stomach wall, and grow to a stubby grub 0.5 in (12 mm) long. In the late spring they are passed out in the droppings and in a few weeks hatch out as adult flies.

They are a fairly harmless parasite except that in large numbers they can cause stomach ulcers and severe digestive trouble during the winter. The autumn dose of dichlorvos suggested for worm treatment deals satisfactorily with any bots there may be attached to the stomach wall. Combing off the eggs from the coat once every week is another quite effective method of reducing the numbers significantly.

Lice

Lice are wingless insects 0.10 in (2.5 mm) long. They sometimes appear in horses' coats, usually in the winter, especially in the mane and tail but sometimes also on the legs. They increase in numbers in the warmer weather of spring and become noticeable because the horses rub the itchy areas against any convenient object. The lice may be seen moving at the base of the hairs on which their eggs are attached. They can cause anaemia if they are present in large numbers as some of them are blood suckers, and they can also damage the appearance of the coat where the horse has rubbed itself raw.

Treatment is by washing the horse over with a solution of louse-dressing. Most are based on gamma benzene hexa-chloride. This should be repeated once a week for at least three weeks to make sure that any young lice hatching from the dressing-resistant eggs are dealt with.

Ticks

Ticks are grey shiny parasites that may grow to 0.5 in (12 mm) in length and nearly as much in width. They live in rough vegetation and attach themselves to animals (usually about the head and neck), living by sucking blood. Small numbers of ticks may be snipped off with scissors. Large numbers are dealt with by dressing the affected parts of the horse with the same dressing as suggested for lice – a solution of gamma benzene hexachloride. Rough grazing should be avoided in the summer if ticks are a trouble.

Mange

Mange is caused by mange mites which are about 0.05 in (1.2 mm) in length. They burrow into the skin causing itchiness and the horse is liable to rub the affected areas raw. These are usually the head, neck, mane and tail, but the legs may also be affected especially in horses with hairy pasterns.

Mange is treated by weekly applications of solutions of gamma benzene hexachloride. It may take a long time to clear up and careful hygiene is required to remove the infestation from the stables, fittings and harness where the mites may persist, as well as from the affected horses.

Ringworm

Ringworm occurs as small circles of dry crusty skin from which the hair falls out. The common sites are on the neck and by the girth. The condition is irritating but if left alone it disappears after a month or two, with new hair growth spreading from the centre of the circle.

Ringworm is infectious and can easily be spread from horse to horse by harness and by grooming tools. If an outbreak occurs grooming should be stopped temporarily for this reason.

Treatment is by ringworm ointment or by spraying the skin. The disease can also be dealt with by feeding tablets of griseofulvin to infected cases, and to horses in contact with them for protection.

Heart trouble

Horses' hearts have tremendous reserves and it is not likely that they will give trouble because of overwork. Hearts are more likely to be damaged by the effects of illness or serious worm infestation.

The signs of heart trouble are that a horse seems unwilling to face up to normal work, gets short of breath (especially going uphill) and, when given a rest, takes a long time to settle down to normal breathing again. More advanced indications of poor circulation are swellings occurring around the fetlock joints, under the jaw, between the fore legs and under the belly. Horses showing these signs are likely to be losing their appetite and getting thinner every day.

Unless some reason can be found, such as worms or poisoning or a growth that can be operated on, there is not much hope for such cases and it is merciful to have them put down.

Old age and disposal

Many horses live happily into their thirties and are all the better for being kept in light work. Complete retirement is not usually a success. A horse needs a lot of attention even when turned out to grass – care of the feet, shelter from flies in summer and autumn, regular worming, attention to the teeth (especially with age) and extra feeding in winter. All these items take time and money and there comes a time when the old horse is either cruelly alone or is unmercifully bullied by its companions. Every horse owner must decide when the time has come for various reasons – common sense, economy or kindness – to put the horse down. Nobody enjoys doing it but the knacker will carry out his task with speed and efficiency.

The salvage value is worth considering and may make a considerable contribution towards the cost of another horse or pony. The horse slaughterer will not pay much for a horse that has died because he can only use its flesh as food for dogs or cats, but if he is given the opportunity of carrying out humane slaughter and professionally preparing the carcass there may be a good deal of money in it for him if he has an arrangement for the meat to be accepted for human consumption. In many countries horse meat is a highly regarded food and horses are bracketed with cattle, sheep and pigs as meat-producing animals.

Asses or donkeys

These useful animals are found in great variety in all tropical countries and many temperate ones. They were known as asses until a couple of centuries ago when donkey began to be used as a pet name for the little grey-brown animal. A lot of asses are not little or grey-brown. The Hamadan is a large white riding ass widely used in the USSR and Iran; The Leon-Zamora of Eastern Europe has long black hair, and the Kassala, the Sudanese riding ass, long grey hair; the Italian Martina Franca is black with white underparts, and the French Poitou is a very large animal over fifteen hands (1.5 m) high, classed in the Americas, where most of these breeds are extensively used, as mammoth in size. The little asses, introduced to Western Europe by the Romans around the beginning of the Christian era, and often known as Irish donkeys, fall into the classification of small asses of ten to twelve hands (1–1.2 m) or even miniatures which do not exceed nine hands (0.9 m), these last being of little use for work but popular as pets.

As farm working animals asses are strong and amenable. Pedigree cattle breeders have even been known to train their bull calves to be led at shows and sales by linking them to a donkey's halter. Donkeys are very adaptable and work well in rough hill country where their loads are often carried in panniers slung on either side of the body. They are gregarious and very willing to work in groups following a led or ridden ass. Their reputation for stubbornness is almost entirely the result of ignorance on the part of people who expect cooperation in activities for which the animals have received no training. They accept a routine even more willingly than horses.

Buying a donkey

The kind of donkey you will buy as a working animal depends on the variety available in your area and the type of work you want the animal for. Donkeys are less demanding than horses, thrive on less nutritious food and are less subject to disease.

This is a good example of a typical small donkey. The only fault of conformation is that the forelegs are slightly bent back at the knee. Its good points include the short head with a soft muzzle, broad forehead, wide-set eyes and fairly long pricked ears. The neck is short, leading to a broad, slightly dipped back and strong loins, and the chest is deep.

Conformation

The general points of conformation described for horses apply equally in choosing a donkey. Asses and donkeys have heads that vary from short ones tapering to a soft muzzle to great bony ones with Roman noses. The only important thing about the head is that it should match and balance the body. Neat little heads for Irish donkeys, great hammer heads for the mammoth Poitou asses. Whatever the variety the ribs should be well sprung, giving a deep rounded chest. The neck is shorter than in the horse and the shoulder more upright. A short back is desirable and it should blend in with strong, well rounded quarters which is the part of the beast that many Middle Eastern riders sit on. The legs should resemble those of a well proportioned horse. The feet tend to be small and narrow. Foot care is directed to encouraging the hoofs to resemble those of a horse as far as possible.

Common defects in donkeys are ewe-necks, which are necks with a concave upper profile; legs too close together where they join the body; knock knees, bandy legs and cow hocks (the points of the hocks turned in and the feet, in consequence, turned out). A donkey's pasterns are normally rather more upright than those of a horse.

Donkeys should move freely with a low, straight action of the feet. They can be trained to trot, canter and gallop under saddle or to trot in harness but they are usually content to work steadily at a mundane pace.

Feeding

Donkeys like a variety of food with a fair proportion of roughage. While horses do very well on grass pastures to which they add a number of variants as already explained, donkeys prefer common land or neglected fields where there is a wide choice of herbage – grass, thorns, nettles, brambles, hedgerows and bushes. In a garden they are nearly as destructive as goats.

If turned out in well fenced fields donkeys will graze like horses but they do not thrive on green grass alone and should be fed some hay even in spring and summer. When grazing is scarce or they need extra feeding because of extreme cold or because they are working hard, their food can be supplemented with crushed oats, corn flakes or horse nuts. The ration should be adjusted to their actual requirements. As explained in the section on feeding horses in relation to condition, it is easy to feed animals so that they get too fat and become ill in consequence, but it is difficult to get that fat off again. Supplementary feeding other than hay should be given in very small quantities until the effects of the ration on the donkey's condition have been carefully observed and then only increased, if at all, by very small amounts.

Condition

The assessment of condition may not be easy by eye alone as many donkeys carry a heavy coat especially in winter. Running a hand along the top of the neck and down the spine will reveal prominent bones in thin donkeys and rolls of fat on the neck and on the shoulders and sides of overfed animals. Assuming that their teeth are in order and that worming has been attended to, the thin ones need fattening up with more or better food and the bulging ones need fining down by giving them more exercise and less food. A hard working adult donkey can usually be kept in condition in winter with rough grazing and 5 or 6lb (2.7 or 3kg) of hay and 1–2lb (0.5–1kg) of crushed oats or pony cubes, but rationing must be by the feeder and by results, not by the book.

Survival

Donkeys do well in a warm, dry climate. In temperate countries the foals need shelter and extra food to get them through their first winter. They also pine without company. Other donkeys are best but a lone donkey will make friends with a horse or a dog or a sheep. They delight in human companionship but few humans have time enough to keep a young donkey from feeling lonely.

Breeding

Donkey mares come into season as horse mares do at intervals of three weeks, and they should be presented to the stallion daily to see if they will accept mating. When a mare does allow the jack donkey to mate with her she should be presented to him again for mating two days later and this should continue at intervals of two days until she no longer accepts him. Mares are in season for about five or six days so two or three matings are usual. If the mare conceives she does not come into season again. If the mating was not successful she will accept the jack again about a fortnight later.

Donkey mares carry their foals for a little over a year. The impending signs are a slight filling up of the udder a month before the foal is due, most noticeable in the morning. Later the ligaments on either side of the tail go slack and the vulva lengthens and its lining is inflamed. Anticipating the day or time the foal will arrive is quite impossible. Foaling is usually uneventful and follows the same course as described for horses and the same management procedure applies.

Breeding to improve a donkey type

Conformation and colour are inherited so, because breeding will tend to be only from donkeys of a locally approved shape and colour, all donkeys in a given area are likely to resemble each other. The quality of a group of donkeys can be improved by breeding only from stallions and mares that reach a set standard.

If you want to change the type or colour this can be brought about by introducing stallion donkeys of the required type, and breeding only from them. This introduction of new blood will produce a new generation showing hybrid vigour and some of their father's characteristics. To establish the new type the half-bred colts should be castrated and the fillies bred to the introduced stallions for several generations. After this time the best colts may be kept as stallions and the newly developed and improved type of donkey may be interbred without risk of reverting to the original, less desirable form or colour.

Mules

Mules are the progeny of an ass or a donkey father and a horse or pony mother. In some parts of the world mules are bred from large mares to produce animals up to fifteen hands (1.5m) high for especially hard work in mountain country. Smaller mules are useful for general duties, their size depending on the more or pony mother whose height they approach but seldom reach. Whether deliberately produced or not (because mules are often the result of an accident) they are tough and hardy animals capable of going anywhere a man can go except up a tree. They have a remarkable capacity for survival.

Hinnies, which are the offspring of an ass or a donkey mare and a horse father, are seldom bred deliberately. They resemble mules in appearance but lack some of their qualities of determination and tirelessness. Mules and hinnies are born male or female and develop normal sexual organs and characteristics but, in spite of travellers' tales to the contrary, they are always sterile and cannot reproduce when mated together or to a horse or donkey.

Mules are intelligent and respond to training as well as horses do and they share with horses a dislike of the unusual and the unexpected, though they are far less sensitive and excitable. 'Don't worry' would appear to be their reaction to signs of trouble and they calmly stand their ground in circumstances that would cause horses to stampede. Mules have exceptional physical abilities. They are strong in the body with a short, slightly roached, back and are sufficiently short in the leg to give them compact strength for carrying or pulling loads. They are straight in the shoulder making their action rather jarring as riding animals. While a horse's reaction to being struck is to kick out, the mule's response is likely to be delayed, not from mental sluggishness but because the striker was out of range at the time. When the offender moves a little nearer the mule kicks hard and effectively. On established terms of mutual respect mules are good to work with. Treat them badly and they become stubborn and spiteful. Treat them well and they will give of their extraordinary best.

Health and disease

It is possible for asses and donkeys to suffer from nearly all the conditions that affect horses but, on the whole, they remain in good health because far fewer demands are made on them than are made on horses. Horses suffer from sprains and injuries and arthritis because of their athletic activities and their large size. The smaller donkeys benefit from their light weight and modest performance and, as they are seldom encouraged to take highly nutritious feeds to produce energy for competing against each other, they are not so often exposed to the problems of colic and other digestive upsets.

Donkeys need worm treatment, just as horses do, at least four times a year. Their lungs are often affected with lung worms even though they show no sign of the disease by coughing, and this lung-worm disease can spread from them to horses in which it is likely to cause much more serious trouble. Regular worm dosing should keep the donkeys free of this problem.

Training donkeys to work

Training a foal

If you have a foal from a donkey mare its training should commence in the first few days. Put a head-collar on and lead the foal when the mare is being led. The foal will obviously comply because it wants to follow its mother, and this helps teach it to be led without resisting. Follow this up with a few minutes' handling and leading every day and by the time weaning comes in six months or so your young donkey is quiet to lead and handle and accustomed to having its feet picked up – and that is half way to being a trained donkey. From then on it is just a question of accustoming the youngster to having a saddle and rider on its back while somebody leads it about. By two years of age you will have a nice quiet donkey fit to go into simple exercises in riding or driving.

Training an older donkey

If the donkey is already two years old and has never had any proper handling you have a very different problem. First you have to gain the donkey's confidence by giving small, regular feeds. Handling all over is important, and to avoid being kicked you may need a broom handle – not to beat the donkey with, but to accustom it to being touched. Quite soon you can use a short stick and then your hand. So far you have caused the donkey not to be frightened of being touched and to like you because you bring food. Now put on a head-collar and lead the donkey behind another led one and continue as for the day-old foal.

You gradually transfer your personality to the donkey which, over the months, becomes confident and even affectionate! But remember that donkeys are intelligent, resentful of the unexpected and have a peculiar sense of humour that you may not appreciate – such as avoiding a patch of mud themselves but making sure you have to slosh through it or, when you have just done a wonderful grooming job in preparation for a special occasion, rolling in that same patch of mud and plastering as much of it as possible on your special occasion clothes. Could affection go further!

GLOSSARY

Addled egg: An egg which has begun to go bad.

Ark: A small house with ridged roof, often used for grazing for a small number of poultry or rabbits.

Baffles: Pieces of material, usually wood or plastic, which are placed in an air vent to prevent direct draught without impeding the air flow.

Bantam: Strictly speaking, a naturally occurring, small-sized fowl which has no large counterpart, but as there has been so much selective breeding to scale down standard-sized birds, the term is generally taken to mean any small-sized fowl.

Blocky: A horse with blocky foot has a noticeably more upright slope to the hoof than is normal.

Bougies: Medicated columns of wax resembling very thin candles that are passed into the teat in the treatment of blocked teats.

Bran mash: A feed prepared by pouring boiling water over bran and then covering the bucket with a sack for half an hour. It is fed whilst still warm.

Bran poultice: Bran soaked in boiling water and applied warm to the foot of a dressing of sacking or cloth, secured with a bandage.

Breast collar: An alternative to a neck collar for a horse. The breast collar is a wide horizontal strap around the lower part of the neck to which traces are attached for traction.

Breech: Presentation of a foetus at birth with hindquarters foremost.

Breeching: A strap similar to the breast collar which is hung around the horse's hindquarters. It is attached by straps or chains to the shafts and prevents the vehicle over-running the horse.

Broiler: A young bird specifically raised for meat.

Broken mouth: A condition where a sheep has lost one or more teeth.

Brooding: Raising newly-hatched chicks in a protected environment.

Broody: A bird manifesting a desire to sit on a clutch of eggs.

Buck: Male goat over two years of age.

Buckling: Male goat between one and two years of age.

Bulling: Presenting signs of being ready for mating (heifers and cows).

Cage battery: System of keeping livestock in cages in a controlled environment.

Calf nuts: Nutritive food mixture supplied by merchants in pellet form.

Candling: Examining an egg against a bright light in order to see the internal structure.

Capon: A neutered cockerel.

Cell count: A laboratory method of examining milk for health protection. Normal milk contains a number of cells from the cow's tissues. If the cell count increases above normal levels, that is an indication of disease and the farmer is informed.

Cloaca: End section of a hen's food canal, from which waste matter is ejected.

Closed breed: A breed where registration is restricted to progeny of animals themselves registered in that breed.

Coccidiostat: An additive sometimes added to propriatry food as a means of protecting livestock against the disease coccidiosis.

Cock: A male bird that has completed one breeding season (American rooster).

Cockerel: A male bird in its first breeding season.

Colostrum: The milk secreted by the udder for forty-eight hours after birth of the young. It contains a concentration of antibodies which are passed to the young and act as a valuable protection against a number of infectious conditions. The young animal can only absorb these antibodies for a matter of hours (possibly only twenty-four) so it is crucial that it should suckle colostrum from its mother frequently during that short period.

Concentrates: Highly nutritious cereal food, fed in order to maintain milk or meat production at a profitable level.

Conformation: The shape of an animal's frame which only varies with growth, as distinct from its general condition (i.e. whether it is fat or thin).

Convection downdraught: Downwards movement of cold air in a building caused by warm air rising by convection and being cooled by contact with uninsulated surfaces.

Creep: Area where young piglets spend most of their time, and which has openings too small to allow the sow to enter.

Crop: Portion of the hen's food canal where food is stored until it is ready for moving on into the stomach.

Cross-bred: Progeny resulting from the mating of two different breeds.

Cudding: Chewing over the small portions of food that have been regurgitated from the cow's rumen. (See rumination).

Culling: Killing any surplus or diseased stock; maintaining good health or productivity in a herd by killing such animals.

Deadweight: The weight of an animal after slaughter and when all the offal has been removed (virtually the weight of all the saleable meat).

Deep-litter: Floor covering in indoor poultry houses, made up of a build-up of successive layers.

Dewlap: A loose pouch of skin on a duck's throat.

Dip bath: A bath containing insecticide in which sheep are dipped to kill external parasites.

Doe: A mature female goat.

Drafting race: A race through which sheep can be driven in order to divide them into two or three groups.

Draining pen: Pen in which sheep dry off after being dipped.

Drench: Liquid medication poured down the throat from a bottle.

Drying-off: The gradual cessation of milk production by the mother after her young have been weaned.

Elastrator: Implement used to apply rubber rings for castrating lambs.

Entropion: A condition where the eyelids are turned in, usually at birth.

Fallowing: Leaving a piece of land idle between crops, usually for weed control.

Farrowing: Giving birth (pigs).

Feed conversion ratio: The amount of feed consumed in order to produce a given body weight.

Feed hopper: A container which allows food to drop down gradually as more is consumed.

Fetlock: The obvious joint between a horse's knee (or hock) and foot.

Finishing: The last few weeks before an animal goes to slaughter, during which period it is given the best attention so that it will arrive for slaughter in the peak of condition.

Flushing: Improving the condition of ewes before mating.

Fold unit: A small house, complete with covered-in run, for the controlled grazing of poultry.

Forage: Feed that is growing (e.g. grass, leaves, kale, etc.) as opposed to manufactured feed (e.g. meal).

Fore-milk cup: A metal cup fitted with a dark shelf. The first milk at each milking is drawn into the cup. If there is udder trouble (mastitis) clots of milk will be seen on the dark shelf.

Free-marten: A female calf that was twin to a male calf. In the majority of cases, the shared pregnancy so affects her that she is unable to conceive.

Free-range: Traditional method of allowing birds to graze over a wide area of land.

Gander: Adult male goose.

Gilt: A name for a young female pig until it produces its first offspring, when it becomes a sow.

Gizzard: A strong, muscular bag within the hen's digestive system, for grinding up food.

Goatling: Female goat between one and two years of age that has not borne a kid.

Grading up: The process where, by the successive use of pedigree sires, the ultimate progeny of unregistered stock become eligible for registration in an open breed.

Green hide: Skin that has been cleaned, scraped and dried, but has not yet been permanently tanned (rabbits).

Heat: The period when the female is ovulating and is ready to receive the male. It is the only time that she can become pregnant.

Heavy breed: Bird which has a high meat to bone ratio and is therefore suitable for the table.

Heifer: A young female cow that has not yet produced a calf.

Hinny: The offspring of a horse father and a donkey mother.

Hock: The large joint half way down the hind leg of a horse.

Hogg: A young yearling sheep before shearing.

Holding pen: A large pen in which sheep are held prior to being handled.

Hybrid: Offspring produced by breeding from two or more distinct lines.
Hypochlorite: Sodium hypochlorite solution used for sterilizing milking equipment by release of free chlorine.

Immunoglobulins: Protein-like materials produced in the body which inactivate or destroy antigens.
Inbreeding: Mating of closely related animals.
Incubation: The process of development of a fertile poultry egg within the shell.

Jack: An entire male donkey.

Keet: Young guinea fowl.
Kibbled grain: Grain that has been chopped into pieces rather than ground.
Kid: A male or female goat from birth to one year of age.
Killing-out percentage: The percentage of saleable or edible carcass after slaughter and butchering. This varies between 70–77% of the liveweight.
Kindling: Giving birth to young (rabbits).

Lactation: The period of milk production after the female has given birth.
Lamb bar: Apparatus for communal feeding of lambs and kids consisting of a container for milk with several teats set into it.
Lambing percentage: The number of lambs produced per 100 ewes.
Let-down: Release of milk from the udder.
Ley: Temporary pasture specially sown by the farmer.
Light breed: A hen which is less suitable for the table, but usually has a tendency to lay more eggs, than a heavy breed.
Litter: Any material used as flooring, e.g. straw, woodshavings.

Mash feed: Food, such as bran and boiled barley, mixed with hot water.
Meconium: Brown/black material passed from an animal's bowel shortly after birth.
Mite: Minute parasite related to the spider.
Moored: Moorit (peat-brown colour).
Morant grazing: System of controlled grazing for rabbits devised by Major Morant in the last century.
Musk glands: Glands which secrete an odour as a secondary sexual characteristic, and in the male goat are mainly situated just in front of the horns.
Mule: The offspring of a donkey father and a horse mother.

Notifiable (reportable) disease: Any disease that must be reported to the government health authorities.
Nursling: A calf that is still suckling the cow.
Oestrus: The period during which the female shows desire for the male.
Open breed: A breed in which entry is not restricted to progeny of animals registered in that breed (i.e. opposite of closed breed).
Overshot jaw: A condition where the lower jaw protrudes beyond the upper one.

Oviduct: Passage from the hen's ovary to the vent.

Pastern: A small joint in a horse's leg between the fetlock and the foot.
Paunching: Removing the entrails from a carcass.
Pelt: Skin.
Pessary: Vaginal suppository for administration of drugs.
Pipping: The process of breaking the egg shell by a chick before hatching.
Point of lay: Age at which pullets begin to lay, usually between 20 and 22 weeks.
Polled: Naturally hornless.
Pop-hole: A hole small enough to allow piglets to creep through to reach their mother.
Pork, Cutter, Baconer and Heavy: Classifications of slaughter weight pigs according to their weight and the purpose for which they will be used.
Poult: Young turkey.
Poultice: A hot, wet dressing applied to an injury or swelling for its softening and soothing properties.
Primary feathers: Main flight feathers.
Pullet: A young hen which has not started to lay.

Raddle: The colouring smeared on the chest of a ram to mark ewes when they are mated.
Rearer or grower ration: Mash formulated for young stock where rapid growth for the table is required.
Ridge ventilation: Gaps left at the apex of the roof to assist ventilation.
Rooster: American equivalent of cock.
Rumination: The process of digestion in cattle whereby food is swallowed to the first stomach, the rumen. Later it is regurgitated into the mouth and chewed over again to be swallowed for further processing by the second, third and fourth stomachs.
Running through: Lactation extended beyond 365 days.

Scouring: Diarrhoea.
Scur: Horny protuberance produced in some cases in apparently naturally polled goats.
Separator: Mechanical device for separating cream from milk by centrifugal force.
Service: The act of mating.
Shearling: A sheep after being shorn for the first time.
Shelly: Fine-boned and thin fleshed (of sheep).
Shelter belt: A small plantation sited to provide shelter from the prevailing wind.
Silage: An alternative to hay as a means of preserving summer food for winter use. It is made from grass and other green crops such as clover, kale, peas and green corn, which are cut while young and are compacted in airtight concrete towers (silos), or in stacks, or pits covered with plastic sheeting, so that they undergo a controlled fermentation to a readily digestible and nutritious feed.
Sitting breed: A breed of poultry which retains the natural tendency to become

broody once or twice a year.
Slurry: The thick liquid formed by the mixing of dung and urine.
Solids-not-fat: The solids dissolved or suspended in milk other than butterfat, composed principally of protein (casein) lactose and minerals.
Space-boarding: Narrow gaps left between vertical boards on the sides of a building to assist ventilation.
Sport: A random mutation.
Springing: Applied to a heifer or cow showing signs of approaching calving.
Squab: Young pigeon raised for meat.
Stag: Adult male turkey (Britain).
Stall-feeding: A system of management where animals are housed more or less continuously (except for exercise) and forage crops are cut and carried to them.
Staring coat: Hairs standing on end indicative of illness or poor condition.
Starter crumbs: Mash formulated for young poultry and made into a crumb structure.
Steaming up: Increased feeding, particularly of concentrates during the latter part of pregnancy.
Steer: Castrated male cattle being reared for beef.
Store: An animal not yet ready for slaughter.
Strain: Type of stock which will produce the same characteristics from one generation to another.
Stratification: A method of cross-breeding sheep of different types.
Stripping out: Removing the final amount of milk from the udder.

Thriftiness: The capacity to make good use of food.
Tom: Adult male turkey (USA).
Tongue-and-groove boarding: Boarding where the panels interlock with each other.
Traces: Two straps, chains or ropes, one on each side of the horse, attached to the collar in front and the vehicle behind, by which the horse draws the vehicle.
Trocar and cannula: An instrument used to relieve cattle suffering from indigestion with accumulations of wind (gas).
Tubbing: Standing a horse's foot in a bucket of hot water with washing soda dissolved in it to soften the hoof.

Urinari calculi: Stones or concretions containing salts found in the urinary system.

Veal: Calf meat.
Vermifuge: Drug or other preparation for killing parasitic worms.

Wattle hurdle: Hurdle constructed of woven, split hazel.
Weanling: A calf weaned from suckling.
Wether: Castrated sheep.
Wool blindness: A condition where wool on the face impedes the vision.

Yearling: A calf between one and two years of age.

INDEX

Acknowledgements

The publishers would like to thank Brian Hale for taking the black and white photographs. We would also like to thank the following for supplying additional photographs of several of the breeds described in the book:

Australia Information Service, London – Corriedale, Merino, Droughtmaster and the photograph of a cow and calf on page 184;
Blonde d'Aquitaine Breeders' Society;
British Duck Advisory Bureau – Pekin;
British Wool Marketing Board – Lincoln, Scottish Blackface, Swaledale, Cheviot, Welsh Mountain, Herdwick, Shetland, Oxford Down and Suffolk;
Brown Swiss Cattle Breeders Society, Wisconsin;
The Commonwealth Bureau of Animal Breeding and Genetics, Edinburgh – Chios, Awassi (courtesy M. Finci), Swedish Red and White;
Devon Cattle Breeders' Society;
Farmers Weekly – Merino (photograph Charles Topham), Romney Soay, Jacob (photograph Peter Adams), Whitefaced Woodland, North Country Cheviot, Teeswater, Border Leicester, Bluefaced Leicester, Dorset Horn, Clun Forest, Southdown and Texel sheep; Berkshire, Landrace, Large White and Pietrain pigs; Kerry and Meuse-Rhine-Issel cows;
Mr D. Hall – British Toggenburg (Balaam's Nimo);
Hereford Herd Book Society;
Luing Cattle Society;
Mr R.T. Martin – Saanen (Gay Jewel);
Milk Marketing Board – Aberdeen Angus, Ayrshire, Charolais, Dexter, Friesian, Guernsey, Jersey, Jamaica Hope (courtesy Mr. J. Friend), Limousin, Shorthorn, Simmental, South Devon and Welsh Black;
National Farmers Union, London – the second photograph of pig housing on page 143 and the photograph of a farrowing house on page 147;
National Pig Breeders Association – Tamworth, Gloucester Old Spot; Welsh (copyright Commercial Camera Craft), British Saddleback and Hampshire;
Optima Graphics – Manx;
Pig Improvement Company – Duroc;
Poultry World – Barnevelder, Plymouth Rock, White Leghorn and Ancona;
Miss E. Rochford – Anglo Nubian (Berkham Butterlady) and British Alpine (Berkham Allegra);
Mrs Rosenburg – Golden Guernsey (Novington Bovis, born 1973, bred by Mrs. R. Karney);
Bill Wilcox – British Milksheep.

The following people kindly allowed their livestock and premises to be photographed for the book:

Mrs Awdry; Christopher Brown; Brian Buckler; Alan Duck; Mr and Mrs B. French; Godfreys Turkey Farm; Jean Gordon-Macleod; Elizabeth and Trevor Gore; Robert Lawton; Mr Leach; Francis May; Morton's Commercial Rabbits; Frank Nutley; Tony Palmer; Bill Price; Mr and Mrs P. Wilson.

The colour photographs were provided by the following:

Bavaria Verlag, Germany (photographs opposite pages 144, 161 and 200);
British Tourist Authority (third photograph between pages 200 and 201);
Paul Gilchrist, Australia (photograph opposite page 32);
Brian Hale (photograph opposite page 113);
Lisa LeGuay, Australia (photographs opposite pages 48, 49, 96, 97 and 145);
Nordisk Pressefoto, Denmark (photographs opposite pages 38, 112, 160 and 209);
Galen Rowell, California (second photograph between pages 200 and 201).

Our thanks are also due to the three consultant editors:

Australia –   Paul Gilchrist (Special Veterinary Officer for Poultry Health in New South Wales) and David Austin.
North America – John Skinner (Poultry and Small Animal Specialist, University of Wisconsin).